FLOODSCAPES

Frédéric L.M. Rossano

FLO
SCA

nai010 publishers

OOD
PES

CONTEMPORARY LANDSCAPE STRATEGIES IN TIMES OF CLIMATE CHANGE

1

2

3

4

5

The title of this book could be seen as a 'contradiction in terms': after all, flood spreads wherever gravity leads it, covering the familiar topography with a dark, grey, and uniform blanket. In that regard, flood is amorphous, as it can distort and temporarily erase forms and features of the visible landscape—nothing that could be described as a 'scape' in the sense of articulated and meaningful scenery. But when the boundaries of a flood are not only defined by the quantity or the velocity of water but also by land forms and structures carefully designed to shape and mitigate the 'disaster', the result can be considered a floodscape, physically and culturally defined.

Existing defence infrastructures designed to tame and control dynamic streams and rivers over the past centuries, are increasingly challenged by storms and high waters. The combination of rising sea levels, increasing climatic fluctuations, and urban sprawl accentuate the risks incurred by flood plains. Since the end of the twentieth century, however, a long-awaited transition has been taking shape. Rivers and floodplains are transformed to make space for floods again, and a necessary cohabitation with natural fluctuations is reinvented. Focusing on both ends of the river network, this book discusses a selection of contemporary projects of floodplain restoration in Western Europe, currently implemented upstream in the Swiss Canton of Valais, the French Isère Département, the German State of Bavaria, and downstream in the Dutch Groningen Province and the Lower Rivers Region.

Today, these projects provide multiple sources of inspiration, themselves inspired by rich cultures of flood adaptation that were applied in the past centuries. These reappropriations widen the range of possible landscape strategies and make it possible to conceive the necessary adaptations outside of the classical oppositions. The issue is no longer to tame and control nature, nor to set it free again and obey its fluctuations, but to reconcile the natural variations of rivers with human activities. These combinatory approaches bring new challenges: beyond hydraulic and civil engineering, flood management becomes a territorial issue involving many stakeholders again. These engage into complex negotiations, together with the designers who are mobilized to sketch out new possibilities and facilitate consensual solutions.

This integrated approach, born out of necessity, is also a sign of profound cultural changes in our relationship to river space and fluctuations. Navigating between mobilization and a reassuring discourse, public policies today must create space for adapting to these increasing variations, rather than aim at a feeling of pseudo-permanent safety. Bringing flexibility into landscapes petrified by centuries of hard engineering is a difficult task, as these portraits of rivers show. Yet I hope that this book will contribute to the understanding of the past and the present of our rivers, and help designers, experts, and decision-makers provide our landscapes with the elasticity needed to face upcoming challenges. It will also be of interest to all those who, living on former marshes, estuaries or floodplains, aim at a safe living environment within a dynamic and fluctuating nature.

1

GOD, THE RIVER AND THE ENGINEER

Coenraet Decker, *Noah's ark in Mount Ararat*, illustration for: *Arca Noe*, Athanasius Kircher, 1675

> 'A man is always a teller of tales, he lives somehow surrounded by his stories and the stories of others, he sees everything that happens to him through them; and he tries to live his own life as if he were telling a story.'
> (Jean-Paul Sartre, 1938, *Nausea*)

Flood prevention plans are more than univocal solutions or intervention proposals, but they still need clear concepts to ensure a large and lasting support. While the Swiss ongoing 'Third Rhône Correction' seemingly announced more coercive measures against an objectified river, the Dutch 'Room for the River' programme suggested a spatial negotiation between man and a personified river. Both programmes were, in fact, of comparable amplitude and undertook a similar task with similar means. How can they differ so radically in leitmotif? Which story is being told, how does it relate to pre-existing narratives, and what future does it propose?

Vast, expensive, long-term enterprises need collective, broad, and lasting support. 'Dear Lord, give us our daily bread and a flood once in a while', Dutch water managers like to say to stress the perseverance and vigilance required for efficient flood prevention, as most flood disasters are the result of maintenance shortcomings or lack of foresight rather than exceptional natural conditions. The projects investigated in this book, as well as their predecessors, show that all projects, beyond offering a set of practical transformations, are embedded in complex and multi-layered narratives that differ in their ability to intertwine practical proposals with cultural elements, and create a meaningful ensemble that can be altogether manipulative, inspiring, and mobilizing.

Various myths and figures are brought into play to oppose or promote flood-prevention measures—with a technical rationale being paradoxically one of the many narrative tools. Furthermore, narration answers the individual need for each one of us to participate in a meaningful epic that is beyond the erratic manifestations of natural processes, their happy or dramatic consequences, and the subsequent human reactions. A man 'tries to live his own life as if he were telling a story': This is true also of collective enterprises. The various planning narrations we encountered differ greatly, depending on the cultural framework used to elaborate and communicate the plan, and the reflexive positioning of humankind and Nature within the context of the project. In a Western European context, they also contain recurrent figures that can be tracked from millenarian myths. Understanding the success and the conflicts behind great river transformations implies uncovering those hidden prejudices that affect both the aim, the process, and the result. From the Flood sent by omnipotent gods, through the absolute protection promised by Promethean engineers, and on to the hopes of a final reconciliation with Nature, myths evolve and accumulate, and are worth investigating in order to better understand tomorrow's hopes, conflicts, pitfalls, and success factors.

THE DELUGE: TRACES, MYTHS AND VARIATIONS

LOCAL FLOOD TURNED GLOBAL MYTH?

Although flooding was associated with wealth and fertility along the Nile river, and despite various examples of flood-resilient societies in Europe, the Near Eastern and European traditions focus on its catastrophic rather than its nourishing aspects. The devastating Flood and the following rebirth of humanity is one of the most widespread myths, found in similar forms in a period stretching over 5000 years, starting from the Mesopotamian epics, told and later written on clay tablets, and spreading to Western Europe and possibly to the Indian sub-continent. The geographical origin of the Flood myth, first recorded in ancient Mesopotamia, is all but incidental, since there is enough evidence of past floods and dramatic changes in lake and sea levels in the Near East regions that could explain its creation, its persistence and possibly its dissemination. Dating the origin of the Flood myth is however more complex: based on written sources, the oldest written account traces the myth back to about 2800 BCE, with the reign of Ziusudra, last King prior to the deluge. However, based on actual flood events of large amplitude, the myth could be 5000 to 12,000 years old, provided these events have left a lasting trace through oral transmission up to the birth of writing.

Already in 1928–1929, two archaeological campaigns carried out in the cities of Kish and Ur, in today's Iraq, found sedimentary flood deposits in the ground, attesting to a major inundation in the area. The excavations made at Ur suggested that a flood had submerged the area around 3500 BCE. Excavations in Kish produced evidence of two more recent floods, dated around 3000 to 2900 BCE (Mac Donald, 1988: 16). Excavations in Shuruppak also uncovered a flood stratum. These findings were quickly related to the biblical Flood by the archaeologists themselves, immediately adding a theological dimension to the archaeological debate. Mac Donald noted that the flood attested to in Shurrupak is of particular significance, since according to the Sumerian Kings list the city is the home of Ziusudra, and a central character in the oldest known account of the Flood. Furthermore, this date would match that of the flood mentioned in various Mesopotamian epics, which is roughly datable thanks to the post-diluvian genealogy found in the Kings List, from which it has been deduced that the Mesopotamian Flood took place around 2900 to 2800 BCE (Mallowan, 1964: 68–70).

Further back in (pre)history, events of an even larger amplitude can be found that would resemble a universal flood to human populations, then confined to a limited number of refuge areas located in the warmer regions and close to fresh water resources, such as the shores of the Black Sea and the Persian Gulf basin. After the Last Glacial Maximum, c. 21,000 BP, each warming episode saw precipitation increase, ice melt, and sea levels rise: three factors of massive flooding. Large quantities of water imprisoned in the ice were progressively released, and sea levels rose again with 120 metres, mostly over the period between 15,000 and 5,000 BP (Gornitz, 2009: 889), with adjustments still occurring after this date. With an average speed of about one metre/century, close to what we could experience in the twenty-first century, this long period of sea level rise does however hide episodes of accelerated variations and violent floods. Three theories are of particular interest in regard to flood myths: the so-called 'Black Sea hypothesis' (11,000 to 10,000 BP and around 8400 BP), the 'Gulf Oasis Model' (8000 BP) and the 'Caspian Sea Flood hypothesis' (from 17,000 to 10,000 BP) (Ryan et al., 1997: 65; Chepalyga, 2007: 119; Rose, 2010: 850). All three envision vast marine transgressions around enclosed fresh water lakes, with fertile shores flooded by the rising waters of, respectively, the Mediterranean Sea, the Indian Ocean, or of the Eurasian basins that were draining the Scandinavian ice sheet and the melting permafrost from a wide catchment area including central Asia. In all cases, human habitats were permanently lost through submersion and/or salinization, causing destruction, multidirectional migrations and, possibly, a cultural 'big bang' of crucial importance in human history.

The question remains as to whether the global post-glacial hydrological variations were catastrophic, gradual, or oscillating (Yanko-Hombrach et al., 2007). In the first instance, these hypotheses could give the mythological and biblical Flood descriptions a basis of authenticity, and shed new light on the history of humankind and its transmission before the development of writing. An oscillating rise in water levels would imply a succession of more moderate floods, while a gradual rise would relativize the accounts of a nearly final Flood to that of one at a local level, or to the mythological domain, and thus dash the hopes of many who wish to see the biblical description confirmed by scientific evidence.

However, even if progressively spreading over several centuries, the radical processes that took place after the Younger Dryas glacial episode would attest to effects on natural habitats and therefore on human communities, whether nomadic or sedentary. Changes in precipitation, the rise and fall of sea levels, shifting coasts and deltas, salinization of lakes and ground water, coastal and river floods, these are factors sufficient to attest to the decisive influence of water dynamics on the early history of human civilization, enough to explain the presence of a flood at the very centre of the oldest cosmogonies. In these unstable times, flood was probably no more lethal then drought, cold, famine, or pandemics; it is, though, the most dramatic of all catastrophes in its suddenness, the most horrific in its visual manifestation, and could still have found its way into oral culture and mythological accounts, elaborated upon by generations of storytellers and priests to remind their contemporaries of the provisional nature of their habitat and their existence, jolted by fluctuating and erratic environmental conditions. This is a situation one could easily find reminiscent of the current challenges to our sedentary civilization posed by accelerating climate change.

FIRST ACCOUNTS OF A FLOOD

Various epics described the occurrence of a mighty flood that would have nearly annihilated humankind, leading to a historical 'reboot' thanks to a handful of survivors. This mythological line, first found in Mesopotamia, written in Sumerian and later translated into various languages used in the Near East, is clearly traceable through Greek and Roman literature, in the Jewish Torah and biblical Old Testament. In most of these cultural traditions, the Flood is a story of destruction and grief, and reveals a deep and collective anxiety about natural fluctuations, their origins, and their possible anticipation. Within a continuous narrative framework, each tradition, however, suggested a particular relationship between humankind and nature, or its divine representatives, and divergent motives regarding the origins, the meaning, and the consequences of the Flood—as it appears not always negative.

Waters and the flood played an important role in Mesopotamian stories, the flood being often presented as an ultimate weapon. The Babylonian poem *Enuma Elis* depicted the origin of the world, and told how the god Marduk defeated chaos: he wielded the deluge, his great weapon, and climbed into his chariot, a frightening storm. Victorious, he became the supreme deity, receiving among other powers the titles of manager of the wells, lord of the canals, inspector of sky and earth channels, who drew the levees, the canals and the furrows (Talon, Anthonioz, 2019).

The oldest traces of the canonical flood myth were found on Sumerian clay tablets excavated in Mesopotamia, where the character of Ziusudra (later Atrahasis in the Akkadian epics), King of Shuruppak, already emerged on tablets dating from the early third millennium BCE. The character was found in various texts and languages spread over vast time distances. This account formed the canon of a story that would be told and written again and again in various Indo European cultures, using the same canvas in different religious or philosophical contexts.

The Old Babylonian version of the Atrahasis epic was written on clay tablets by Ipiq-Aya around 1700 BCE. At the time, Mesopotamia was a 'largely urban society' (Dalley, 2000: xv); its fertile landscape was intensively populated and exploited thanks to an extensive irrigation system. This irrigation system itself was believed to have been made by the gods, before even the first human was created. Rivers and canals together were strikingly described as one creation. They were built by the gods, not by means of a divine miracle but through hard labour, labour so hard that the gods one day burned their tools and revolted against the authority of counsellor warrior Enlil. Enlil found a way to unburden the gods and avoid war: 'Let the womb-goddess create offspring, / And let man bear the load of the gods!' (Dalley, Atrahasis: 1.4.) And so the first humans were created out of blood and clay, receiving, together with life, the endless task of shaping, working, and irrigating the land.

After 1200 years, humankind had multiplied to such an extent that Earth was threatened by chaos. Enlil convinced the Council of Gods to regulate the human population through the use of 'natural' hazards, to end this chaos and clamour—and allow him to sleep. He subsequently sent various catastrophes down upon humankind to remedy the situation. Several similar episodes followed, repeating this same passage, a characteristic typical of oral poetry (Sandars, 1972: 41), where Enlil again reports chaos and overpopulation, sending first a plague, and later drought and famine. It was during these dark times that the human character of Atrahasis came in contact with the god Enki/Ea, to whom he prayed and eventually sent a message, which was carried by the irrigation canals and rivers. Every time humankind threatened to disappear, Atrahasis succeeded, with Enki's support, in appeasing the gods' anger, and returned humankind to health and good harvests—leaving Enlil angry and dissatisfied.

The third catastrophe sent by Enlil upon humankind is the Great Flood. This event, related in the third tablet of the Old Babylonian Atrahasis epic, is later included in the Epic of Gilgamesh (id.: 38), which informs us about the content of the missing parts. After complaining again about the excessive number and chaotic existence of humans, Enlil proposed to the Council of Gods that humankind be wiped out by releasing an almighty and final flood. Several of the gods were reluctant to destroy what they had created, but the powerful and influential Enlil eventually convinced them to follow his plan, and demanded that all the gods swear an oath to keep the plan secret. However, his brother Enki was opposed to the proposal, wondering, 'Why should I use my power against my people?' And so, he secretly warned one righteous human, Atrahasis. In a dream, Atrahasis was instructed to 'Dismantle the house, build a boat, / Reject possessions, and save living things'. He was taught precisely how to build the boat and to 'put aboard the seed of all living things'. Enki instructs Atrahasis on how he was to justify his absence to his fellow citizens, and how to keep them unaware of the coming catastrophe. Atrahasis left the city, built and loaded the boat, and, once he sealed the door, Enki began the Flood. 'For seven days and seven nights / The torrent, storm and flood came on.' The Flood was of such devastating proportions that '[e]ven the gods were afraid of the flood-weapon'. Horrified by the damage caused, the goddess Mami lamented the destruction, and, together with the other gods and goddesses, regretted having obeyed Enlil's command. Enlil, in his turn, was furious when he realized that Enki had helped Atrahasis escape the flood together with his family and representatives of all other living creatures. Enlil and Enki reconciled, however, and Atrahasis was eventually allowed into the assembly of the gods and granted everlasting life (id.: 79).

The tale of Atrahasis and the story of Noah, known from the Hebraic Torah and the Christian Bible share various narrative aspects. The origin of this flood is similar: in both the Mesopotamian Epic and Genesis, the Flood is commanded and monitored by Enlil/God; in both accounts, the god is angry at humankind and initially resolves to wipe it out. Both tales however differ on several aspects, significant in regard to their geographical context, the perception of floods and natural hazards. The first intrinsic characteristic that differentiates the Mesopotamian tale from that of the

Old Testament is the motive for the Flood. Noah's Flood puts God's perfection, infallibility, and omnipotence in opposition to the weakness and wickedness of humans; the Flood is explicitly a unique occurrence, a one-time absolute punishment for moral failure: 'Every inclination of the thoughts of the human heart was only evil all the time. / The Lord regretted that he had made human beings on the earth' (Holy Bible, Genesis 6.5); a punishment that ends, in the Koran, even more abruptly than it started: 'And it was said: O earth, swallow down your water, and O cloud, clear away; and the water was made to abate and the affair was decided' (The Holy Qur'an, 11.46).

In the Mesopotamian tale, however, the Flood released upon humankind is not primarily motivated by moral judgment but rather is an ad hoc measure aimed at taming the uncontrolled growth of humanity. Dalley notes that, according to Mesopotamian mythology, the first humans had no predefined lifespan, and the gods thus needed 'to regulate their population intermittently with plague, famine, or flood', with the last of these being just 'one method by which the gods tried to reduce overpopulation' (Dalley, 2000: 8, 43). More tangible, the Mesopotamian myth evokes the geography and the culture of its place of origins. The size and the density of Mesopotamian cities was indeed an unprecedented phenomenon, and Enlil, complaining about his sleepless nights troubled by humankind's uproar that only the great flood could halt, probably echoes the very 'earthly' experience of his human contemporaries. Ancient Mesopotamia, considered to be the oldest urban civilization, one that was made possible by productive, fully irrigated agriculture, most probably experienced a demographic boom comparable to other developing communities when shifting from foraging, with its self-regulating survival mechanisms, to an organized collective and productive existence. For the dense Mesopotamian population, the main condition for survival was its capacity to control and redirect the waters brought by the fluctuating Tigris and Euphrates Rivers, to exploit regular river floods, and to sustain its agricultural production level, which depended on groundwater and irrigation more than on direct precipitation. This early Flood tale was brought to us by a flourishing hydraulic civilization, whose cosmogony logically placed the creation of rivers and canals at the very beginning of history. This dependency upon irrigation, water storage, and flood control suggests a high level of social stability and administrative coordination, two vital aspects of hydraulic civilizations. When Enlil sees how 'the world bellows like a wild bull', he might also be pointing out a level of dangerous disorder or social unrest, which could ultimately lead to flooding and famine, caused by absent or deficient control and maintenance of the vital water infrastructure. Seen in this perspective, the tale of Atrahasis may offer as much of a religious metaphor as a political warning or an 'allegory of the bad government'. The Mesopotamian epics were primarily multi-facetted tales told orally, probably for various purposes: recreational, historical, philosophical, or religious. Dalley notes that 'some of the most famous author-scribes were also incantation-priests' (Dalley, 2000: xviii), suggesting that literature and religion were not strictly differentiated. Although the narrative sequence of the Mesopotamian tale obviously forms the canvas of later Flood stories, it comprises a rather different content: a call for controlled growth to prevent chaos, and secure lasting wealth and peace in a fertile but vulnerable flood-prone region. This dimension of the Mesopotamian story was unfortunately lost.

NOAH'S FLOOD: THE END AND THE ORIGIN

The biblical Flood, which most certainly influenced today's Western culture more directly than the Mesopotamian, Greek, and Roman tales together, undoubtedly integrated many aspects of the story of Atrahasis, which seemingly encountered 'international' success even before it took on a religious meaning. Dalley points out that 'Akkadian myths and epics were universally known during antiquity, and they were not restricted to the Akkadian language. Some were definitely told in Sumerian, Hittite, Hurrian, and Hebrew' (Dalley, 2000: xviii). She furthermore suggests that an abbreviation of (Uta)-na'ish(tim) was pronounced 'Noah' in Palestine from very early times (id.: 2), which would confirm—if necessary—a direct affiliation between Mesopotamian mythology and the biblical Genesis, and thus make the Flood one of the oldest living myths, bridging pagan mythology and Abrahamic religions over more than five thousand years. These did not, however, retain in their narratives the divine doubts and hesitations, nor the implicit references to a territory put under pressure.

The context plays no significant role in the biblical Flood, since no reference is made to social/territorial organization, its shortcomings and possible remedies—as radical as they may be. From the very beginning, the Flood aims at a near-destruction, selection, and renaissance. In the warning sent to Noah, no doubt is left either about the violent character of God's plan: 'I am now bringing the floodwaters over the earth to destroy everything under the sky that breathes.' With this deliberate disaster, 'God wiped away every living thing that was on the fertile land—from human beings to livestock to crawling things to birds in the sky' (Genesis, 6:17; 7:4, 10–12, 23). At no time can the origin, the progress, or the effects of the Flood be questioned or hindered: it is the proof of God's omnipotence and the exemplary punishment for doubting it. There is no ambiguity about the religious and moral nature of flooding: the Flood is intrinsically a resolution to remedy the moral degeneracy of the creation, and reassert the divine power and his mercilessness, for God 'did not spare the ancient world when He brought a flood on the world of ungodly people' (2 Peter 2: 5).

Yet besides being a destructive process, the Flood is also a process of regeneration, which destroys the bad but spares the good, and prepares the rebirth, here resembling the great fertility myths. While 'all the people on earth had corrupted their ways' and perished in the rising waters (Genesis 6:12), the Ark contained enough survivors to populate the world anew, allowing a new start after an almost total destruction of life. The New Testament draws a parallel between collective and individual regeneration by comparing the Flood to the effect of baptismal waters: 'Those flood waters were like baptism that now saves you' (1 Peter 3: 21). On both scales, flood puts man/humanity through a destructive and regenerative process, leaving no choice but to accept the loss, abandon the past, and construct a new life on new and fertile ground.

Le Déluge, 1571 (Gallica Bnf, author unknown)

The Flood can also be seen as an instance of revelation; it is mentioned in the New Testament when Matthew compares the ignorance of men before the Flood to the indifference witnessed upon the arrival of Jesus: 'As it was in the time of Noah, so it will be at the coming of the Human One. When the Son of Man appears, things will be just as they were when Noah lived.' 'In those days before the flood, people were eating and drinking, marrying and giving in marriage, until the day Noah entered the ark. They did not know what was happening until the flood came and swept them all away. The coming of the Human One will be like that' (Matthew, 24:37–39).

Indeed, in a period of evangelization, the Biblical Flood, just like actual flooding, offered a persuasive argument in converting pagan populations. Throughout the Christianized Middle Ages, floods were seen as divine manifestations, if not as divine punishments. The disastrous flood that ravaged the Rhine Delta in 838 was successfully exploited by the Christian Franks in that way to convert the pagan Frisians: ecclesiastics claimed that the disaster was foretold to Bishop Hungerius and brought by God to punish the Frisian polytheists, whose own divinities failed to protect them, thus showing the superiority of the Christian God to Germanic paganism (Rooijendijk, 2009: 25). The Christian God could also turn into a protector against flood, as attested by Saint Willibrord, a Catholic archbishop who actively spread Frankish influence and the Christian faith in the Frisian lowlands. Willibrord described, in the early eighth century, how God had erected a row of dunes after the flood of 700, which resisted the waves 'like a strong wall', one quite powerful argument in favour of the Christian God for a population regularly threatened by sea and river floods (id.: 25).

Populations affected by floods in Catholic France were also likely to attribute their misery to God's anger rather than to environmental factors or human failure. Floods in the Middle Ages were often countered by carrying around holy relics in religious processions, followed by monks and the inhabitants. In 1206, a thorn from the Holy Crown of Thorns, brought from Constantinople to the Basilica of Saint-Denis, was carried in procession to plead for an end to the flooding in the Seine Valley (Rigord, 1885: 413). That same year, an arm of Saint Simeon was taken out from the same Basilica for the same purpose (id.: 500). Concerning the 1196 flood, the historian Mezeray related that 'prayers, alms and processions were the only remedy against these ills; and when after all this, cross signs were made over the waters, they miraculously crawled back into their regular bed' (Mezeray, 1668, quoted in Champion, 1858–64, vol. 3: 192). The editor Delaborde, who published the Rigord oeuvre in 1882, ironically points out in a footnote that different churches could claim to have repelled the same catastrophe: 'The decrease of the inundation that Rigord, monk at Saint-Denis, says to be a result of his convent's prayers, is attributed by a canon from Sainte-Geneviève to the patron saint of the Parisians' (Rigord, 1882: 165).

Unsurprisingly, the medieval representations of the Flood focused on the religious symbolism, showing the wooden Ark (the cross, the Church) of Noah (prophetic figure, prefiguration of Christ) floating on the waters (the baptism) that covered the old world. They celebrated the beginning of a new era (Wuhrmann, Cariel, 2006: 10) rather than empathized with the sinners, in a world where every natural catastrophe was meaningful, explained, and legitimized as the divine plan. They catechized more than they diverted, and they stressed the magic rather than demonstrated the authenticity of tales, which were not open to doubt or public discussion: the Flood itself made clear what awaited those who lost faith.

FLOOD DIGRESSIONS IN ANCIENT ROME AND GREECE

Long before the development of climatology, the presence of fossilized seashells in mountainous areas already suggested that the Earth's climate and morphology were not all that fixed and stable. Already in the sixth century BCE, Xenophanes suggested that their presence attests that vast parts of the Earth had been submerged by the sea. Xenophanes is credited 'with a theory of alternating periods of worldwide flood and drought that was inspired, at least in part, by the discovery of fossilized remains of sea creatures at inland locations' (Lesher, 2014). The hypothesis is transmitted by various pre-Socratic thinkers who tried to envision the hydrological mechanism that could explain such findings—with Anaximender and Diogenes of Apollonia suggesting that the Earth's waters were gradually sinking and disappearing. Herodotus believed that sea levels fluctuated whenever one sea would communicate with another (Duhem, 1958–59, IX, 247)—a hypothesis reminiscent to those that today link the biblical Flood to great marine transgressions which would have taken place during the last episodes of de-glaciation, around the eighth millennium BCE.

In the fourth century BCE, Aristotle, however, refutes the theory of a flood of universal amplitude. In his Meteorology, the mythical flood is downplayed to a local event rather than the global disaster of divine origin described in Mesopotamian, Hebraic, and Greek texts. Envisioning a fluctuating continuity rather than a string of catastrophes, Aristotle described natural conditions as permanently changing, and permanently re-balancing: 'The same parts of the earth are not always moist or dry, but they change according as rivers come into existence and dry up. And so the relation of land to sea changes too and a place does not always remain land or sea throughout all time, but where there was dry land there comes to be sea, and where there is now sea, there one day comes to be dry land' (Aristotle, 1931). He suggested an asynchronous reading of the Earth's history, consisting of a perceptible and registered history, made up of dramatic and abrupt events, and a hidden history, made up of mighty but gradual processes. The latter could only be known by reasoning, based on the signs they left behind, making it difficult if not impossible for humans and nations to relate to this parallel history: 'The whole vital process of the earth takes place so gradually and in periods of time which are so immense compared with the length of our life, that these changes are not observed, and before their course can be recorded from beginning to end whole nations perish and are destroyed' (idem). While human history involved creation and destruction, the Earth's history was for Aristotle fundamentally cyclical, and no phenomenon could be observed that was not compensated by its opposite—for 'some parts remain moist for a certain time, and then dry up and grow old, while other parts in their turn are filled with life and moisture', Aristotle noted. There was thus no reason to fear the end of the world, 'since there is necessarily some change in the whole world, but not in the way of coming into existence or perishing (for the universe is permanent)'.

Aristotle's successor Theophrastus also studied climatic phenomena and referred to what could be a manifestation of climate variability. In his treaty 'On winds', he reports the case of elevated regions of Crete that used to be cultivated and populated and were later abandoned as a consequence of a general cooling. The story remains merely an anecdote in the treatise, and an uncertain and unremarked trace of climate variability.

Ancient Greek cultures showed a great porosity between poetic, scientific, or theological texts and presented the flood both as an agent of destruction and as a fertilizing or regeneration process. 'From the flooding of the river abundant crops spring up' noted Apollonius, referring to the Nile (Apollonius, 1919 ed.)—but the flood was also the most dangerous weapon used by Hercules to defeat the Mynians, and by Zeus to destroy humankind in a similar way as Enlil did before him. The story line of the Deucalion Deluge found in various sources followed the Mesopotamian canon quite closely, although its scale was subject to discussion: Aristotle considered that 'the deluge in the time of Deucalion ... took place chiefly in the Greek world and in it especially about ancient Hellas, the country about Dodona and the Achelous, a river which has often changed its course' (Aristotle, 1931).

According to Dalley, Prometheus, whose name literally means 'forethinker', may be an approximate Greek translation of Atrahasis, 'extra-wise'. The parallel between the two mythological figures goes further, as Prometheus is also father of Atrahasis' equivalent, Deucalion, who is saved from a destructive flood released by Zeus to punish humankind. Ovid describes in his Metamorphoses how the mightiest god, 'not thinking the punishment of Lycaon sufficient to strike terror into the rest of mankind, resolves, on account of the universal corruption, to extirpate them by a universal deluge'. Jupiter intends to punish humankind for its arrogance and restore his authority after Lycaon has dared to test his omniscience by serving him human flesh as a dish. Furious, the mighty and merciless god decides to wipe out the whole of humankind. Lycaon is not only famous for being a wicked man: he dares to doubt Jupiter's divine nature when he is visited by the god, and the deluge comes as an absolute demonstration of power and as a final punishment for humans who dare question their divinities. But in Sumerian polytheist mythology, as well as in later Greek and Roman mythologies, the decision is initiated by one of many gods, and it is not equally appreciated by his peers, thus creating a source of discussion and conflict. As was the case for Enlil's decision in the Epic of Gilgamesh, Jupiter's lacks consensus among the gods, who wonder who will worship them then. 'Yet the entire destruction of the human race is a cause of grief to them all, and they inquire what is to be the form of the earth in future, when destitute of mankind?' Jupiter, like Enlil, silences his critics by 'promising, that from a wondrous source he will raise a generation unlike the preceding race'. From a moment of destruction, the Deluge becomes a necessary step in the creation process, a repeat with better conditions and components.

Ovid's description of this deluge is particularly rich and evocative. It mixes mythological elements with descriptions of altogether dramatic and absurd scenes unique to flooded landscapes. Jupiter, in one of his most impressive transfigurations, shows 'his terrible face covered with pitchy darkness; his beard is loaded with showers, the water streams down from his hoary locks, clouds gather upon his forehead, his wings and the folds of his robe drip with wet' (Ovid, 1893: 1:8). The face of the Earth is dramatically transformed, as this great flood uniformly but also equally covers familiar landscapes and all recognizable features, returning the world to its original nothingness. 'Sea and land had no mark of distinction; everything now was ocean' (idem). Even the laws of the animal realm were abrogated, since the 'wolf swims among the sheep' and the 'wandering bird ... falls down into the sea'.

Deucalion, warned by his father Prometheus, is the only survivor of the catastrophe, together with his wife Pyrrha. After nine days and nine nights, their ark lands on Mount Parnassus, according to Ovid (Ovid, 1893, 1.9) and Apollodorus (Apollodorus, 1921: 1.7.2). Together they 'remain as mere samples of mankind' (Ovid). After this deluge, Deucalion and Pyrrha pray to the gods, and are instructed to repopulate the world. Like his father Prometheus who formed men out of clay, Deucalion repopulates the Earth by throwing stones over his shoulders; each stone falling onto the rich mud gives birth to a man, while Pyrrha does likewise, except that from each stone she throws, a woman is born. But contrary to Prometheus,

Deucalion and Pyrrha repopulate the world after the deluge sent by Zeus (in J.-W. Baur, *Deucalion and Pyrrha*, ca. 1639, Harvard Art Museums)

who brought life to his creation with fire he had stolen from heaven, Deucalion obediently follows the instructions of the gods, and turned the destructive event into a rebirth of humankind.

Besides this Deluge, flooding has indeed more than just a catastrophic side in ancient mythology. In the Deluge tale, Ovid goes on to illustrate the nourishing virtue of this flood, since 'the Earth, covered with mud by the late deluge, was thoroughly heated by the ethereal sunshine and a penetrating warmth, it produced species of creatures innumerable; and partly restored the former shapes, and partly gave birth to new monsters' (Ovid, 1893: 1.11).

Flood is also mentioned in ancient literature as a useful ally for Hercules in his battle against the Minyans. The story told by Pausanias in his Description of Greece (9.38.6–8) and by Diodorus Siculus, shows Hercules eager to avenge the Thebans enslaved by the Minyans, known for their formidable cavalry. In order to weaken his opponents, Hercules unleashes a massive flood over the plain: 'By damming the stream which flowed near the Minyan city of Orchomenus he turned the country into a lake and caused the ruin of that whole region' (Diodorus Siculus, 4.18.7).

The story is described in more detail by the second-century author Polyaenus in his treaty on stratagems of war, later printed in Basel and spread in Western Europe from the sixteenth century on: 'This river [Cephisus] flows by the two mountains Parnassus and Hedylium, and directs its course through the middle of Boeotia; but before it reaches the sea, it discharges its stream into a large subterranean chasm, and disappears. Heracles filled this chasm with great stones, and diverted the river into the plain where the Minyan cavalry was stationed. The plain soon became a lake, and the Minyan cavalry were rendered useless. After he had conquered the Minyans, Heracles opened the chasm again, and the Cephisus returned to its formal channel' (Polyaenus, 1793: 1.3.5).

For Heracles, who is also said to have reclaimed the marshes of Tempe by digging a drainage ditch to the Peneus River (Diodorus Siculus, 4.18.6), flood is a reversible tool that can turn a peaceful plain into an impassable marsh, and back into fertile land again, a strategy since then used in more plains and estuaries (the last time being in the Netherlands during the Second World War, by both German and Allied Forces).

FROM MYTH TO PHENOMENON

In all periods of time, myth and rationale, however sporadically, meet in the texts of Greek thinkers and Christian theologians. Plato's Laws nuanced the idea, putting the myth in a geographical and historical perspective. Ancient traditions telling that 'the world of men has often been destroyed by floods, plagues, and many other things, in such a way that only a small portion of the human race has survived', seem credible to the protagonists of Plato's dialogs (Plato, Laws III: 3. 677). They argue that 'the cities situated in the plains and near the sea were totally destroyed at the time', and that, with the destruction of cities, 'everything in the way of important arts or inventions that they may have had—whether concerned with politics or other sciences—perished at that time' (idem), implying a dramatic cultural 'reset'. Only scarce populations of shepherds must have survived the flood, together with 'the few herds of oxen and flocks of goats that happened to survive' (idem). The past flood(s) would thus have brought the human population back to a certain state of innocence, which Plato associates with the simple life of small communities ignorant of writing, 'unskilled in the arts generally, and especially in such contrivances as men use against one another in cities for purposes of greed and rivalry', preceding a slow reconstitution of human knowledge and more complex social organizations.

Aristotle presented the mythological deluge ordered by Zeus as a historical fact to support his own theory of hydrodynamics (Aristotle, 1931: part 14). In the third century, the Christian theologian Origen attempted to demonstrate the authenticity of the Flood by reconstructing in detail how the Ark was fabricated. He discussed among other things the origin of the wood, the form of the planks, the insulation of its door, and the capacity of the vessel, and targets his authentication 'against those who endeavour to impugn the Scriptures of the Old Testament as containing certain things which are impossible and irrational' (Origen, 1982: 77). Origen hereby attests that discussions concerning the veracity of the biblical account were not that rare, at least in the Alexandria where he was writing, and that the Flood already raised doubts and questions in the early centuries of Christianity.

It is only in the late Middle Ages that the oldest observed and documented climate change begins: The Little Ice Age, currently considered as lasting from the end of the thirteenth century to the mid-nineteenth century, with a high point reached during the fourteenth century (Holzhauser and Zumbühl, 1999: 223–4). This 'modest cooling of the Northern Hemisphere' had significant effects locally, bringing, among other things, 'unusually cold, dry winters in Central Europe' (Folland et al., 135). During this period, increased variability and extreme weather events put European communities and organization under stress (Behringer, 2010: 85–86),

Coenraet Decker, *Noah's ark in Mount Ararat*, illustration for: *Arca Noe*, Athanasius Kircher, 1675

and are widely noted and documented by the contemporaries. Around the North Sea, in Germany, and in France down to Marseilles, canals and rivers remained frozen in the winter, inspiring the famous images of icy landscapes left by Averkamp and Brueghel the Younger in the seventeenth century and Van Hove in the eighteenth. Borders are challenged both by the sea and by the rivers, resulting in major transformations such as the drowning of the Grote Waard region between Holland and Brabant, and the drowning of the Dollard region between Frisia and Saxony (Meyer, 2010: 16). The newly colonized Greenland had to be abandoned again by its Scandinavian settlers. Alpine populations saw glaciers irresistibly grow and expand, sometimes obstructing valleys and creating new glacial lakes, which could suddenly empty and flood entire valleys. Such outbursts were attested in the Swiss Valais region already from the late sixteenth century. They lasted until the nineteenth century (Holzhauser and Zumbühl, 1999: 226), when Ignace Venetz tried to release the waters captured by a dam of ice that eventually flooded the Val de Bagnes in 1818, killing 44 inhabitants.

In the Christian Middle Ages, however, the debates around floods, late frosts, and poor harvests long focused on finding a moral explanation to these catastrophes—whether God's punishment or witches' curses—rather than on understanding their physical origins. The Universe's permanence was not questioned, and seashell fossils found on high land were still interpreted as a proof of the biblical Flood, a thesis defended by, among others, the Catholic monk Restoro d'Arezzo (Duhem, 1958: 281). Any measurement of the climate that would make a long-term perspective possible was still lacking. Furthermore, investigating the origins and the transformation of the Earth could quickly lead to questioning Genesis itself, a dangerous exercise until the Renaissance. Although the 'Little Ice Age' offered a living example of climate variability, Aristotle's lessons seemed forgotten.

The close coexistence of science and theology in flood studies stretched further through the Renaissance. While Galileo sought to take liberties with Catholic dogma, Athanasius Kircher dedicated a whole book to the detailed reconstruction of the Flood and Noah's ark. *Arca Noe*, beautifully illustrated by Coenraet Decker for the 1675 edition, was simultaneously the devoted enterprise of a pious Christian, who accepted the most supernatural as historical fact and tried to demonstrate its veracity, and as a work of science, which by means of rational inquiry tried to measure and decipher phenomena that were, in 1675, not yet fully understood. Decker's representation of the Flood covering the Earth, showing the Ark floating over the highest mountains before landing on Mount Ararat, is particularly striking in its topographical realism. Vertical geographical stages—mountains, hills, and valleys—are carefully represented.

Leonardo Da Vinci, *Deluge*, ca. 1515 (Royal Library, Windsor Castle, London)

The representation of water itself is free from all cataclysmic exaggeration but instead is cleverly rendered through a fine hatching superimposed on a still landscape, producing an image of great eloquence and clarity that could illustrate a treatise on pre-glacial eras. Kirchner's essay still included groundless beliefs common in his time—springs were still believed to flow from lakes hidden under the mountains, and all waters to eventually disappear into submarine whirlpools—but he actually used the Flood to construct a theory of hydrological transfers, thus furthering the work on the systemic vision of the world proposed by Aristotle's Meteorology. Mythical representation and physical manifestation of hydrological processes were mixed, in an attempt to understand simultaneously the origins of humankind and its physical world.

With the Enlightenment new perspectives emerged concerning the Earth's history. Along with scientific emancipation from the Christian canon, geology became a full-fledged discipline, and the formation and evolution of the universe received attention from various scientists. Almost 2000 years after Aristotle, seventeenth-century geographers dared again to question the hypothesis of a flood covering the whole planet. Scholar and geographer Bernhardus Varenius shared his doubts about the Great Flood: 'But to that question, whether there may be a Deluge that shall cover the whole, even the very Islands: we answer the way how such a thing may happen, may be conceived and explained, yet can scarce ever happen, the Earth being to compactly joined and the Mountains too high' (Varenius, 1650, 1734 ed., 2: 416). Shortly after Varenius published his treatise, Nicholas Steno was the first to envision the long process of stratification and the formation of mountains—yet without questioning the biblical account of the creation. Putting 'Scripture' and 'Nature' in dialogue, he argued that 'Nature proves that unevenness was great, while Scripture makes mention of mountains at the time of the flood. But when those mountains, of which Scripture in this connection makes mention, were formed, whether they were identical to the mountains of the present days, whether at the beginning of the Deluge there was the same depths of valleys as there is to-day, or whether new strata opened new chasm to lower the surface of the rising waters, neither Scripture nor Nature declares' (Steno, 1916: 264–65). Challenging the immediate character of the Genesis, Steno succeeded in creating a new perspective on the planet's history and morphology, basing his research on the visible stratification of mountains.

The artists of the Renaissance—Ucello, Michelangelo, Hans Baldung Grien—abandoned the medieval canon to explore new directions in the visual transcription of the Flood, developing its dramatic, figurative, historical, and scientific dimensions.

Da Vinci showed more interest in nephology and hydrodynamics than in the religious message. The various sketches of floods found in his sketchbooks show nothing of the human or the divine story but try to decipher the morphology of storms and sea surges in a kinetic figuration that evokes the photographic work of Edward Muybridge more than Noah's

Nicolas Poussin, *Winter* or *The Deluge*, 1660–64

Flood. In his sketches, Da Vinci investigated the dynamics of flooding rather than seeking any grand meaning, as he searched for beauty and order in the morphology of waves and whirlpools, not in the magic or the divine.

In terms of dramaturgy, paintings unfolded in complex narrations that stressed more and more the earthly catastrophe rather than the miraculous survival of Noah. Instead of illustrating the biblical message, painters represented animal and human creatures in desperate situations: fighting for survival, praying to the sky, trying to save their families from the waves, dragging their belongings onto the last dry hills. Figuration evolved strongly, as flood representations became increasingly theatrical and complex, announcing the nineteenth-century Sublime. Painted at the dawn of the Enlightenment, Nicolas Poussin's gloomy representation of Winter—one of his Four Seasons painting series—took the Déluge/Flood as a central narrative motif, and used his great sense of composition to assemble the individual, intimate tragedy and the horrific but grandiose catastrophe, turning this Flood into a rich and complex drama. In this dramatic composition, dark clouds soak the surface of Earth, turned into one gigantic lake, lit by dramatic lightning that forecasts long-lasting rain. Attempts to rescue wives and children seem hopeless; a boat carrying four men is about to sink, one of them still imploring the heavens in a last effort to escape drowning. On the left of the picture, the Snake who brought God's curse down upon man, contemplates the scene from a distance, as if no more harm could be done. Of the inundated farms, only the roofs remain visible, while Noah's Ark is about to disappear behind the rocks and leave the dark scenery, taking away all hopes of survival. Again, sin and unpreparedness lead here to inexorable death and destruction, affecting without mercy all creatures except those God has intended to save. Nevertheless, as Noah's Ark disappears on the horizon, the human scene remains in the foreground, leaving the viewer behind, forced to share the tragedy of those who are left to their fate.

Religiosity, dramaturgy and rationality coexisted in the various accounts of flooding given by writers and chroniclers. This cautious shift from a religious to a rational perspective is manifest in the way the 1469 flood in Grenoble was depicted in a note written by a contemporary, a note reported by the nineteenth-century historian Champion: 'A note written by a contemporary on a manuscript of that time is made as such: "The year 1469, eighth day of August, the Isère River, from the morning until the tenth hour before midnight, coming from the Mount Cenis, destroyed and submerged the low fields and green areas; seeing this, ecclesiastics from the [cathedral of] la Bienheureuse Marie in Grenoble, made a beautiful procession around bell's time, and water stopped rising right after that." This inundation was doubtless the result of a storm, which explains the sudden swelling of the Isère, a river subject to these spontaneous floods, as well as the waters rapid discharge' (Champion, 1858–64, vol. 3: 205–206). The fifteenth-century chronicler rather accurately describes the flood but only to attest of its divine nature and the efficacy of prayer as means of mitigation. The nineteenth-century historian, without contradicting the note, analyses the event in order to understand the concrete origins of the inundation, thus implicitly refuting its divine nature.

Along the same Isère River, in between the Middle Ages and the Disenchantment, a poetic description of the devastating flood in 1733 carefully mixed pragmatism, Christian religiosity, and paganism. In *Grenoblo Malhérou*, the grocer and poet François Blanc described how a sudden panic overtook the city, as 'one runs away without knowing where to go / Another, made captive, fears for his future' (Blanc, 1733: chap. X). Blanc attributed the catastrophe to various origins, and first evoked religious metaphors and apocalyptic visions: 'Waters pour from a sky turned into hell / The moon, the sun suspend their journey / The Drac and Isère [rivers] climb back up to their sources.' Blanc later blamed a river personified into a furious and profane divinity: 'Isère, everywhere your fury appears / You should have spared the Lord's temples. Up onto the altar and in sanctuaries / You dared to violate our sacred mysteries.' In a no less dramatic but more rational vision, the author eventually pointed to the city's founders for their poor understanding of basic topography: 'Grenoble, you are lost! The monster swallows you / Fool he who founded you so low!' Although Blanc wrote long before the Isère River was channelled, he showed, in addition to his faith in God's power and evocation of the river's evil Spirit, a clear understanding of the role played by topography in the amplitude of the inundation. This last verse expresses, on the one hand, the physical process—water harms most in the lowest areas—and, on the other, the human responsibility and, in this case, lack of foresight: natural disasters might be caused by fate, but their consequences can be aggravated or limited by human choice, right from the very founding of new cities.

Religious interpretations and rational explanations of flood events begun to collide at the end of the eighteenth century. The fight had already lasted for centuries, opposing a vision of a meaningful world obeying an

Diodore Rahoult, *Grenoblo Malherou*, 1860 (illustration for the Dardelet edition, text François Blanc)

omnipresent God to the vision of a world created by God but explainable by human reason, defended by Galileo, Descartes, Newton, and some movements within the Christian Churches (Jansenists, Calvinists). But while religious interpretations of predestination still dominated at the middle of the eighteenth century, they progressively lost ground in favour of the notion of scientifically measurable risk (Picon, 2006, 38, 39). The Enlightenment and the European Revolutions accelerated the secularization of European societies, and parallel scientific and technological progress brought new means of controlling nature's dynamics. The 'Disenchantment of the World,' depicted a century later by Max Weber, gradually freed flood from mystical interpretations and human guilt, and brought it back to the realm of tangible phenomena. The 1791 flood in Arles is representative of this historical turn, as a local and later national quarrel opposed the bishop of Arles, who organized processions to calm God's anger against heretics, to the revolutionary and anticlerical Mayor of the City (ibid.).

As religious myths started to lose their power, several philosophers of the Enlightenment however still placed floods at the origin of early human migrations and development. Jean-Jacques Rousseau imagined how natural catastrophes might have spread human language: 'We may form a conjecture also concerning the manner in which various causes may have extended and accelerated the progress of language, by making it more and more necessary. Floods or earthquakes surrounded inhabited districts with precipices or waters: revolutions of the globe tore off portions from the continent, and made them islands' (Rousseau, 1754, 1: 211).

The marriage between science and religion around the Flood question persisted among others in the work of Nicolas Antoine Boulanger. Boulanger was an engineer at the prestigious French Corps des Ponts et Chaussées, founded in 1716 to develop France's road and canal network, and was a contributor to Diderot's Encyclopaedia. He compiled numerous flood and catastrophe myths, which he then related to geological observations in order to create hypotheses, at a time where the formation of the Earth's landscape was still mysterious. In 'L'Antiquité dévoilée par ses usages', Boulanger put the biblical Flood at the centre of geology and anthropology (Wuhrmann, Cariel, 2006: 31), using the great Flood as the key to an understanding of the origins of the planet and human societies. In the introduction to Book 6, 'Description of the physical and moral effects of the Deluge' Boulanger wrote: 'We must consider it as the fecund source of all human institutions ... the memory of it is preserved in all nations of the world and the naturalist finds its ravages written in legible and indelible characters in all parts of the globe' (Boulanger, 1772, VI: 242). To stress the universal character of the deluge and the veracity of his deductions, Boulanger noted that even the 'wild people' of the Caribbean believed that a Flood created the topographical features of their islands, such as mountains and cliffs. He partly succeeded in reconstructing the de-glaciation and its effects on sea levels and topographical distortions, using the Flood as a central agent: 'All seas could have surged instantly over our continents and destroyed all nations in a flash; they then later went back to their regular basin, to be brought back again onto the land, which they frequently and repetitively assaulted. Thereby the waters could change the surface of the terrestrial globe, form new valleys, shred mountain ranges, dig new gulfs, overthrow ancient heights, create new ones and cover the ruins of the old world with sand, mud and other substances they were able to carry in their extraordinary agitation.'

Boulanger, in his turn, inspired Buffon's *Histoire Naturelle*, which still generously mixed scientific deductions and dramatic description to depict the cataclysmic origins of Earth's surface. Buffon stated that man had 'conserved a durable and almost eternal memory of these misfortunes of the world: the idea that one must perish by a universal deluge or by a general conflagration; the respect for certain mountains where they were saved from inundations' (Buffon, 2018 ed.: 227). He did not literally use Genesis to construct his geological theory, but he still thanked 'the Creator for not letting man witness these horrific and terrible scenes which preceded, and so to speak prefigured the birth of intelligent and sensible Nature'. A Nature with its own history and dynamics, which seemed to eclipse the genesis of humankind, merely an epiphenomenon within the grandiose geological process.

Buffon furthered emancipated science from the religious canon, by stating that the Earth had to be older than the Great Flood, as 'the epoch of the separation of these two great continents, and even that of the rupture of the barriers to the ocean and to the Black Sea, seem to have been much older than the date of the deluges conserved in human memory' (id.: 107). Antique floods are thus products of a 'particular cause' such as an earthquake, and as regards Noah's flood, it should be considered a miracle beyond scientific considerations: 'To add facts to the Holy Writ, to draw consequences from such facts, is only a presumptuous attempt to measure the power of the Most High. The natural wonders which his benevolent hand performs in a uniform and regular manner, are incomprehensible; and by the strongest reason, these wonderful operations and miracles ought to hold us in awful wonder, and in silent adoration' (Buffon, 1797: 153).

This new fascination for Nature's powers upon fragile human lives, together with the dramatic floods taking place in Western Europe, inspired new artistic representations of the Flood, which figured prominently in the

Gustave Doré, *Le Déluge*, illustration for the Bible, 1866

Paris Salons. The Flood was even chosen as subject for the 1872 Prix de Rome competition, the most prestigious art prize of nineteenth-century Paris. Artists competed with ever larger and more dramatic compositions to represent and transfigure the most epic and horrifying storms. Following the humanistic path opened by Poussin, a new generation of painters seized the Flood theme to explore the limits of horror and drama: Girolet, Gustave Doré, Leon Comerre, Francis Denby, and Merwart magnified the emotional tension, focusing even more on the distorted faces, bulging eyes, and floating corpses of the victims. In Great Britain, Turner, Westall, and Danby transfigured the violence of natural elements, bringing painting a step closer to abstraction. More than any other theme the Flood, whether biblical or factual, could reach the Sublime, when constructed by a combination of opposites: power and fragility, innocence and pointless destruction, bold materiality and spiritual elevation.

FROM ANGRY GODS TO HEROIC SAVIOURS

A strong link between power over waters and men is established by historian Denis Coeur about the development of a flood defence policy in the Alpine region of Grenoble: 'The return of destructive high waters, starting from the end of the sixteenth century, coincides with the affirmation, and even more specifically, with the territorial embedding ['enracinement'] of the royal power' (Coeur, 2008: 196). The physical transformation of rivers and wetlands was thus directly related to the political structuring of national powers, whose investment capacity and reflexive legitimacy in the field of water management withdrew the power of initiative and the prerogatives of churches and local communities, which until then had been fully responsible for the protection of their own land in most Western European regions.

With the rise of secular states, the use of flooding for religious propaganda could eventually be denounced freely by the liberal press, as seen in the coverage of the massive 1856 floods in France, here seen from neighbouring Switzerland: 'The recent floods give the Chroniqueur the opportunity to bring up again his old theme of "the hand of God that weights on a world guilty of so many violations of the most sacred laws". Yet the place hurt the most [by the floods] is the Catholic city of Lyon, seat of an archbishopric ... from where the ultra-montane propaganda extends its influence until Switzerland.' The journalist reaffirms his faith in God, but advocates a clear distinction between God's will and natural processes: 'The rest of us who believe in God's omnipotence, but also in the power of man upon the elements through science, we see purely natural origins to these disorders.... When shall we stop blaming God for the evil that men do?' (*Courrier du Valais*, 10.06.1856: 4).

William Bouguereau, *L'Empereur visitant les inondés de Tarascon*, 1856 (Tarascon City Hall)

Although natural hazards finally left the divine and magical realm, they were, however, not freed from cultural narratives and Christian motifs. Nature might no longer be the expression of God's will or substance for enlightened minds but it still had to be characterized in moral terms to fit into a coherent cosmogony and participate in the collective narratives, then defined by newly empowered European nations rather than religious authorities. If the God figure tended to disappear from flood accounts (the violent, merciless God that unleashed the Flood), the Promethean character of the Engineer gained in importance in the political drama, with Frederick the Great and the Emperors Napoleon I and III as mighty representatives. The growing fascination for the Wild ran parallel to territorial transformation of an unknown scale and rapidity along most European rivers and floodplains, a metamorphosis only possible thanks to the development and the rapid transmission of new technological knowledge throughout the continent, concomitant with strong demographic and economic growth. Flood protection integrated the domain of the land surveyor and the water engineer with the techniques then used to reinforce river banks and reclaim new land, showing strong similarities in Holland, Italy, Germany, and France, from the deltas up to the alpine valleys. Within decades, valleys, plains, and deltas changed from fluctuating landscapes of meandering rivers, alluvial woods, and marshlands into highly productive farmland ready to feed booming populations, increasing the tax base of the young states. Empowered nations could then, in turn, carry out bigger and more ambitious reclamation projects, made possible by abundant labour, and facilitated by their centralized organization and their growing authority over local communities.

These great enterprises needed knowledge, and financial and political power, if not military, as this 'progress' implied a certain violence to local ecologies and to the economy of the local community, depicted as unhealthy and unproductive. Referring to the achievements of Prussian

King Frederick II in the Oderbruch lowlands in the eighteenth century, historian David Blackbourn notes that 'Violence was an essential part of the transformation: without the use of soldiers and military coercion, the project would not have been completed when it was' (Blackbourn, 2007). King Frederick associated himself with reclamation works to a more global enterprise of (forced) civilization and policing, held for the good of a barbarian population: 'In Prussia I abolished serfdom, reformed barbaric laws, promulgated more reasonable ones, dug a canal which joined the Vistula, the Nètre [sic], the Vaste, the Oder, and the Elbe; rebuilt cities destroyed in the plague of 1709; drained twenty miles of swamp; and established a police force in a countryside where the name itself was unknown….' (Frederick II and Voltaire, 1889 ed.: 24.10.1778).

As loving counterpart of the river 'corrections' and reclamation 'campaigns', the compassionate visit became a new figure in flood narratives. With the development of printed media, regional inundations recurrent in Western Europe throughout the nineteenth century could quickly reach national newspapers. The individualization of post-revolutionary societies, parallel to the integration of regional powers into bigger ensembles and the development of mass media, gave a new status to natural catastrophes. Regional floods soon became national dramas and political issues. For young European nations, flood defence was a tool of territorial integration *par excellence*, as it crossed regional boundaries and transcended local interests, offering a stage for national bonding with distant leaders.

As widespread information called for a rapid, political reaction at the national and local level, carefully staged compassionate visits became a central element in political narratives, offering a public demonstration of national concern and personal commitment from the highest representatives of the state. The well-orchestrated communication of Emperor Napoléon III after the Loire and Rhône floods showed his personal engagement in the rescue of flood victims—not to mention that it paved the way for the subsequent territorial transformations planned from Paris. In the fight to structure and unify the country behind its self-proclaimed Emperor, the visit became a technical inspection, a rescue operation, and a propaganda tour, used to frame the fight against floods into specific narratives that reinforced the central power's legitimacy. In the year 1856 alone, Napoléon III was pictured in three large paintings comforting the victims of the Loire and Rhône floods. In William Bougerau's *L'Empereur visitant les inondés de Tarascon*, the Emperor was shown on a boat inspecting the flooded Rhône Valley and comforting the victims. That same year, the Emperor was painted by Lazergues helping the victims of Lyon in a devastated city still covered with mud. In both paintings, the Emperor was depicted as a man of action and compassion, dressed in military uniform, reaching his hand out to the victims. The dense and popular crowd showed simultaneously that the Emperor was close to his fellow citizens, no matter their social level, and was willing to experience and understand their situation.

In Hippolyte Beauvais's *Le Prince-Président visitant les inondés d'Angers*, Napoléon stands in the middle of the picture, holding out his left hand towards an imploring woman, his right hand towards young children.

H. Beauvais, *Le Prince-Président visitant les inondés d'Angers*, détail, 1856 (Musée National du Château de Compiègne) • H. J. Flandrin, *Jésus Christ et les petits enfants*, 1837 (Musée d'Art et d'Histoire de Lisieux)

An animated crowd, escaping the flooded plain shown in the upper right corner, gathers behind him. The Emperor, while comforting the victims, turns a resolute look towards the spectator, standing in the middle of his compatriots, compassionate but ready to take the lead. Merging religious and secular iconography, the painting's narration and its composition are strongly reminiscent of the contemporary representations of Christ Blessing, such as the large *Jésus Christ et les petits enfants* from H. J. Flandrin, exhibited at the Salon of 1839. On the political stage, the theme was probably not a coincidence either: Angers had a strategic value for Napoléon, as a year before, the city had almost fallen into the hands of an insurrectionary movement that protested against low wages and defied the power of the new Emperor.

PRE-INDUSTRIAL FLOODPLAINS

Although modern water management policies primarily focus on controlling and banning natural fluctuations from the living environment, traces of societies that were used to living with floods can be found along Western European rivers and coastal plains, indifferent to the biblical prejudice that associates flood with punishment and destruction. Several example of positive coexistence with natural fluctuations have been found. The fertilizing effects of flood—at least the 'good flood' that brought fine alluvium—was well known and welcomed by early farming populations. Adapted ways of living, such as elevated houses and villages, used to be commonplace in flood-prone regions. Hydraulic techniques were already developed in the Middle Ages to take advantage of natural silting and to benefit from seasonal flooding, from the North Sea along river valleys up to the Alps, and they continued in use until the nineteenth century. Today's binary landscapes, which restrict and isolate dynamic nature areas from petrified inhabited environments, are to a large extent the result of a gradual shift from decentralized, rural, and adaptive societies to an urban, centralized, and capitalized economy, willing to secure and upscale navigation and farming, eventually at the cost of local adaptive mores. Millenary 'flood-compatible' traditions, however, are now being rediscovered in light of necessary adaptation to rapidly changing hydrological conditions.

EARLY ADAPTIVE LANDSCAPES

Along the North Sea's undefined and ever-changing south-eastern coast, a vast coastal plain hosted a population that had long learned to live with tidal floods. Before dikes and windmills could turn the Netherlands into 'hollow lands', the Frisians had learned to secure their homes and benefit from natural fluctuations in the regions located between the estuaries of the Rhine and Ems Rivers, an area colonized around 1000 BCE from the overpopulated sand ridges of the Drenthe region. In this wide coastal plain, sand and clay banks formed by the combined movements of rivers and tides offered vast open pastures, first used temporarily and later permanently inhabited.

Few sources are available to picture this long-gone landscape, the oldest known being the description of Pliny the Elder written in the year AD 47. The Roman naval commander was sent to today's Northwest

Germany to lead a punitive campaign against the Chauci people, and brought back one of the first precise descriptions of the North Sea landscape. Pliny dedicated a chapter of his *Natural History* to the 'Countries that have no trees' (Pliny the Elder: XVI, 1), describing the Chauci coastal regions as particularly inhospitable. In his account, he first doubted 'whether these regions are to be looked upon as belonging to the land, or whether as forming a portion of the sea'. He described how a 'wretched race' survived harsh natural conditions, settling on naturally elevated spots of land or onto artificial mounds (*exstructa manibus*), erected in the flat and endless landscape. These 'eminences artificially constructed, and of a height to which they know by experience that the highest tides will never reach', were found in the Frisian territories as well, the oldest specimen known dating from about 600 years BCE, 1500 years before the first dikes were built to connect mounds to one another (Ter Haar et al., 2004: 28). They could measure up to seven metres in height (Lebecq, 1997: 365) and eighteen hectares in area (id.: 26). All along the North Sea, from today's Flanders to the Jutland peninsula, the specific words used to designate these mounds showed great similarity: in Dutch *wierde*, in old Frisian *werd* or *wird*, in German *warft*, *wurt*, or *wierde*, in Danish *værft*, *varft*, or *verft* (Boersma, 1972); this attests to the existence of a well-developed 'flood-resistant' civilization stretching along 500 kilometres of fluctuating coastline. The word *terp*, village in the Frisian language, is nowadays abusively used in Dutch and English publications to designate these mounds, although not all mounds had a full village, and not all Frisian villages were built on an artificial mound.

Like most ancient writers evoking the edges of the Empire, Pliny saw foremost a barbarian people surviving in a hostile landscape. But ancient authors were prompt to describe the populations living beyond Roman influence as brutal and underdeveloped, rather than stressing their knowledge and skills (Lendering, 2002, Lendering and Bosman, 2012). Compared to the structured and civilized coasts of the Mediterranean sea and the flourishing cities of the Roman Empire, the endless and fluctuating landscape of the North Sea coastal plains, punctuated only by terpen and huts, must have seemed inconsistent and sterile to Roman generals and sailors, especially if they measured the value of the Frisian and Chauci people in terms of their capacity to structure their environment. Here, control and geometrical structuration seemed less relevant than resilience and adaptation to a watery, flood-prone, unstable, and salty environment.

Tidal movements and sea storms made it necessary to secure homes and cattle in times of high water, although the absence of artificial obstacles such as dikes and levees allowed these fluctuations to spread over vast areas and lose part of their energy. Few natural elevations could be found in the region, and the construction of artificial mounds with local materials—clay, animal refuse, construction materials—was thus a simple way to create elevated shelters and provide the basis for a unique civilization living on a part-land part-sea archipelago. The extensive coastal plain, crisscrossed by tidal creeks between sea and ridges, provided vast pastures. Cattle was kept on the prairies surrounding the inhabited mounds, where stalls were built that could shelter up to forty animals, and brought milk, meat, and hides—undersized hides, according to Tacitus' *Annals*, that were seized by the Roman army heading towards the North Sea.

Botanical research has since shown that saline vegetation formed the natural habitat around these artificial mounds, from their construction attested already around 600 BCE to their being abandoned around the tenth century. Nevertheless, the Frisian people succeeded in turning saline soils into productive land, while sustaining fishing and hunting activities. Although salt was found everywhere in various concentrations, various salt-tolerant crops were cultivated on and around the mounds in clayey soils, more fertile than swampy forests and sandy ground. Archaeological excavations of ancient mounds have attested to the presence of *Hordeum vulgare* (six-row barley), *Linum usitatissimum* (linseed) used for oil and fibres, and Camelina sativa (false flax). Fruit trees and legumes were found as well. The use of agrarian tools is confirmed, including the use of ploughs revealed by traces of grooves found during archaeological excavations conducted on the Feddersen Wierde in Northern Germany (Lebecq, 1997: 365).

The terpen were initially divided according to a radial structure, with a circular field left open at the top of the mound, which would be occupied by Christian churches centuries later. Depending on its size, a terp could host one or several families, and small terpen could be gradually heightened, expended, and eventually connected to form a village. The land was initially not privately owned but periodically distributed among the inhabitants. Only later would the terpen be associated with and owned by families and privileged individuals, who held special rights and duties. Among the duties came later the 'dijkplicht' (dike-service) that appeared with the first dikes and overruled all other duties, as the servicer was required to remain near the dikes he was in charge of (Ter Haar et al., 2004: 28). Although life on a terp would appear to be isolated, the ubiquitous waters offered a large network of transport and communication, of particular importance for a country designed more for boats than horses. The first written Frisian law, therefore, severely punished the obstruction of 'a public passage in a river' with a fine of 12 solidi—identical to what one had to pay for wounding someone in the stomach (Lex Frisionum, ca 790: De Flumine obtruso).

In the High Middle Ages, Frisia was still open to the sea, and formed a rich and populated country of fishermen, farmers, and traders. In his 1963 inventory, Halbertsma counts 1170 terpen in the northern regions of the Netherlands alone. Around the year 900, the core regions of Frisia (Oostergo and Westergo) had an estimated density of twenty inhabitants per hectare, similar to the density then found in the urbanized basins of Paris or Cologne (Lebecq, 1997: 366), this despite the constant risk of flooding, the saline soils, and the lack of wood, but thanks to the Frisians' remarkable capacity for adaptation. Although not totally protected from flooding, the Frisians were able to secure people, goods, and livestock on the terpen during high waters. In return, the Wadden Sea, the estuaries, and the creeks offered rich fishing resources, and the Frisians learned to use tidal movements to maximize their catch. They used the wide coastal plain for extensive pasture, hay production, and for crop farming; the terpen provided fertile soil for intensive agriculture close to the house; the house itself had a limited life of a few decades, depending on the moisture of the ground and the type of wood, but could be recycled and rebuilt together with raising the height of the terp, when necessary (Boersma, 1972: 69, 87); these scattered terpen individually offered flood-proof, divisible, and expandable lots, and together formed a resilient decentralized spatial structure, maximizing accessibility by sea and by land.

The Frisian tidal civilization was not destroyed by the sea but more probably by competing Viking plunderers attracted by its wealth. During the ninth and tenth centuries, repeated attacks weakened Frisian society. In the second half of the tenth century, the first polders and dike rings announced the end of the terpen, and the progressive incorporation of the Frisian archipelago with the mainland (Ter Haar et al., 2004: 31). As the sea was increasingly closed off, the utility of the terpen diminished—only to be recalled by dramatic floods that spared the highest villages—and their construction stopped with the advancing generalization of the levee. Farmhouses and barns moved to the lowlands, close to their farmland. Many terpen were demolished in the nineteenth century, their rich soil excavated and sold by the ton to farmers willing to fertilize sandy lands (Boersma, 1972: 100). Paradoxically, the interest in terpen was rekindled during the same period, and for various reasons: the abundance of objects found during the excavation and the transport of the terp's soil when sold as fertilizer; the rise of regionalism and nationalist movements, for which archaeology became a matter of identity; the international interest in prehistoric archaeology, aroused by discovery of the pile dwellings in the Alps, that archaeologists wrongly associate with the Frisian terpen, together with the Italian 'terramare'; the dramatic 1825 flood, that hit large parts of Frisia and Holland, calling the national flood defence system into question (id.: 93–94).

A renewed interest has been seen in the last few decades in Germany and in the Netherlands, partly driven by the increasing salinization of coastal farmland and by the need to create space for temporary flooding while keeping settlements secure, characteristics inherent in the ancient Frisian landscape. Experiments conducted in the 1960s by the University of Groningen showed that the plants found on the excavated 'wierden' could survive a spring tide, whereas modern cultivated plants died. Archeobotanist Udelgard Körber-Grohne showed that these species could grow on brackish ground and produce satisfactory harvests if no inundation took place during the growing period (Boersma, 1972: 33). In the Netherlands, where 125,000 hectares of coastal area was expected to be unsuitable for regular agriculture before 2020 due to salinization, farmers and research institutes have been testing the resistance of various vegetables, with 'saline' potatoes commercially available nationwide since 2014. One of the most commented projects of the Room for the River programme is the transformation of the Overdiepse polder into a temporary retention basin bordered by a series of elevated mounds connected by a dike—clearly inspired by the Frisian terpen and often presented as a return to ancient Dutch adaptive strategies—although no terpen were ever found in this part of the Netherlands. The long ignored and forgotten Frisian terp, key feature of a remarkable landscape built for flood-prone conditions, has now become a reference for new adaptation projects that seek to achieve gradual safety and flexible solutions.

Traces of the terpen in today's land elevation around the Lauwers River in today's northern Netherlands

LIVING WITH THE FLOW

Examples of culture adapted to flood dynamics can be found not only along coastal areas but also along continental wetlands and riverine floodplains. The vertical distribution of function and submersible flood defences seen in Frisia could also be found in the Elbe basin, in the Oderbruch region. Before its reclamation by the King of Prussia, the area 'contained a scatter of villages built on higher sandy mounds' (Blackbourn, 2000). Historian David Blackbourn gives a description of the mores of this region that very much recalls what is known from the Frisian coastal plains and the Loire Valley: 'Their inhabitants were amphibious. They lived primarily as fishermen, from the rich stocks of carp, perch, pike.... But they also produced hay and pastured animals when water-levels were lower, using animal dung mixed with mud and bundles of twigs to construct protective walls against floods, and on those walls they grew vegetables. For much of the year, except during lower water and winter ice, communication was by water' (Blackbourn, 2000).

The Loire River area, historically known for its overwhelming floods, was, however, inhabited and exploited long before the first levees were built at the end of the Middle Ages. Ancient settlements were carefully located on the riverbanks (*chantiers* in local dialect), on natural ridges or artificial mounds, whereas today about 300,000 people live in the river's natural expansion area (Guillou, Maurin, 2005: 37). The elevated mounds called 'tertres' were rather similar in functionality to the mounds found around the North Sea, located just a few metres above flood levels, which were much lower when the river was able to spread over vast areas. Unlike

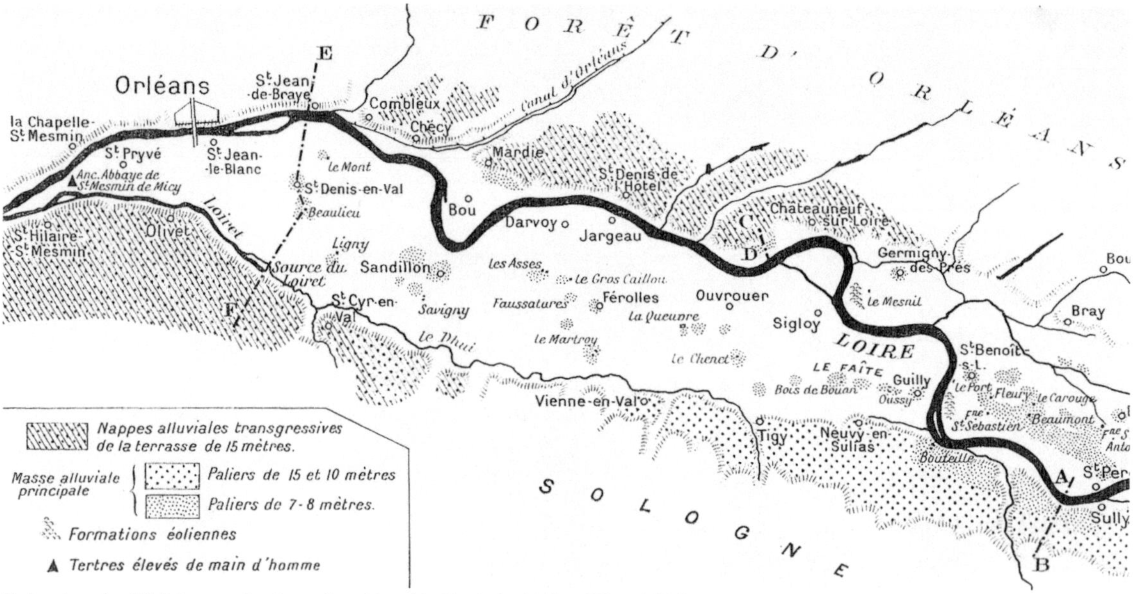

Natural and artificial mounds above flood level in the Loire Valley (Dion, 1961)

the burial tumuli found in various countries, the tertres of the Loire Valley were places to live and work, large enough for families, small farming communities, and sometimes for a full abbey. As flexible as the Frisian terpen, these tertres could also be expanded and agglomerated into circular or linear ensembles, and eventually could grow into villages (Oudot, 1989: 212). Such constructed topographies were also found in the estuary of the Rhône river, and in the Camargue region, more limited in size but similar in function: during a flood, livestock and equipment could be brought there in the safety of the *recati*, a small artificial knoll built close to the farm (Allard, Labeur, 2009: 180).

For the early inhabitants of the low Loire valley, directing and making use of floods was a better strategy than trying to block it off and lose the benefit of its rich alluvium. The first artisanal levees, called 'turcies', were in use at least since the eighth century. They consisted of low and rudimentary ridges made of a mix of earth and faggots, and could be used to connect the inhabited tertres, to create continuous paths along the river arms, or to protect side areas of the floodplain from erosive floods. The turcies formed soft and submersible barriers against the Loire's powerful stream, which would not guarantee safe ground for habitation but could efficiently slow down the stream in times of high water and mitigate its effects. The energy of the flow could be thus attuned and used to catch and spread fertile sediments onto the inundated land (Fournier, 2008: 9).

Pressured by big landowners wanting to secure their farmland, King Henri II of England, Count of Anjou, organized the installation of permanent dwellers on the turcies. According to the 1160 Charter, the new inhabitants of the turcies were made responsible for their maintenance—for their own safety and that of the farmland behind. Strangers in the river's domain, the settlers were given no other choice than to fight against the

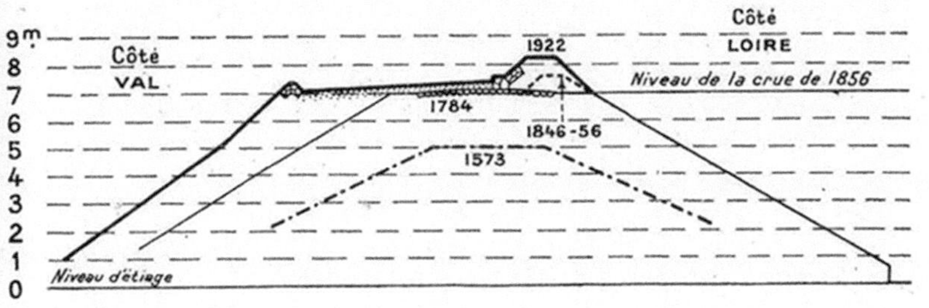

Ever-growing levees along the Loire: the Authion levee in Angers, evolution since the 16th century (Dion, 1934)

flow and defend their land in order to survive, being, in exchange, exempted from certain military duties.

This forced colonization and the pioneers themselves were placed under the control of the Count's army officers, thus attesting to the coordinated and political character of the campaign. As water transport intensified on the Loire River, pressure grew on local authorities to further regulate the river's profile in order to control its level. The turcies were then gradually replaced by stronger, permanent, and continuous levees, and for two centuries this development was not challenged by major floods, possibly due to a mild hydrology and the mitigating effect of the tributaries' still open floodplains (Guillou et al., 2005: 31). The fixed and strong levee temporarily won the battle against the submersible turcie. But by allowing more floodplain to be exploited, and by raising the level of the river, the levee made every transgression more dramatic, and opened a 'frantic chase for insubmersibility' that lasted until the Plan Comoy in the mid-nineteenth century (id.: 39). Furthermore, the shift from submersible turcies to fixed levees was not so much the result of technical advancement but rather the consequence of economic developments that saw the economic interests of traders and big landowners prevail over local interests and small-scale agriculture. Thanks to the levees, navigation intensified and enabled the empowerment of urban trade centres over local communities, which underwent the often-negative consequences of the channelization implemented in and around cities (Fournier, 2008), and the downsides of an early globalization.

With the return of major floods in the fifteenth century, many dikes and goods were destroyed by the river. Following the paradox common to most modern Western European basins, the narrowed riverbed and high levees led to more dramatic floods that led, in turn, to raising the same levees even more, and so on. No one then questioned the basic principle itself, and royal authorities, financially supported by the local trade bourgeoisie, carried on with the canalization of the river, foremost in order to secure river shipping (Guillou et al., 2005: 32). Despite this priority given to canalization, already in 1584 a first spillway is mentioned, thought to be able to release the pressure on the levee and so prevent dike breaches in Blois. A plan made by the royal council of Louis XIII in 1629 proposed the creation of six of these spillways, and even the destruction of existing levees. Local authorities however succeeded in hindering their implementation, and instead the levees were heightened again at the end of the century (idem).

After four catastrophic floods in the early eighteenth century, submersible levees were again suggested and this time implemented between Gien and Tours in the shape of lowered dike sections of 200 metres or longer, which were kept at the old level of fifteen 'feet' (4.88 metres) to allow local overspill (id.: 33). After the 1733 flood, however, these spillways caused an outcry in the flooded areas and failed to prevent the breach of several higher levees, leading to another campaign of dike raising and extension. But the dramatic floods of 1846, 1856, and 1866 again showed the failure of ad hoc raising of the levees and the risks brought by the systematic channelling of the riverbed. More and more engineers estimated that finding space for expansion and mitigation was inevitable.

This shift led to the creation of a commission headed by the engineer Comoy, explicitly tasked with studying the controlled flooding of side valleys in order to mitigate the effects of future flooding. Particularly interesting here is the role played by Emperor Napoléon III in the emergence of adaptive river management, in a century that saw most of Europe's rivers disappear behind levees. As imperial leader, Napoléon III favoured order and reliability over chaos: 'I am honour-bound that in France rivers, like revolution, shall return to their beds and rise no more', he declared in his opening speech at the 1857 parliamentary session, a few months after the 1856 floods.

The Emperor, however, saw with his own eyes the destruction caused by the Loire and the Rhône Rivers, both corseted by high and continuous levees to secure shipping and trade. His personal commitment to territorial development and civil works went far beyond Paris, and his letters show a great concern and a broad perspective on the management of great French rivers. Like many of his engineers, Napoléon III was frustrated by the poor results of the massive investments made after the 1846 floods, mostly directed to the heightening of portions of the levees without any broader perspective, which 'only made the effects of the last tragedy even more disastrous' (Bonaparte, 1856). The Emperor called for a scientific and systemic approach, and took a critical stance about the levee system, 'merely a ruinous palliative'. About the reservoir-dams planned in the upper Loire basin, Louis-Napoleon wrote: 'Even if these dams were causing any wrong to the cultures of the valleys, one would have to come to terms with this, even if it implies compensating the owners, for one should accept giving water some space [faire la part de l'eau] just like one has to cut one's losses [faire la part du feu]' (idem).

Unfortunately, the areas designated to be used for preventive overspill were increasingly built upon and exploited, making it more difficult to implement the project of the Comoy Commission every year (Guillou et al., 2005); only seven out of the nineteen planned overspill sections were eventually built. Still the 'Plan Comoy' remains known to this day as one of the earliest flood mitigation programmes implemented on a grand scale that tried to reconcile variability, safety, and navigability—prefiguring the contemporary Room for the River programmes.

In the third century BC in AD, Apollonius of Rhodes evokes the majestic river that waters a 'fertile land, Egypt, mother of the first humans', that Nile whose waters spread fertility over regions where 'never does the rain from Zeus moisten the earth' (Apollonius, 1919: 313). Flood was always known to erode and destroy, but in many places and under certain conditions it was also welcomed as a free and natural fertilizing agent, ever since the first agricultural societies developed along the great rivers of the Fertile Crescent. The quality of the soil found on river banks after a flood was also known along European rivers, as well as the value of the water that was brought by the flooding of the large rivers, accurately named 'vetwater' in Dutch (Bink, 2011: 62). Isaac Newton used the same expression in its English translation of Varenius's *Compleat System of General Geography*: 'the Seyne, with it's fat and soft Waters, makes the Land fertile'. The origin of this particularity, however, remained obscure for Varenius, who suggested that 'the reason why all these Deluges make the Fields fertile, is because the Water that overflows them is either melted Snow or Rain, which being light and spirituous, and containing sulphureous Matter mixed with it in the Air, is more prevalent to make the Ground fruitful' (Varenius, 1734 ed., 2: 344).

Spontaneous flooding had the disadvantage of being erratic in its distribution of material—a wild stream could bring more gravel or sand than fertile alluvium, or erode more than it would silt. Various attempts were thus made to guide and make use of this natural process, in order to heighten and fertilize unproductive land. Controlled flooding for fertilization was a practice seen in much of Europe from at least the thirteenth century (Cook et al., 2003), both in monastic domains and secular practices. These techniques could vary from artificially slowing down the stream to facilitate the accumulation of alluvium, practices seen in the Dutch Delta or along the Loire River, to deliberately rerouting these river waters towards poor land in order to raise their level and improve their soil, making use of gravity or tidal movements to redirect the rich waters.

Besides fertilization, flooding was also used to make land through warping techniques, known in the sixteenth century in Tuscany under the name *colmata* (Girel, 2008: 84). Colmata techniques, literally meaning 'filled up', aroused international curiosity in the early nineteenth century, as booming European populations put pressure on the demand for arable land. In his *Journey in Carniola, Italy and France* published in Edinburgh in 1820, Scottish traveller W. A. Cadell relates how 'almost the whole of the Val di Chiana has been raised by the process of colmata' (Cadell, 1820: 266). He describes the process of warping as it was applied in the eighteenth century to the marshlands surrounding the river, that were first transformed into a closed compartment: 'The field is surrounded in an embankment to confine the water; the dike of the rivulet is broken down so as to admit the muddy water of the high floods.' The Chiana River itself was too powerful to be allowed into the new compartment, thus only water from the tributaries was used. 'This water is allowed to settle and deposit its mud on the field. The water is then let off into the river at the lower end of the field by a discharging course called *scolo*, and, in French, *canal d'écoulment.*' Cadell estimated the duration of the process to take ten years for an impressive 1.7 to 2.30 metre land rise. He specified, however, that various factors could influence the thickness and the quality of the new soil: 'If the dike is broken down to the bottom, the field will be raised the same height in seven years; but then, in this case, gravel is also carried in along with the mud.... The water which comes off cultivated land completes the process sooner than that at which it comes off hill and woodlands' (idem).

The success booked in the Val-di-Chiana, after centuries of vain attempts to drain the wetlands, received a great deal of attention and influenced ensuing reclamation projects in Italy and elsewhere. 'In 1817 laws were passed in Piedmont and Central Italy (under the respective titles of *Regie patenti* and *motu proprio*) for government-sponsored land reclamation (*Enciclopedia Italiana*, 1933–42), and in 1839 and 1855 in the Kingdom of Naples. Following Unification, the Piedmontese Regie patenti were adopted throughout Italy; the colmata principle was incorporated into new laws in 1873, and the period 1870–1912 involved much legislation dealing with river regulation, slope protection, and reforestation' (Alexander, 1984: 547).

Following upon the Val di Chiana, the warping technique was further developed in Tuscany, on the coastal plain of Maremma, and in the neighbouring Kingdom of Piedmont-Sardinia, which engaged in the late eighteenth century in a large programme of land reclamation in the Isère Valley (today located in the French Département de Savoie). In the mountainous regions of the Kingdom, arable land was scarce, and pasture had reached its maximum intensity, causing deforestation, erosion, and more frequent floods on hillsides and in valleys, where rivers regularly changed bed. Although these fluctuating floodplains still provided the local population with construction wood, brushwood, and grass used by farmers as feed and fertilizer (Girel, 2010: 42), their reclamation was seen as an opportunity to fix populations in a border area that needed to be in a fit state to resist possible military threats, a region where demographic and economic development was hindered by the lack of arable land. With the planned reclamation, the Sardinian government hoped to empower its borders and facilitate the integration of Savoy into a 'united, organized and modern' Kingdom, with the transformed river symbolizing the power of Man over a wild and hostile Nature (idem). The channelling project, prepared by French engineers before the Revolution, was thus developed by the Sardinian government and completed between 1829 and 1854. The braided stream was confined to a single channel bordered by so-called insubmersible levees, and the vegetation of the floodplain, adapted to frequent fluctuations, lost about ninety percent of its habitat (id.: 52).

The land isolated from the river was ready for *bonification*, drainage, and exploitation. Along the Isère River, 44 hectares of former riverbed were turned into farmland through *colmatage* (Choron, 1871: 359). Two warping techniques were actually used: the 'Italian system' that led waters from the main bed directly into a succession of compartments, each one feeding the next until the water is sent back into the main bed; and the 'French system', where waters were distributed into all the compartments by a single canal laid parallel to the river, and evacuated by a

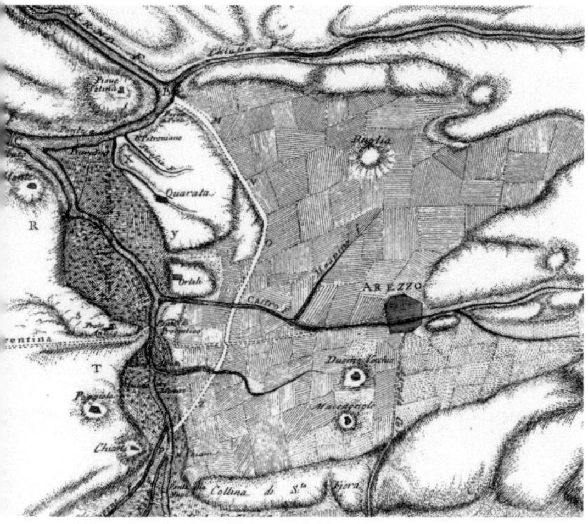

Warping in the Chiana Valley, Tuscany (Fossombroni, 1835)

ditch dug on the opposite side of the reclaimed land. The French system ensured a homogeneous soil quality, while the Italian technique generated a high variety in soil composition. The warping operation would take five to twenty years, bringing even more variation in soil quality. The work was not yet finished when the *Annales des Ponts et Chaussées*, a yearly journal published by the School of the same name, published two essays on the warping experiment, in 1868 and 1871. Both articles praised the achievements of the engineers, Choron stating confidently in 1871 that the 'insalubrious regions that are constitutive of state properties in our valley are not unproductive, and that with enough perseverance, the grains sowed onto these grounds will one day produce a rich harvest' (idem).

In neighbouring Switzerland, engineer Ignace Venetz, who envisioned the channelization of the river Rhône in the Cantons of Vaud and Valais, also promoted the warping technique for the reclamation of the river delta upstream from Lake Geneva. The principle of warping had already been known in the Rhône valley since the fifteenth century, when submersible levees made of wood and stones were built across secondary river arms to slow down the stream, accelerate silting, and create new farmland (Borgeat-Theler et al., 2011: 31). Insufficiently drained by open ditches, in 1840 the Rhône delta consisted of wetlands, pastures, and *flachères*, wet meadows where farmers produced a forage mostly composed of sedge, locally called 'flats' (Schoeneich, 2009: 152). In order to raise the plain and increase its productivity, the 1843 preliminary project included the creation of two *canaux de colmate* (warping canals) running parallel to the central drainage canal, meant to bring the fine sediments carried by the river onto the low parts of the delta. Fifteen years later, the municipalities of Riddes and Saxon together decide to launch a study on the drainage and warping of the plain between these two villages (*Gazette du Valais*, 16, 1859: 1–2). J. A. Raby, *ingénieur civil, membre correspondant de la société des sciences chimiques, physiques, arts agricoles et industriels de France, etc.*, promoted the warping technique over the construction of levees alone, in a series of articles published by the local newspaper. In his ambitious proposal for the transformation of the Rhône valley, Raby estimated that 19 million *toises* (about 7600 hectares) could be protected from flooding and partly warped (*Gazette du Valais*, 7.02.1861: 1.3). He also proposed very symbolically to locate a new school for agronomy on the land that would be reclaimed through warping from the marshes of Conthey, 'a good and useful example'; he admitted that the first warping attempt in Granges had not been conclusive—but 'one

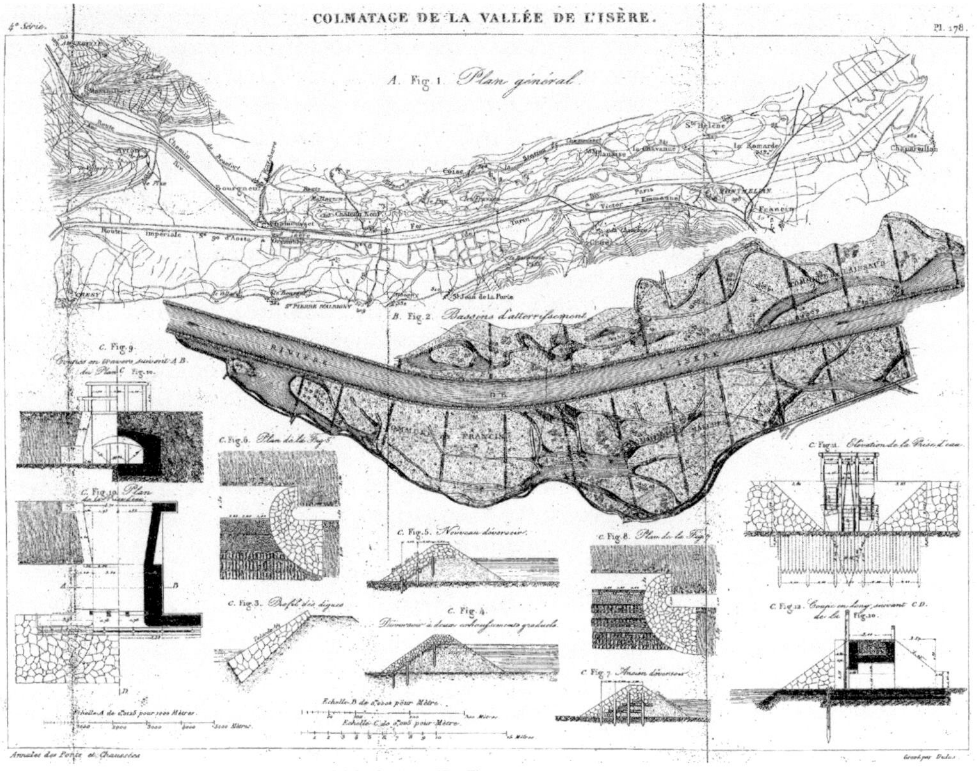

Warping in the Isère Valley (Drizard, 1868, Gallica/BnF)

had applied the means of warping and diking adapted to the nature of the river and the land. Many similar enterprises have been made in other countries. All were totally successful, regarding the completed works as well as the profit obtained' (*Courrier du Valais*, 3 July 1856, 74.1).

The technique was not limited to alpine rivers but also received serious attention in England, where it was widely practiced from the 1730s to the 1880s (Smith, 2012: 18). Warping was used onto several estuarine wetlands (Somerset Levels, Humber area, Gainsborough), and seemed common along the River Ouse according to Loudon's 1826 *Encyclopedia of Agriculture* (Loudon, 1826: 324). In the second edition, completed in 1831, Loudon used the description of the colmata given by Cadell in 1820 (Loudon, 1871: 330), adding however that 'the Italian process called colmata is nothing more than a variety of the British process called warping' (Loudon, 1871: 330). Warping was for him an ancient and obvious technique that 'has long been practised, and is an evident imitation of the overflowing of alluvial land, whether in meadow or aration' (idem). Initially, the warping implemented in England seems indeed not to be related to a specific tradition, but rather be an attempt to reproduce the benefits of spontaneous river floods, since 'superior quality pastures that were regularly flooded pointed the way to controlled flooding or warpings' (Williams, 1963: 179). It is also sometimes combined with various techniques of soil improvement, such as burning and claying (idem).

A description of the warping process as applied in Yorkshire was published in 1845 in the British *Farmer's Magazine*. In a detailed article, Ralph Creyke presented his warping technique and promoted his own experience with warping and draining the South Yorkshire peat bogs (Creyke, 1845). The company of Dutch engineer Cornelis Vermuyden (famous for introducing the land reclamation to England in the seventeenth century) had already been active in the area, building the aptly named Dutch River, a diversion of the River Don, that allowed the drainage of 30,000 hectares of land. Ralph Creyke went further in land creation and developed the warping techniques for use on the Yorkshire peat bogs. A landowner himself, Creyke closed a deal with three townships, offering to reclaim 648 hectares in an area known as Thorn Waste, a bog that had long been open to the tidal movements of the Humber estuary. The levees built along the Ouse River and the drains crossing the land had caused the peat bog to subside below high tide levels. Using the same warping process that he had previously applied to raise his own land, Creyke told how he created a channel to convey the rich waters of the Ouse River through cultivated land all the way to the bog, which had first been divided into three compartments through the construction of new embankments. These compartments were then gradually raised by the deposits brought in by spring tides, which were powerful enough to keep a sufficient flow in the channel and prevent it from becoming clogged with alluvium. Looking back at the warping process completed from 1821 to 1826, Creyke estimated in 1845 that ten to twelve tidal movements could warp four to six hectares with a 30- to 90-centimetre thick layer of fertile alluvium, although the speed depended on many factors such as the distance to the river, the depth of the area, and so on (id.: 33). In total, only five years had been necessary to warp the initial area, and the process was so successful that an additional 800 hectares was eventually reclaimed with the same technique in the surrounding townships by the time Creyke published his article. More land followed after a second Act for the Improvement of Thorne Waste in 1861, but the central part of the area remained a peat bog, and some warped area was not reclaimed (Smart et al., 1986: 739). Ironically, the effort to create new land by raising the peat bogs was followed by a period of intensive peat extraction in the second half of the nineteenth century: after succeeding in creating large areas of land out of muddy waters, vast areas were excavated and returned to wetlands.

DREAMS AND ACHIEVEMENTS OF THE GREAT RECLAMATION WORKS

Wetland reclamation took place and levees were built several millennia before the Industrial Revolution and various plains and marshes of Italy, England, or the Low Lands had already been reclaimed during the Middle Ages, as monasteries spread hydraulic technologies over the continent. From the eighteenth century onward, however, converging evolutions in technology, demography, agriculture, and politics led to a strong acceleration in large canalization and reclamation projects, which together addressed flood risks, pandemics, food safety, the stabilization of local populations, and the empowerment of young nations. Economic development and political integration merged into river management and land-making that turned into national projects, sometimes personally orchestrated by King-Engineers. These vast landscape enterprises were supported by a new discourse of progress: projects of *bonification*, *Flusskorrektion*, *Rektifizierung*, *correction*, *melioration*, or *assainissement* all implied the promise of a better, safer, and wealthier future for all, and were often supported by the denigration of pre-existing landscapes and pre-industrial economies, and the downplaying of the corrections' downsides that already manifested themselves during construction work. Opponents criticized the impoverishment of the landscape and also doubted the long-term efficiency of canalization measures. Economically profitable, the 'taming of rivers' indeed soon appeared to be an endless task, as each canalized river allowed for more land to be reclaimed, yet also called for still more correction work in order to remedy its side-effects and in order to protect newly exposed assets.

A EUROPEAN SELF-COLONIZATION

To a large extent, the origins of land reclamation techniques in Western Europe can be traced back to the practices found in ancient Egypt, Greece, Etruria, and the Roman Empire. Contemporary landscapes, however, owe as much to the medieval development of feudal and ecclesiastical domains, and more particularly to the remarkable expansion of the Cistercian Order in the late Middle Ages, although the myth of a pioneer order colonizing new land from scratch has now been widely revised (Van Acker, 2014; Curtis and Campopiano, 2013). More than just a spiritual project, the

Raphaël Ritz, *Ingénieurs dans la montagne*, ca. 1870 (Musée d'art du Valais, Sion)

Cistercians gave a new dimension to the Benedictine adage *ora et labora* (pray and work), laying more stress on the second term and including in their credo a territorial and economic model. Each new abbey had to be self-sufficient but remain closely connected to the overall organization through the yearly gatherings in Cîteaux, an organization particularly well adapted to the instability of medieval borders. The apparently contradictory combination of productivism, collective capitalization, and individual poverty, presaging the strict protestant ethics of work, wealth, and austerity, were soon shown to be successful as the new monastic order became expert in structuring and exploiting the land, including barren marshlands, remote valleys, or tidal mudflats.

The development of the first Cistercian Abbey in Cîteaux was itself closely related to not only the great hydrological works undertaken to drain, irrigate, and power the domain but also to its central location within the European navigation network. Following the Rhône, Seine, Loire, and Rhine, the Cistercian Order became a master in the control of water and the exploitation of waterways, creating a wide spiritual and commercial network that expanded towards Flanders, to the East along Alpine valleys, and to the South in the Camargue, Spain, and Italy. Everywhere, the Cistercians developed specialized knowledge that could be put into practice on 'wastes' donated by pious landowners. Thanks to the Cistercian centralism and its expansionary drift, information and knowledge was regularly exchanged and spread again over Western Europe. Thanks to the autonomy of abbeys, the flexible system of granges, and the free labour of lay brothers, this knowledge was turned into profitable enterprises. The rapid expansion of the order from the twelfth century on was largely fuelled by this unique strategy associating pioneer development of neglected land, diversified production, and international trading—all encouraged by the Christian authorities and supported by numerous land donations (Van Acker, 2014: 15). In 1129, thirty years after the foundation of the Cîteaux Abbey, thirty-five monasteries were active, stretching from the Lucedio monastery in Piedmont through France and Switzerland to the Saxon monastery of Walkenried (Cooke, 1893: 640), with these latter two domains having again been donated to the Order, to be reclaimed from wetlands and floodplains. In 1153, at the death of Bernard de Clairvaux, 343 monasteries could be counted (id.: 639).

The Flemish domain of the Abbey of Duinen illustrate this spread empire conquered from the water. Founded in the twelfth century, it covered circa 10,000 hectares in the year 1300, becoming one of the largest landowners in the Low Countries (Lehouck et al., 2011: 2).

Its filial abbey, Abbey Ter Doest, came to play a crucial role in the history of water management with the transformation of a brackish intertidal area into a fertile polder: the Land of Saeftinghe in Flemish Zeeland. Building dikes, draining and reclaiming new land, maintaining locks, and in some cases heading local water boards, the Cistercian abbeys expanded their reputation further to the North, into Holland, Frisia, Northern Germany, and England (Rooijendijk: 51–2), where more abbeys were founded in coastal plains and estuarine areas.

The great medieval colonization cannot be reduced to one single model, and the techniques differed according to the local conditions: tidal mud flats and continental marshes required different drainage and diking techniques. Furthermore, the social and economic schemes used for pioneering reclamations differed from country to country. The Cistercian expansion of monastic domains run by monks, and Granges supported by the labour of lay brothers, was successful, especially in France and Flanders, and lasted several centuries. In Medieval Holland, aristocrats, lords, and clergy instead encouraged colonists by granting them full ownership and by setting them free from serfdom (Curtis, Campopiano, 2013: 6), thus involving the pioneers in the maintenance of the water system they depended on. Contrary to the ancient Nile fertilization systems, or to the later colmata techniques, the diking of the North Sea lowlands was not based on an open system but on a permanent isolation of the land from the surrounding waters. In Flanders, Zeeland and Holland, a new culture of fighting and containment competed with the adaptive mores developed by the Frisians. As the population grew and trade increased, the diking of new polders was followed by extensive peat extraction for salt and fuel. Peat mining and land subsidence caused by grazing and draining rapidly worsened the position of the new lands in regard to sea and river levels, calling for more radical measures in order to keep waters out of profitable but fragile properties. In this increasingly dual system, floods turned into considerable threats against the very existence of the hard-won domains. A dramatic example of this vulnerable success was the completion of the Grote Waard in 1283 and its relatively short life before it was swallowed by the 1421 Saint Elizabeth flood. This large area won from the floodplain of the Meuse River remained 'one of the biggest waterworks in Dutch History': 130 kilometres of dikes, 540 square kilometres of land, which disappeared after less than 150 years of intensive drainage and peat mining (Bink et al., 2011: 14–16). The total disappearance of the Land of Saeftinghe, together with many more Zeeland villages, provided another dramatic example of the fragility of the new system, protected on its perimeter but forever exposed in elevation, creating a land that appeared to be borrowed rather than won from the waters of the estuary.

Despite these major losses, reclamation still gained ground on sea, rivers, and lakes. Unfinished and not published until the twentieth century, Andries Vierlingh's treatise written in 1577 described the successful interaction of technological, economic, and political developments that allowed for the reclamation of ever-larger territories, and the structuration of a 'water-repellent' landscape. Vierlingh sketched a time when dike-building had become a flourishing industry powered by the *polderjongens* (polder boys) who alternatively worked on the levees and in the

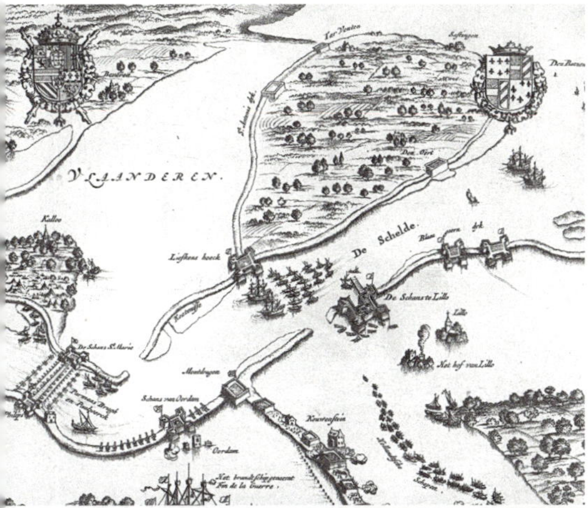

The Cistercian Saeftinghe Polder in the 16th century (Jan Luyken, *Siege of Antwerp in 1584–1585*, 1679, Amsterdams Historisch Museum) • Current topography of the Saeftinghe nature reserve

fields, and dike masters who sometimes 'loosely go diking and damming at all cost, throwing someone else's money to keep the honour and the profit to themselves' (Vierlingh, 1920 ed.: XVIII). As Holland gained more and more territory over water but lost some again after each big flood, Vierlingh warned his contemporaries against the sea but also against division. 'Your enemy Oceanus does not sleep or rest, days nor night, but like a roaring lion seeks to destroy all that surrounds him.' He insisted this enemy should be fought together: 'In dike-making, the external enemy must be resisted with joined means [*beurs*] and joined capacity [*macht*]', as 'this should be considered: give way to the sea, the more it will want to flow in' (id.: 396). From an agglomeration of autonomous domains independently gained on marshes, lakes, or sea, where each local negligence could have dramatic consequences for all, the Low Lands evolved towards a more coherent organization, which Vierlingh wanted to further professionalize and detach from the local interests of the dike-builders and land owners, who lacked coherent vision, and tolerated division, corruption, and incompetence for the sake of individual and short-term profit.

The collectivization of flood defences accelerated from the seventeenth century onward, and allowed for more massive and coherent reclamation projects to be planned by qualified engineers. The development and the communication of knowledge intensified throughout Europe, among countries but also among scientific disciplines that were not clearly differentiated. At the time Dutch water expert Cornelis Vermuyden was working in England, Swiss-born mathematician Leonhard Euler, friend of Daniel Bernouilli, was called upon by King Frederick II to work on the transformation of the Oder in Prussia, together with German engineer Simon Leonhard von Haerlem, son of a Dutch dike master (Blackbourn, 2007). Before studying the Swiss Reuss and Linth, Jean-Samuel Guisan first worked at the reclamation of marshlands in Dutch and French Guyana. Army officer, engineer, and painter Jan-Pawel Lelewel fled from Warsaw to France before working in Switzerland, among various projects, on the correction of the Jura waters (Zuercher, 2007).

The up-scaling of European reclamations was also furthered by the progress made in technologies essential to landscape architecture: surveying, mapping, and hydrology, along with mechanical engineering (Blackbourn, 2000). For a while, the power and imagination of the engineers seemed limitless, as nature became increasingly considered merely as a resource that man was meant to shape and exploit. More than flood defence, these large transformations were the materialization of a *Peuplierungspolitik* that increased population and food supply for the large urban centres (idem).

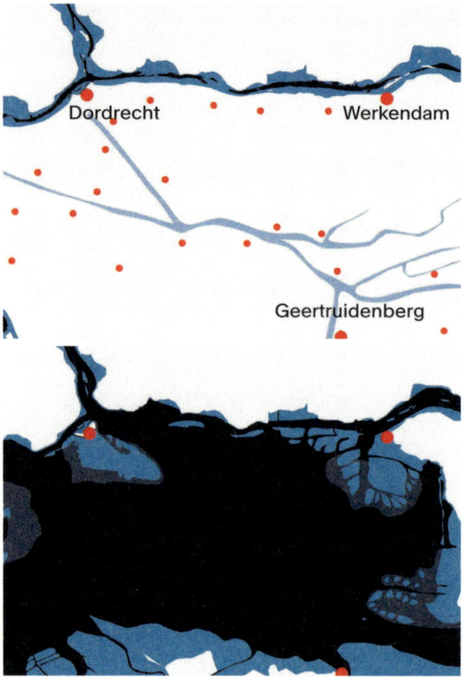

The Grote Waard polder, in ca. 1420 and 1560 after the Saint Elizabeth Flood

Large drainage works were undertaken where the conditions for these enterprises were met: a strong demand for arable land, an abundant labour force, capitalistic entrepreneurship, and a strong support if not leadership from political authorities. The seventeenth and eighteenth centuries saw large areas diked and/or drained in North Western Europe: the Fens and Somerset Levels in England, the Watergraafsmeer and Beemster Polders (72 km^2) in the Netherlands, the more than 600 km^2 Oderbruch in Prussia, the Moëres Polder between France and Flanders, not to overlook a bog reclamation of modest size but of great symbolic power: the drainage of the Versailles marshes.

From the nineteenth to the late twentieth century, flood and river management definitively shifted from local adaptive measures to centralized defence engineering, leaving for good the religious and emotional domain to enter the field of rational choices, whether technical or political. But it soon became apparent that technical choices could be as dogmatic as religious beliefs. In 1812, Johann Gottfried Tulla, 'the man who tamed the wild Rhine' (Blackbourn, 2007), concisely summed up the prevailing vision of river management: 'Every river or stream needs only one bed, and if it has several arms, one should work towards a narrowed stream. It should be kept as straight as possible to facilitate a constant run-off by high waters in order to preserve the banks and so that the flow lowers its own bed and does not flood the surrounding area. The old river arms should be filled and the reclaimed land should be planted.' Following Tulla's rigid prescription, almost all the principal Western European rivers were channelled between levees and their floodplains were subsequently drained to feed the ongoing industrial revolution: The French lower Rhône, the Saône, the Seine between Paris and Rouen, and the Isère; the German Rhine, the Elbe, the Emscher, and the Isar; and the Swiss Linth, the Reuss, the Jura, the Aare, the Upper-Rhône and Rhine rivers, to name only the largest canalized rivers.

These big and long-term projects also reflexively determined the new relationships established between local and national powers. The English reclamation was generally the result of private entrepreneurship supported by the monarchy. The reclamation of the Oderbruch in Prussia was largely the work of Frederick the Great, his personal views on territorial development, and the power of his army and administration. The French river corrections were also mostly orchestrated and financed by the central royal and later imperial authority, supported by its increasingly influential corps of engineer. Even in federal Switzerland, the central government played a key role in the promotion of canalizations, though with a different balance of powers. The First Rhône Correction was, for example, initiated but also long disputed within the Valais Canton's borders. After the massive 1860 floods, the Federal Council took a clear stance on how the plain should evolve, setting strict conditions on its contribution. The Council refused to fund the reconstruction of the Canton, arguing that 'we could not positively answer this demand, for its potential consequences and because the Federal Constitution does not allow such subsidies'. The Council also feared that other cantons hit by the floods would soon follow with similar claims (*Feuille Fédérale Suisse*, 14.02.1863: 273). A different answer would be made if only 'the Valais Canton, in order to prevent similar catastrophes, would consider a correction of the Rhône through diking', a project that the Confederation offered to finance at least one third of, while expecting the railway company to fund about 40 percent of the total cost (id.: 357). To the Canton of Uri, also hit by severe floods, the Confederation 'gives a similar answer by noting that we could only consider the request if it includes a correction project with plans and an estimate of the costs' (Conseil fédéral, 1860, p. 316).

In both cases direct support for the victims of flooding was ruled out by constitutional law, but massive federal funding was made available to finance the diking of rivers thought to solve the problem for good, and to be in the interest of the country as a whole, since the Upper Rhône Valley was a major transportation axis.

Not only was the Confederation a crucial supporter of marsh reclamation and river canalization but the corrections also needed to be radical, as recommended for the Rhine River in 1861: 'The Confederation declares being prepared to support the correction of the Rhine.... However, no subsidies will be granted to favour the enterprise unless a plan is adopted for the correction to be as complete as possible' (Dapples et al., 1861: 664). In return, the Confederation covered a large part of the investment, provided a financial safety net that private companies often failed to secure in the long term, and let the Canton lead the construction works. Through this strict policy, the central authorities stimulated the systematic canalization of Swiss rivers all through the nineteenth century, propagating a unified vision of flood defence similar to Tulla's vision: rivers should be totally channelled, and priority be given to the development of agriculture and infrastructure. The corrections and reclamation projects thus both revealed and shaped a conditional relationship between the Federal and Cantonal powers, made up of a careful balance of solidarity and self-interest, central control and local autonomy, a balance that is sought again today in all large water management enterprises.

THE HYGIENE OF DRAINAGE

With cholera, water was already long known as a vector of deadly diseases. Until the origin of malaria was found in the late nineteenth century, freshwater wetlands furthermore remained synonymous with dangerous fevers. People 'located the origin of all medical evil in the swamps. Their smell was designated as 'mal'aria' [bad air], in which people figured a sickening substance' (Speich, 2002a: 13). The etymology of the word itself and its various translations all closely link the dangerous disease to wetlands, swamps, and their unpleasant smell, rather than to the long unknown parasite or its carrier, the mosquito. The Latin origin *mal'aria*, a contraction of *mala aria*, refers to the swamps' 'bad air'; the Dutch *moeraskoorts* and the German *Sumpffieber* both literally mean swamp fever; the French *paludisme* originates from the Latin *paludis*, swamp. Stagnant waters were commonly seen as unhealthy, but this bad reputation did not necessarily extend to flooding. As Rousseau had noticed, flood was not the enemy as long as water remained in motion: 'Although the Areuse [river] crossing the glen frequently overflows, making its banks a sort of marshland, it does not smell like marsh, the air is not humid or unhealthy, as the vivacity generated by its elevation prevents filthy fumes to last' (Rousseau, 1763a).

Malaria became a growing concern in the eighteenth and nineteenth centuries, since it became endemic in many coastal and alpine plains (Pautou et al., 1995: 37). The Rhine and Rhône Rivers and their alpine tributaries (Limmat, Reuss, Arve, Isère...) saw the disease expand, favoured by frequent inundations and anthropogenic disruptions that created new habitats for mosquitoes to develop. Malaria became a collective cause, and its representation was eventually incorporated within the disaster genre thanks to Ernest Hébert, a painter born in the French Alps, whose work *La Mal'aria* composed in Italy depicted an afflicted family escaping the infected area on a boat. Attesting of the subject's relevancy, the painting brought him his first big success in the Salon of 1850, and was shortly thereafter bought by the state to be exhibited in the Musée du Luxembourg.

With the knowledge of their time, the promoters of diking and reclamation could quite rightfully stress the health benefits that could be attained with the eradication of wetlands, as the recurrent outbreaks of epidemics after big floods seemed to confirm the assumption that water and humidity were at the origin of the plague. Officials did not hesitate to directly blame wetlands and the smell of decomposing organic materials for the spreading of the disease, occasionally to support their own ambitions. 'The emanation from these inundated or desiccated wetlands are contrary to the human race', boldly stated French Engineer and Deputy M.-A. Puvis in a technical treatise on the drainage of ponds (Puvis, 1844: 185). The remedy was thus easily found: 'In the Dombes region, the ponds are the main and almost unique source of insalubrity, but it will disappear with all the swamps when we release the waters.'

Upstream in Switzerland, the State Commission for the Rhône Correction concluded that malaria was spread by the 'noxious action of miasmas originating from the Rhône inundations'—another argument in favour of a full diking of the river, which was praised by both state and federal authorities (Schenk, et al., 1863: 38). An official report from the same Commission saw one more argument for the Confederation to fund the costly river correction: the fight against 'cretinism', a case of mental and physical debility later attributed to iodine deficiency. The reporters argued that 'the Valais has, as we know, a large number of cretins, and according to various experts, this sad impairment is directly related to the immediate effects of the Rhône's floods' (id.: 38).

It appeared, however, that not only did floods bring malaria but also the construction of flood defence itself. A recent analysis of the history of alpine rivers in relation to the development of malaria confirmed a direct correlation between the construction of new dikes and the recurrent outbreaks of the disease: the civil works proposed by engineers to eradicate the disease appeared to be 'simultaneously responsible for the expansion and the regression of malaria' (Pautou et al., 1995: 45). Channelization hindered the drainage of the surrounding plains, creating larger areas of stagnant waters. Warping basins offered new breeding grounds for mosquitoes, since they remained wet for several years. Furthermore, these great civil works put even more individuals at risk: besides residents and farm workers, state employees, ditch diggers, and levee builders were also exposed, increasing the total amount of victims of the 'fevers'—that historically include Grand-Duke Ferdinand III of Tuscany, involved in the drainage of the Maremma plains, and later Engineer Tulla, famous for the diking of the Rhine between Basel and Worms. Yet the correlation between diking works and malaria was soon used as an additional argument by the supporters of drainage and warping techniques: the full drainage of the valleys became a matter of public health. Antoine Chiron, Architect of the City of Chambéry in Savoy, noted in 1844 that the population of the valley rightfully blamed the new dikes for the regular epidemics of 'fever': He recommended using warping technique to heighten the valley floor, so that 'with cleverly and patiently completed works, vast areas will be cultivated and the initial cause of fevers will disappear' (Chiron, quoted in Lortet, 1846: 732).

FROM RIVER SPACE TO BORDER RIVER

As fluctuating space, the floodplain resisted enclosure and land monetization. Its boundaries were vague, changing according to climatic shifts, seasonal floods, and variations in the ground water table, which could rapidly turn a pasture into marshland or pond. The river itself often failed to provide steady borders between communities, and conflicts emerged, when a stream would change its bed and redistribute the land among neighbouring municipalities or countries.

In the fluctuating Rhine-Meuse Delta, flooding regularly blurred political borders and user rights. In 1560, a commission was designated to find back the Meuse riverbed, covered after the inundation of the Grote Waard in 1421 and eroded by 140 years of tides and river flow. The issue: finding the border between the belongings of the Prince of Orange, south of the Meuse, and the Count of Holland to the North, with all implications on navigation and fishery rights. Using prods to test the resistance of the sunken land, the surveyors tried to locate the old riverbed, but could find only two-thirds of its old course. The rest was arbitrarily drawn on the map and materialized with markers—that were maliciously displaced by fishermen from one or the other country, and ultimately had to be anchored by sunken millstones (Van Wijk, 2012: 14–15).

In the medieval Alps, with the first attempts to resist the flow, there also came the protest of those who suffered from the counter-effects. It was first as a border between village communities that the Rhône became a source of conflict. The plain became the platform for continuous negotiation and arbitrages requested of the feudal and religious authorities, which also varied in the fourteenth and fifteenth centuries as the Rhône formed a disputed border. As the population grew and agriculture flourished in the floodplain, conflicts among users and commoners from different communities emerged frequently. Milestones and crosses of larch wood were used to indicate reference points and delimit the different pastures, but the placement (and displacement) of these milestones could become a source of conflict as well. The topographical description found in a 1409 arbitrage showed the difficulty of agreeing on the borders and the location of pastures in a fluctuating landscape that lacked fixed elements, a difficulty that the arbitrators sometimes tried to resolve by using the surrounding relief and villages as landmarks in their descriptions (Borgeat-Theler et al., 2011: 9).

Stabilizing the Rhône thus became a question of flood defence, border protection, and social peace-making together, a question that intensified when the hydrological activity became more intense and exacerbated competing interests. Various devices were used to stabilize the borders and the riverbanks, such as *barrières* (levees made of branches, stones, and gravel), groins, *arches* (chests), or *fourrage*, a coating of stone and branches anchored with wood stakes sunk into the riverbank, a technique already known in Provence.

However, these techniques often had the effect of pushing the stream in the opposite direction, causing fear and anger on the opposite side of the river. Medieval arbitrages thus tried to differentiate between offensive and defensive installations. Longitudinal levees were generally seen as defensive, whereas transversal structures were considered as offensive. Elements placed inside the riverbed were also seen as a threat, since they restricted flow in the high-water season. The right to install defence structures was a disputed privilege: no community was allowed to intervene with the river course beyond its territorial borders—a double challenge, since these borders were not all that clear and stable, and since it was then already understood that flood defence issues went far beyond administrative borders. In a trial that began in 1489 in Valais, the community of Fully accused the people of Martigny of setting their wooden levees on fire, exposing their territory to dangerous floods, to which the witnesses who were heard for Martigny replied that the levees were illegitimately built on their territory, threatened their possessions, and altered the usual river flow. The underlying motivation seemed, however, to be the orientation of the Rhône waters: by building the barrière, Fouilly redirected flood risks towards its neighbours from Martigny, who saw this as a direct threat (Borgeat-Theler et al., 2011: 44–47). The verdict, more than an arbitrage between right and wrong, clearly tried to reach a balance between what should be protected and what should remain open to flooding in order to protect the interests of both communities, thereby showing a global understanding of the river's dynamic and the need for a balanced approach in order to share the plain with a fluctuating river. It also presaged the first coordinated plans for the Rhône in this region, plans that were made in the mid-sixteenth century by the Diet and aimed at coordinating the different structures built along and within the riverbed, until then built without an overall picture in mind. It is only with the map of the Rhône made in 1768 that the border between the Vaud and Valais Cantons was agreed, a map that also represented the width within which the river should be kept. The actual diking was only considered in the 1836

Ernest Hébert, *La Mal'aria*, 1848–49 (Musée d'Orsay, Paris)

convention signed by the two cantons (De Torrenté, 1964: 114). This long process shows that the evolution of the border, the planning of the canalization, and the actual shifts of the riverbed were not simultaneous—the river obviously resisting the fixity and the permanence desired by the human administrations.

Not only did the river's fluctuation hinder the stability of local borders but transnational dissensions also hindered water management. The management of the Rhine River in particular was a source of conflict and discussions during the great reclamation era, since it formed a sensitive border between Switzerland, Liechtenstein, Austria, Germany, France, and the Netherlands. In 1861, the Swiss Confederation reported having trouble implementing the correction of the river in the St.-Gall Canton, since negotiations with their Austrian neighbour had not made any progress in more than forty years (Dapples, et al., 1861: 666). The Commission subsequently recommended dividing the works in two segments: an upstream segment fully located on Swiss territory, and a downstream segment shared between the two countries. The Rhine Commission further suggested that Switzerland could start with the diking of the upstream segment, and noted maliciously that 'one can even anticipate that these works will wake Austria from its current apathy, as when the river would be narrowed upstream its waters would run even more violently into the [Austrian] Vorarlberg territory'. The proposal, addressed to the Federal Council, was at the least undiplomatic, since it suggested no less than the deliberate flooding of foreign territories to force the neighbouring country into joint river management. 'One cannot deny there is something specious to this proposal', admitted the commission, but 'not being able to do as good as we wish we could, is no excuse for doing nothing at all' (id.: 667).

Despite the Swiss threat, it would take another thirty years to arrive at a treaty settling the status and the shape of the Rhine between the two countries (Attinger, 1930: 473)—the river being eventually straightened and shortened by ten kilometres, and the border left unchanged.

At the same period, in Prussia, 'reclamation and settlement "secured" the border—for how could you protect or even define a frontier that was under water half the year?' (Blackbourn, 2000). Channelling works not only aimed at stabilizing international borders but also at reinforcing them through a pro-active policy of population supported by new agricultural production and settlements, a policy also applied in the Isère Valley, the Alpine border of the Kingdom of Piedmont Sardinia (Coeur, 2008).

FROM COMMON GOOD TO REAL ESTATE

Alluvial pastures, woodlands, marshes, and peat bogs long remained 'waste' or commons useful to local populations, their unstable nature protecting them from enclosure and appropriation other than collective and informal. Their transformation into permanent and arable land radically changed their status, their access and their use.

Most uncultivated lands were generally burdened with various rights of use and property, applying to hunting preserves but also to commons (Curtis, Campopiano, 2013), and remained the property of abbeys, aristocrats, or lords after their reclamation. However, several exceptions are known. Even before the 1160 treatise encouraging pioneers to settle in the Loire floodplain (see chapter 'Living with the flow'), early reclamation in the Dutch Low Lands was generally carried out by colonists themselves, granted broad autonomy by the religious, aristocratic, or feudal powers. These could, in return, expand the population and increase the production of their jurisdictions, generating higher tax revenues. The pioneers eventually owned their land and were set free from serfdom. They were involved in the maintenance of the water system through dike service. The water boards generally corresponded to a specific reclaimed area, giving them local knowledge and legitimacy, and are to this day run by elected administrators and allowed to raise taxes. This fairly egalitarian distribution of new land remained specific, as land, including wasteland, was generally not 'new' nor free of ownership or rights of use in the rest of Europe.

In medieval Switzerland, alluvial commons generally belonged to feudal lords, who granted their use to the local community in exchange for goods or labour days; the land remained open, offering free access and transit. Only in the eighteenth century and after were agrarian communities increasingly deprived of their right to use wild areas, as restrictions were applied to free pasture in woods and forests, and fishing rights became more and more restricted and codified (Walter, 1990: 31). With the corrections and reclamation works, the reduction of riverbeds to dammed channels further reduced the area of commons and *vaines pâtures* available to local communities, while the nineteenth-century campaign against deforestation diminished their access to wood. The intensification of land exploitation increased global production but intrinsically tended to reinforce inequality between owners and users.

Flood defence and land-making became increasingly connected, as the knowledge and money needed for large diking and reclamation works generally exceeded the means of local communities, which were often hit by floods at the time diking appeared necessary. Royal authorities were commonly called upon for help in order to implement new flood-defence infrastructure, occasionally associated with third parties in order to (co-)finance and carry out construction of the levees, in exchange for ownership or exclusive use of the land taken from the river. The deal implied the loss of large parts of the commons on the part of the local communities in return for a smaller area of reclaimed land.

In seventeenth-century England, the draining of the Fenland around the borders of Norfolk, Cambridgeshire, and especially in Lincolnshire, was only possible thanks to a collusion between royal authorities and wealthy investors-entrepreneurs mandated by the Crown. 'This was carried out by enterprising individuals or "improvers" generally under Royal mandate, who with the help of workers and ideas from the Low Countries drained marshy areas, using straight canals, ditches and embankments. The improved landscape which resulted replaced an economy based on exploitation of the marshland, with one based on arable and pastoral farming.' (Smith, 2012: 18–19). Frequently, private investors on their own approached governments to propose a dike-for-land deal. A Dutch-French group led by Cornelius Vermuyden was reported to have 'contracted with the Crown to drain certain lands south of Thorne and in the parishes of Epworth, Crowle, etc... receiving for their labour and expense one-third of the land so drained' (Creyke, 1845: 29). A similar offer was made for the French Isère river by Lesdiguières, Governor of Grenoble, who at the end of the sixteenth century proposed diking and straightening the river in exchange for the drained areas; the consequences of the straightening for the city downstream seemed too uncertain, however, and Grenoble declined the proposal of its governor. The same offer was repeated in 1670 by a descendant of Lesdiguières, the Duke of Sault, and was again declined for the same reason (Agard, 1942: 711 and Champion, 1858–64, vol. 4: 76).

With the development of capitalism, new forms of funding were called upon to fuel flood defence and reclamation. For the correction of the Alpine Rhine in the Canton of Graubünden, following the dramatic 1807 floods, the engineer Richard La Nicca presented a project to the council and recommended the creation of a joint-stock company to finance the works meant to create 575 hectares of arable land. The company was created in 1832, but could not complete the work, which ended up taking half a century. In 1851 the Canton had to take over, followed by the Confederation in 1871 (Attinger, 1930: 473). Along the Rhône in Valais, the creation of the Simplon railway line and the simultaneous First Rhône Correction also were too much for the investment capacities of local authorities, which had to offer, free of charge, the land that was necessary for both civil works (Bender, 2004: 65). Already when the idea of a general correction of the Rhône emerged, Franz Samuel Wild, commissioner for the Confederation, suggested creating a joint-stock company that could finance the operation in 1800. His idea was this: 'The whole plain is expropriated; councils of experts estimate its value before and after execution of the works; owners who wish to keep their land are authorized to do so, provided they pay the company for the added value realized on this land' (De Torrenté, 1964: 43). This was a system similar to the *Aktiengesellschaft* that was created for the Linth reclamation (1807–1823). Later on, with no correction yet decided, municipalities received various proposals from private companies, eventually backed by local notables: 'Mr. Junod, chemical engineer living in Monthey, in the name of a company that has undertook large agricultural works in Switzerland ... has offered to take care of the drainage and warping of the marshlands spreading on both sides of the Rhône from Riddes to Martigny', reads *La Gazette* in 1859 (*La Gazette du Valais*, 24.02.1859: 1–2). The company, which remains unnamed in the article, offered to build all infrastructure necessary in order to transform the land into an 'arable state' and pay a rent equal to ten years of average profit in exchange for twenty-two years of enjoyment and the perennial property of five percent of the reclaimed land. The municipalities declined the offer but immediately commissioned a reclamation study (Bender, 2004). A year later, with the great 1860 flood, the canalization of the river became the first priority for local and federal authorities, and put an end to piecemeal diking. Eventually, Confederation, canton, and the private railway company joined forces in a public-private enterprise meant to bring about a national transportation system as well as local agricultural development.

Potential conflicts between the common good and private interests were most visible in England, where capitalistic enterprises clashed with local rural societies early on. The 'British Agricultural Revolution', characterized by enclosure and reclamation works, often implied the sale of the commons and/or the expropriation of its users and of small landowners. These transformations were legitimatized by the global rise of arable land and production. John Locke theorized about this new relationship to land and nature in his *Second Treatise of Government*. In Locke's perspective, the value of land was proportional to the amount of goods that are actually gained from it and useful for humankind; property rights could thus directly be derived from the exploitative capacity of the one who claims ownership, as 'he who appropriates land to himself by his labour, does not lessen, but increase the common stock of mankind' (Locke, 1690, V, 37). Nature in the sense of unexploited environment, could be assimilated as 'waste' in the literal English meaning (the unused land) and moral meaning (the worthless, disposable). This utilitarian vision translated into capitalistic terms the new relationship constructed between nature as resource and humankind as its only legitimate ruler and beneficiary.

However, the 'progress' brought by the 'improvers' was not always welcome, and 'opposition took both legal and illegal forms, involving petitions to Parliament and full-scale riots' (idem). In 1845, Creyke, a land developer, relates how, two centuries earlier, Cornelius Vermuyden had received one-third of the land he had drained in South Yorkshire, in return for his financial and technical investment. The Crown of England, which had to agree to the transaction, received another third, with the remaining land going to 'the commoners of the district' (Creyke, 1845: 29–30). Creyke maliciously stresses that the deal was not welcomed by the locals and led to numerous conflicts. 'The Dutch seem, by numerous lawsuits, and by a marked dislike of the inhabitants to those *innovators*, to have been finally driven out of the country, and their property to have been taken from them (although there are many bearing Dutch names living in the neighbourhood to this day)' (id., emphasis in the original). Whether reclamation could be seen as a global improvement or as 'class robbery' (a term used by historian Thompson to define enclosure) remains controversial, as the positive aspects of land reclamation and cultivation, which allowed for economic and demographic growth, went together with drastic changes in agricultural practices and land distribution, increasing social and economic inequalities—without forgetting the devastating impact of the drainage works on river and plains ecology.

TAME OR SET FREE: NATURE AND THE ENLIGHTENMENT

Various ontological attitudes towards the rivers coexisted after the Enlightenment: that of acceptance and contemplation towards a living element to be enjoyed in its primary and unspoiled state, and, at the other end of the spectrum, that of contempt for a chaotic and pointless phenomenon that only became meaningful through its transformation and exploitation, with the latter eventually winning over the former.

In France, from the sixteenth century on, the reclamation of marshes was seen as an untapped resource and as a global answer to poverty. Sully wrote in his *Economies Royales* 'that in many of our provinces and countries, along the seas on either side, in large and small rivers and other places in our kingdom, there are large quantities of wastes and marshes flooded and invaded by water... these wastes and marshes being dried up will serve partly for ploughing and partly as meadows and pastures', farming and grazing being 'the two udders from which France is fed', he added in an expression that remained famous (Sully, 1820 ed., p.97).

Through the lens of productivism a new definition of good and evil emerged, opposing the useless and formless to the productive and formalized. Projects of *bonification*, *correction*, *Rektifizierung*, *melioration*, or *assainissement* were opposed to the waste, the *prés pourris* and the miasmatic swamps. The philosophers of the Enlightenment blessed the civilizing enterprise of the great monarchs: 'From the side of my marshland, I bless you for all the marshes that you have reclaimed: together with my ploughmen, I bless you for the many ploughmen that you have set free from slavery and changed into men', writes Voltaire to Frederick II to celebrate the King's achievement in reclaiming the Oderbruch marshes (Frederick II and Voltaire, 1889 ed.: 8.11.1773).

The advocates of reclamation often denigrated the local and the informal celebrated by the poet. Where J. G. Tulla saw efficiency and progress, J.-J. Rousseau, one of the few defenders of pre-industrial waterscapes along with Theodor Fontane later, saw foremost homogenization of the landscape and cultural impoverishment in the drainage of the valleys. For Rousseau, marshes were a place of retreat from the tumult of urban life, and he criticized the 'draining of marshes as one of the destructive interventions in nature that threatened to obliterate physical distinctions in the face of the land' (Blackbourn, 2007: 71). Furthermore, none of these great technical achievements could be justified by the happiness these works failed to bring to humankind. 'When we consider, on the one hand, the immense labours of mankind, the many sciences brought to perfection, the arts invented, the powers employed, the deeps filled up, the mountains levelled, the rocks shattered, the rivers made navigable, the tracts of land cleared, the lakes emptied, the marshes drained, the enormous structures erected on land, and the teeming vessels that cover the sea; and, on the other hand, estimate with ever so little thought, the real advantages that have accrued from all these works to [the happiness of] mankind, we cannot help being amazed at the vast disproportion there is between these things, and deploring the infatuation of man, which, to gratify his silly pride and vain self-admiration, induces him eagerly to pursue all the miseries he is capable of feeling, though beneficent nature had kindly placed them out of his way.' (Rousseau, 1754: appendix). More specifically, Rousseau considered the reclamation of marshlands, just like deforestation, as an impoverishment of the landscape that was being standardized for the sake of cultivation. This poor landscape led in turn to a cultural impoverishment for populations, which no longer needed to adjust to specific conditions but shared a standardized environment. 'This is why the ancient distinctions of race, the effect of soil and climate, made a greater difference between nation and nation in respect of temperament, looks, manners, and character than can be distinguished in our own time, when the fickleness of Europe leaves no time for natural causes to work, when the forests are cut down and the marshes drained, when the earth is more generally, though less thoroughly, tilled, so that the same differences between country and country can no longer be detected even in purely physical features' (Rousseau: 1762). Furthermore, the forces of nature, whether expressed in terms of human beings, other creatures, or the elements, should not to be constrained but set free. To Rousseau, flood was to rivers what passion was to youth: restrained, it would break free and destroy; set free, it would expand and pull back naturally, a parallel drawn during an epistolary controversy opposing him to Christophe de Beaumont, Archbishop of Paris, in 1763. Beaumont asked: 'Left to itself, into what errors, what excesses would youth not throw itself? It is a torrent that overflows despite the powerful dikes built to contain it. What would happen, then, if no obstacle stopped its flow and broke its force?' To which Rousseau answered: 'Youth never goes astray on its own.... It is a torrent that overflows despite the powerful dikes built to contain it. What would happen, then, if no obstacle stopped its flow and broke its force? I could say: it is a torrent that topples your impotent dikes and breaks everything. Broaden its bed and allow it to run without obstacle. It will never do harm' (Rousseau, 1763b: 33). Although the dispute concerned education rather than hydrology, it revealed both the opposition and the similarity between Rousseau's vision of nature and the prevailing technicism. To Rousseau, the flood was a temporary manifestation that should be accepted as such; to the engineer it was a flaw that had to be corrected. Nature should either be set free and correct itself, or strictly educated and 'corrected'; in both perspectives, it was a child in regard to its human counterpart.

However, Rousseau's was a minority view in a progressive and optimistic eighteenth century. For most philosophers and scientists of the Enlightenment, man's destiny was rather to rule over nature and adapt the world to his needs rather than to adapt himself to specific conditions.

For Buffon, nature's faith and reason was to be helpful to humans, and useless plants or animals could doubtless be disposed of in order to improve natural environments. There was no region where 'the power of man has not assisted that of Nature, either by bringing in or turning away water, or by destroying useless herbs and plants that are harmful or superfluous, or by favouring useful animals' (Buffon, 2018 ed.: 129). Exploiting and transforming Nature into a productive system was for the naturalist an act of humanity and civilization: 'Thus the first trait of man as he starts to become civilized is the empire that he learns to take over animals, and the first trait of his intelligence then becomes the greater character of his power over Nature. Because, it is only after having subjugated them that he has, through their help, changed the face of the Earth' (id.: 130). Similarly, Walckenaer, one of the first *polytechniciens* educated to transform revolutionary France into a modern country, saw humankind as a species chosen among all beings, for 'among the multitude of living creatures spread over the earth by nature's generous hand, it is to man it gave prominence.... Everything announces his superiority and proclaims him master and dominator of the planet he inhabits' (Walckenaer, 1798: 3, 6). This new self-consciousness, drifting further away from the Catholic tradition that saw nature as a divine creation (Boersema, 2011), gave a philosophical justification to ever greater and more distant civil works allowed by technological developments. Rivers could be 'improved' and objectified in writings and in arrangements.

The descriptions made of the Rhine civil works conducted in Germany in the first half of the nineteenth century were representative of the new paradigm, as they mostly showed contempt towards the river's former natural course—for Tulla 'a twisted snake through the German Landscape' (Blackbourn, 2007). The vocabulary used for the canalization of the river was far from innocent: the *Rektifikation des Rheins* (Tulla, 1825) related both to its geometry—the 'snake' losing 75 kilometres of its original curves—and to its morals; the twisted, unproductive, and dangerous river was turned into a channel made 'right', suitable for navigation from Mannheim up to Basel—losing en passant about 80 percent of its riverine ecosystem (Blackbourn, 2007: 71–111).

Besides the global movement of 'disenchantment' that was spreading over Europe, could this objectification also be explained by the abundance of water in North-West Europe? The utilitarian protestant vision emerged in the wet plains of the Rhine and Elbe basins, where agrarian economy was not so much dependent on retention and irrigation but rather on drainage and containment—a precious resource here and a nuisance there. From this perspective, the Alps formed a transitional stage that combined both perceptions, since water was there both a treasure and an enemy, depending on the season, the inclination of the land, and the quality of the soil. This duality could be read between the lines in a notable article from the *Confédéré* newspaper, which asked the architects of the first Correction to consider storing the Rhône waters for irrigation before collecting them in a giant drain. An anonymous letter published on the front-page stated that 'the Rhône could be to Valais what the Nile is in Egypt' instead of 'throwing all the waters of our torrents into the Geneva Lake' (*Le Confédéré du Valais*, 28.02.1864: 1).

REWRITING HISTORY: THE NEGATION OF PRE-INDUSTRIAL ECONOMIES AND THE APOLOGY OF TECHNOLOGY

For public authorities, landowners, engineers, or entrepreneurs promoting great reclamation works, it was crucial to convince local populations of the legitimacy of their enterprise and the radical landscape transformation it implied. The reclamation and subsequent privatization of large parts of the land thus needed to be understood as progress in terms of wealth and public health, before the work started but also retroactively. All arguments that could legitimize the enterprise and a fortiori prove it to be beneficial to all were welcome. The need to secure the land and protect the plain from disastrous floods was naturally emphasized as the first motivation for canalization rather than the profit expected from the operation. Furthermore, the narratives developed to advocate these great enterprises often included some dramatization and frank optimism, which associated 'insubmersible' levees with absolute safety, drainage with productivity and health, on the one hand, and flood with disaster, and marshland with wasteland, on the other.

Since the winner gets to write history, few accounts are left of pre-correction river landscapes, whether textual or graphic, apart from the often apocalyptic reports made by pro-correction engineers. The assumption that pre-correction floodplains intrinsically consisted of wetlands, and that only canalization and drainage could make them exploitable has a posteriori been proved partly wrong if not tendentious, although it still can be found in the current reports made for Third Rhône Correction: 'Until the half of the nineteenth century, the Rhône valley looks like a vast swampy area,' (HydroNat, 2000). Yet, before the first Rhône Correction, commons and marshes hosted an active economy reliant on the pasture of pigs, hunting, and fishing—and marshes were more often than not hunting reserves for the nobility, some of whom opposed the reclamation works (Blackbourn, 2000). In the case of the Upper Rhône, the floodplain before correction was considered all but 'waste'. Medieval writings confirm that the riverine space played a central role in the rural economy of alpine valleys. About the life of the inhabitants of the Upper-Rhône historian Borgeat-Theler noted that 'far from the stereotype of a swampy and unusable valley floor, far from the cliché of a frightening river one should avoid, their existence is punctuated by agro-pastoral practices that perfectly integrate the varied resources of the Rhône plain' (Borgeat-Theler et al., 2011: 7). The numerous conflicts that took place throughout the Middle Ages precisely reflect the importance of these resources for local populations, and medieval documentation sketches an image of the intense agricultural activity that took place in the plain already in the fifteenth century. Areas that could produce grass were turned into pasture; straw and cereals were also produced; wood was cut from the riverbanks, while enough was left to stabilize the slopes. After a flood, inhabitants collected driftwood, sand, and stones, sometimes selling their finds to neighbouring communities (id.: 10). Marshes were exploited as well, for pasture, forage, livestock bedding later used as fertilizer, wicker, or for peat extraction. Ancient documents sometimes named the plain between Sion and Martigny a *campagne* or *champagne*, attesting to its open and fully exploited character (id.: 14, 16).

Unfortunately, few general descriptions of the Upper-Rhône Valley were found before the mid-eighteenth century, since scientists and tourists mainly had eyes for the spectacular mountains, waterfalls, and other *curiosités* (Bender, 2001: 88). The description made in 1800 by Franz S. Wild, federal commissioner in the Valais from 1798 until his death in 1802, evoked the opposite of a rich campagne: 'Close to Naters, a large zone is totally devastated and occupied by the river during high waters season; downstream of the village, the land is marshy and produces bad hay.... Beneath the pleasant village of Glis are wide areas left marshy by laziness and inefficiency.... In the summer, from Loèche to Sierre, the Rhône covers almost the whole plain, besides a few miserable coppices and a few Island of poor profit. From Sierre to Sion, the biggest part of the plain is barren, covered by stones and marshes.... From Sion to Saint-Pierre de Clages, the plain is absolutely marshy and belongs to the Rhône riverbed; from this last place, until the Branson bridge, almost the whole valley is flooded and the stream threatens to wash away more of the few areas that are protected by scarce and meaningless defence works' (Wild, 1800, quoted in de Torrenté, 1964: 42–3). Wild's description can hardly be confirmed from contemporary maps, which were generally drawn for military purposes and were not too precise as to land use. The Swiss Atlas drawn between 1786 and 1802 shows indeed that the Rhône was then meandering and counted several secondary arms, but none of the historical maps can attest to the general devastation he describes (Weiss, 1797).

Furthermore, the objectivity of Wild's description could be doubted, not only because he omitted to describe the parts of the valley that were profitably exploited but foremost because he was himself a passionate promoter of diking, promoting the correction of the Rhône by the Vaud and Valais Councils as well as in Bern, where he presented his 'projet pour le diguement du Rhône' shortly after the Linth Korrektion project was revealed in 1783. To Wild, taming the Upper Rhône was a personal and patriotic mission: 'Which brave-hearted man with the power to do so, would not try joyfully and with all his strength to withdraw the pitiable population of a large part of the Rhône Valley from the evil influence of the devastating river? What friend of his fatherland and of humankind would refuse to contribute with all his heart to let our fellow countrymen from Valais get a taste from the sweet fruits of fraternity after all the pain they endured? May my appeal be echoed by all Swiss hearts; may it be approved by the government, and receive its support and protection!' (F. S. Wild, quoted in *Le Confédéré*, 17.09.1913: 1). Patriotic feelings, Christian charity, trust in progress and technology, Wild's vibrant appeal leaves no room for doubt; the correction was the best and only road to safety and prosperity, and the river, Valais' worst enemy, had to be channelled without delay.

The description of the valley made by Charles Lenthéric a century later in his monograph of the Rhône River expressed once again all the contempt of the progressive engineer for a stream unsuitable for navigation, unusable for industry, and inconvenient for intensive agriculture: 'In the long corridor of Valais ... the Rhône is nothing more than a dangerous and violent torrent. Man can at best use it as driving force. But the region is essentially pastoral and not industrial at all. Everywhere meadows, woods, poor commons [*vaine pâture*]. Merely a few sawmills.' Besides his contempt for the limited industrialization of the Canton, Lenthéric had no good word for the river itself: 'Man nowhere feels the need to take possession of water; it overflows [metaphorically: *elle abonde*]. The river is an enemy more than a support' (Lenthéric, 1892: 498–9).

In 1924, Walter Haenni again described a terrible pre-correction landscape before the audience of the third Rhône Congress, justifying the river correction retroactively. He wrote: 'The Rhône used to flow in countless meanders, causing terrible inundations and every year changing its bed', (Haenni, 1929: 39) probably exaggerating the wilderness and the freedom left to the river. And indeed Haenni wanted to celebrate the correction as a victory of the 'Valaisan People that was able to tame the Rhône and keep it in its bed', as the alleged past chaos had been replaced with a productive landscape: 'Valais, thanks to the works we just described, is now protected from flooding and the plain is transformed into fertile gardens with a bright future' (Haenni, 1929: 43, 45). But did the river change so radically between the late Middle Ages and the Industrial Revolution? Most probably the European Little Ice Age did influence the regime of the Upper Rhône, as it did for the lower Rhône (Pichard, 1995) and for other alpine rivers (Girel, 2005). Although the river surely influenced the landscape and the land use of the valley, it was not a vast and barren marshland (Staüble and Reynard, 2005). Historian Evéquoz-Dayen notes that 'the majority of publications dedicated to the Rhône upstream from Lake Geneva in the nineteenth and twentieth centuries are marked by the negative perception that was used to convince of the necessity to correct the river' (Evéquoz-Dayen, 2009: 47).

A later retrospective depiction of the Rhône before diking was given by Charles de Torrenté in 1964, in his monograph dedicated to the Rhône Correction, and published by the Federal Department of Roads and Levees. Echoing Walckenaer's theory, De Torrenté saw the exploitation of natural resources as the destiny and the duty of humankind: 'As the Rhône obeys immutable natural laws when it carries down its water and gravel and can contribute to this work only with its strength, it is up to man to use this energy in his best interests' (De Torrenté, 1964: 40). The dramatic description of the untamed Rhône made by Franz S. Wild was for De Torrenté an indisputable reference: 'To figure out the state of the Rhône before the correction, there is no better description than the one made in 1800 by M. Wild, director of the Salterns of Bex and Commissary for the Federal Government in a "Proposal for the diking of the Rhône and its arrangement into a navigable river from its mouth in the Lake Geneva to upstream Brigue."' To eliminate any further doubt, De Torrenté added: 'The official mission of the author confers an indisputable veracity to this description'—an utterly doubtful statement as when Wild wrote his own description, he also advocated his own correction project (see above), and did not seem to make use of various sources besides his own occasional observations. De Torrenté further enumerated the inundations that had been remembered in historical documents, mentioned the conflicts between neighbouring states over their fluctuating border, and described the measures taken since the Middle Ages to stabilize the river before a global correction project emerged in the early nineteenth century. However,

Raphael Ritz, *Correction du Rhône à Rarogne*, 1888 (Musée d'Art, Sion)

he fully ignored the activities that for centuries had taken place in the floodplain before the correction, simply noting that 'already [in 580] parts of the plain were cultivated'. But 'Man long remained very weak in front of the fieriness of the young Rhône', he noted, and this power play had finally been turned to man's advantage.

THE PROS AND CONS OF CONTAINMENT

National authorities were more inclined to promote, if not enforce, global and (supposedly) final solutions, rather than local and flexible ones; similarly, predominant rhetoric evoked the needs of the 'people', conceived at the national or regional scale, rather than the complex and sometimes conflicting interests of local communities. However, the very principle of river damming was occasionally doubted by a few renowned experts, such as Vierlingh in his 1577 Treatise (see chapter 'A European Self-colonization'). Pierre du Buat, a respected hydraulic engineer graduated from the French École Royale du Génie, author of a treatise on 'Principles of hydraulics verified by a large number of experiments made by order of the Government' (du Buat, 1786), was also critical. Du Buat was asked to examine the damming project made in 1786–88 for the Isère river upstream from Grenoble, which included eighteen cuts in the river's course and the creation of a second river arm south of the city (Champion, 1858–64, vol. 4: 77). His answer was an urgent warning against channelling: 'I received the document and the plan concerning the straightening of the Isère river.... I examined carefully the objections that could be raised about this project, the consequences it will have, and the imminent danger that would threaten the city of Grenoble. In my treatise published in 1786 (*Principes d'hydraulique*) I exposed in great details the risks of such enterprises; the risks are so great that, knowing the laws of river flows and the terrible effects they can have when one frustrates them, I would see myself as guilty of losing this city if I would in any way seem to approve the straightening of the Isère river upstream Grenoble ... and the construction of levees supposed to contain its swollen waters ... this project might seem right to people with little experience in hydraulics ... but I would advise them to build these dikes with metal, copper or iron, since I wouldn't even trust stone embankments such as those built for sluices and sea levees' (Du Buat, 1789, quoted in Champion, 1858–64, vol. 4: LXIII). Ahead of his time, Du Buat advised instead to make space for the river to fluctuate: 'I thus absolutely believe that one should abandon the straightening, and make do with a widening of the former bed, without damming it, following the old meanders' (id., LXIV). In his treatise, Du Buat conceded that 'there are several cases when nature seems to require the help of art, to moderate the course of rivers', but warns against overenthusiastic river diking, 'as all excess is harmful, it could happen that through straightening a river excessively, this would cause accidents as damageable as the ones to be prevented; for the speed of the flow, accelerated by the increased slope, could excavate its own bed and erode its banks, which would cause various damages' (Du Buat, 1786, 2, notes: LXV, LXVI).

In the nineteenth century, the period of the great European river corrections, local circumspection if not frank opposition was not unusual, as the rural economies still greatly depended on the production of the floodplain (Pautou et al., 1995: 38). The results of completed river canalizations were being discussed among engineers, who had in mind the spectacular diking of the Po River 'which, since its bed was narrowed between powerful levees, has raised so high that it now flows at the level of Ferrare's roofs' (Pigeon et al., 1845: 236). Permanent diking deprived the floodplain of valuable alluvium and made flooding less frequent but potentially more damaging. Ignace Venetz, co-author of the Rhône Correction scheme, former Chief Engineer of Valais and inventor of a new warping technique, wrote in 1850 that 'a weighty question is agitating flood-prone regions: that of knowing whether one should dike rivers and streams in the plain in order to prevent overflow even during high waters, or put no obstacle to this overflow' (Venetz, 1851: 1). He referred more particularly to the controversy generated by a publication by Marc-Antoine Puvis in 1844: *De l'endiguement des fleuves et des rivières* (On river diking). Puvis, graduated from the prestigious École Polytechnique, was at the time a respected agronomist, Deputy, and President of the agriculture society of the French Department of Ain, close to Geneva. His work was also known outside France, translated into English and German (Puvis, 1847), and he was a correspondent for the agriculture societies of Turin and Geneva. In his essay, Puvis praised the fertilizing effect of seasonal floods, which for him largely made up for their negative aspects, and denounced the longitudinal diking of rivers as disastrous, as it hindered natural fertilization, tended to raise riverbeds, and leave 'pestilential wetlands' behind (Venetz, 1851: 5). Puvis noted that he did not oppose levees in general, especially not those built to prevent erosion or secure the navigability of waterways, but he denounced levees 'that aim at creating an artificial bed for the river's high waters, and deny them the basin that they themselves defined as an additional riverbed, especially for these high waters' (id.: 7). He directly opposed the diking scheme made for the Saône River and indirectly criticized the general movement of river channelization that had spread across Europe, becoming an early supporter of mitigation strategies and 'soft engineering' over strict and fixed correction schemes. Learning from his own experience, he tried to generalize a hydro-dynamic model applicable to all rivers in all segments: streams collect sediments from mountains, deforestation increases their volume, diking keeps the sediments inside the main bed, leading to it constantly rising above the surrounding land, portending more dramatic floods.

Criticizing publicly the canalization of the Saône River, Puvis' pamphlet generated vehement reactions in his own Département and beyond. The Société Royale d'Agriculture de Lyon published a fierce letter from chief-engineer Laval, who was in charge of the correction of the Saône River, and a report of a review commission composed of no less than five members of the society. In a tone that left few doubts about his feelings, Laval first acknowledged the author's global understanding and agreed on his general considerations, to further denounce Puvis' arguments against the diking of the Saône, which he considered misleading and based on fantasy calculations. Laval remarked that when the river threatened to flood the land during summer, inhabitants built improvised dikes on the riverbanks, which were naturally heightened by the river's deposits. 'The levees I propose are thus, so to speak, just the regularization of a system already adopted by the inhabitants' (Laval, 1845: 225). The planned submersible dikes would still allow the fertilizing winter floods to overflow onto the fields; erosion during flooding would be avoided by letting the water flow from downstream into compartments; openings in the dike could also let alluvium in, if the dike blocked them out; drainage ditches should finally make sure that no swamp was left behind after the flood. Finally, Laval deplored the effects of Puvis' essay on local opinion, 'spreading suspicion and discouragement among the local population most concerned by their implementation [of the planned levees].'

A dedicated review commission was more nuanced in its comments but ultimately supported Laval's scheme. After a brief summary, the reviewers refuted Puvis' assumption that diking leads to the rising of riverbeds, 'a rare phenomenon only seen close to the estuaries' (Pigeon et al., 1845: 236). The constant threat formed by raised riverbeds such as the Pô could be averted by 'a tireless vigilance and well-directed efforts', the commission argued. Wetlands left behind the constructed levees were similar to those naturally generated by riverbanks raised by spontaneous deposits, and could be improved by warping techniques. Spontaneous floods were often welcomed for their fertilizing virtues but might also cover rich fields with a bed of gravel, ruin a harvest when occurring in the wrong season, or change the riverbed, making permanent land use impossible: In these cases, the authors would favour non-submersible levees. Confident that the range of techniques available could compensate for the downsides of diking, the commission supported Laval's project and believed in its success—if 'the construction [of the dikes] is conscientious', if 'the repair work is done timely', if the warping and drainage works are sufficiently funded, and if the locks are used 'with intelligence and coordination'—a combination that would rarely be achieved.

The proceedings of the special session held to discuss Puvis' book and its review especially revealed the lack of global and dynamic vision that then hindered a full understanding of the river. Various arguments were exchanged, which above all showed the difficulty of drawing general conclusions from fragmentary information, gathered from different contexts, with little hindsight on completed corrections. Contradictory conclusions could thus be drawn from the same examples. For instance, the total volume of carried sediments increased with deforestation and channelling, but the straightening of rivers actually sped up the streams and

The 1856 flood in the Loire Valley (*L'Illustration* no. 28, 3 July 1856)

accelerates erosion, some answered. Puvis himself, in his attempt to defend floodplains and rivers against full channelling, failed to find a universal hydrological model that could apply to all rivers. But, despite his errors, what made Puvis' text visionary was his description of an irresistible chain of decisions, events, and reactions that inevitably led to full and permanent diking: 'One will start with a combined system of submersible dikes, but will soon be forced to raise them, seeing the fields badly preserved and losing their fertility; waters filtrating through the locks will rust the hay and spoil the harvest; these locks will be removed and one will make the dikes insubmersible, then raise them again as the riverbed will rise continuously, like in Italy or on the Rhône River' (Puvis: 1844, quoted by Laval: 1845: 228). To a large extent, Puvis' scenario came true for many European rivers and for various other reasons: partial clogging, increased flood due to systematic channelling upstream, deforestation, soil sealing, disinterest for alluvial fertilization, increased exposure behind the levees.... His message was lost, however, in the collective fear of more damaging floods and the faith in technological perfectibility.

However, four years later, Napoléon III encouraged the reclamation, land clearing, and irrigation of the French plains in a letter to his State Minister Achille Fould: 'Clear forest located in the plains and reforest the mountains, dedicate each year considerable sums to great works of reclamation [*dessèchement*], irrigation, and forest clearance. These works, turning uncultivated commons into cultivated land, would enrich municipalities without impoverishing the State, which will recover its advances by selling part of these land returned to agriculture' (Bonaparte, 1860: 46). The *Journal de l'Agriculture Pratique*, one of the many that published the letter, reacted enthusiastically to the Emperor's programme: 'The cities will no longer be the only ones to benefit from great public works, life will flow through the countryside. France's destiny shall be fulfilled.' However, the French central government has not always appeared so univocal, especially after the 1856 inundations that washed away the levees that had just been heightened or repaired. In a public letter to his Minister for Public Works, Napoléon III appeared very critical of the levee systems pretending to prevent all floods and that ultimately lead to more damage, as seen along the Loire and the Rhône in a series of major inundations (Bonaparte, 1856).

CORRECTING THE CORRECTION: ILLUSION OF PROGRESS AND NEVER-ENDING ADJUSTMENT

Commenting on the optimism and the self-confidence of reclamation defenders, David Blackbourn remarks that 'the reality was that (taking the Oderbruch once again as an example) none of these supposedly definitive solutions ... was able to prevent floods that were now a threat to the work-cycle rather than a part of it' (Blackbourn, 2000). This statement could be extended to just about all reclamation works and river canalizations, which did not all succeed or fail, but did all lead to an endless chain of adjustments and further land transformations that were generally neither announced nor foreseen.

Besides the drowning of the Dutch Grote Waard mentioned earlier, one of the most resounding failures is no doubt the attempt on the part of the Swiss Federation to divert the Kander river in the Berner Oberland region in 1713–14. In order to protect the *Allmende* (commons) of Thun from regular flooding, a tunnel was dug to divert the waters of the waterfall towards Lake Thun. The tunnel collapsed soon after completion; instead of regulating the river's discharge, the diversion led most of the river flow into the lake, rapidly raising its level. The road to Spitz was closed, the city of Thun was flooded several times, and it would take decades and massive defence works to protect the city and the lake area against the irreversible consequences of this failed attempt to correct nature (Walter, 1990: 34).

Even for those correction works that seemed to go as planned, not all their goals were achieved nor all consequences foreseen. A posteriori, it appeared that the global surface of wetlands did not necessary diminish after the channelization of the river, as tributaries had trouble finding their way towards the 'corrected' river, which would locally lead to rising ground water, growing marshes, and sometimes to increased flooding along these tributaries and behind the dikes. The correction of the Linth, envisioned by German engineer Tulla and completed by Swiss geologist Hans Konrad Escher, did not improve production at the expected rate, and 'in fact, to achieve this objective, several additional projects of land drainage had to be carried out, the last of which was not completed until 1964' (Speich, 2002b: 434). The inhabitants of the Isère Valley in Savoy, and those of the Rhône Valley in Switzerland, saw marshland expand and malaria spread locally during and after the diking works meant to achieve just the opposite (Puvis, 1849: 93, and Staüble, 2009: 172–174). Around the Jura lakes, the productivity of meadows decreased after the first correction, completed in 1891, and drainage caused the land to sink rapidly, more than a metre in some places (Walter, 1990: 73). Retrospectively, there is nothing surprising in these unforeseen consequences, considering the scale and complexity of these enterprises and the limited knowledge and modelling possibilities available at the time. More intriguing are the promises of eradicating flood risks and fluctuations that could never and would nowhere be made true, since the assumption that a landscape and a hydrological system could be stabilized, if not petrified, later proved to be totally wrong.

Samuel Bodmer, Plan for a diversion tunnel between the Kander River and the Lake Thun, 1710 (Bern State Archive)

Officially, the canalization of rivers was still celebrated as a triumph of man's technology over nature's needless fluctuations, the achievement of an entire people and its engineers. David Blackbourn notes, however, that the 'great illusion of the "improvers" was the belief that they had found a once-and-for-all technical solution' (Blackbourn, 2007: 74). Despite the progress of cartography and civil engineering techniques, engineers still lacked the knowledge of hydrodynamics and hydromorphology that could help them predict the long-term and large-scale consequences of their interventions. In their enthusiasm and faith in the superiority of man's rationality vis-à-vis nature's purposelessness, engineers often failed to consider rivers as being part of complex and dynamic ecosystems. Often, the canalization of an isolated segment exposed the downstream regions to greater and faster floods. The concentration of deposits in a narrowed profile led to the rise of the river, silting its own bed above the surrounding land, making every dike breach even more devastating. Catching materials upstream in weirs, dams, or retention lakes led to accelerated erosion downstream and, eventually, to the sinking of the riverbed and the subsequent weakening of the levees. The diking of rivers hindered the flow of its tributaries and the natural drainage of the plain. So, each wetland reclamation and river correction rapidly created the problems that would lead to the next correction, a correction of a correction, initiating the spiral of technical intervention feared by Puvis.

In 1929, Swiss Engineer Haenni was still praising the First Rhône correction, quoting an anonymous expert: 'The success of the [Rhône] Correction is resounding and the speciality of the system has proved itself.' Although he went on to acknowledge the multiple negative consequences of the river's channelling and shortening, that simply 'demonstrated that the Rhône Correction had to be completed with the Correction of the (tributary) streams' (Haenni, 1929: 41–42). On the French side of Mont Blanc, the local newspaper *Dauphiné Libéré* used similar words forty years later to praise the opening of a new straight channel cutting off a three-kilometre-long meander of the Isère River, as another final victory over the river's wilderness character: 'There was a time when the Isère would go its own way and be a headache for the agglomeration of Grenoble. Today the hand of man dictates the way, and she surrenders' (*Dauphiné Libéré*, 1968). Despite the successive cycle of disasters and corrections (which to a large extent still goes on today), the belief that a holy battle between rivers and technology was going on and that technology would eventually triumph, could still be found until quite recently. The legitimacy of the engineering as such has, until recently, seldom been questioned. 'How people reacted to these reverses has changed over the 250 years. But for most of that time it was the specific remedies that changed, not the underlying assumptions' (Blackbourn, 2007: 68).

Water, the most malleable of all elements, allowed man to measure his power against the creator's. 'God created Valais, but the Valaisans made it inhabitable by creating the aqueducts (*bisses*) that carry the sacred waters', states Professor Schöter from the Société Valaisanne des Sciences Naturelles (quoted in Haenni, 1929: 42). A perspective brought to a paroxysm by the Dutch with the often heard saying 'God created Earth but the Dutch created Holland', regularly used by Dutch landscape architects to promote and export their collective skills. The world-famous Delta Works completed in Holland in the 1980s and sanctified as one of the 'seven wonders of the modern world' by the American Society of Engineers, fit in the tradition of massive infrastructures meant to provide a final answer to flood risk but rapidly criticized for their lack of flexibility and their underestimated side effects. Yet this Promethean vision of flood defence, still partly supported, is today in contradiction with a growing uncertainty about future hydrological conditions, which should inspire flexible and cautious interventions rather than heroic infrastructures.

The name 'Third Correction', recently chosen to designate the programme currently applied to the upper Rhône river, stresses, ironically enough, that not only Nature's oeuvre but also man's work needs correction, since this new enterprise also involves not only a correction of nature but also a correction of the two previous corrections. The term also shows the difficulty of breaking with centuries of technological triumphalism, despite the adaptive and integrated character of the new project.

ENVIRONMENTAL GROUPS AND THE ENGINEERED LANDSCAPE

As mentioned above, river correction projects were in their time discussed on technical grounds, economic efficiency, and for aesthetic reasons. The environmental aspect, however, did not seem to play a significant role in the discussion until recent times. In Switzerland, the first example of a collective consciousness of destruction of nature was provided by the fight against deforestation, begun in the mid-nineteenth century. The erosion, the increased run-off, landslides, and plain flooding were then identified as the direct consequences of the deforestation of the rivers' catchment areas. But the strong mobilization that followed in Switzerland as well as in France with the *Restauration des Terrains de Montagnes* above all indicated an enlightened self-interest: the massive reforestation campaigns were not begun in order to restore natural habitats but primarily to prevent flooding, the most feared hazard in the Alps, and to secure wood production for fear of the exhaustion of a crucial resource (Walter, 1990: 67, 77).

The *Heimatschutz* and *Naturschutz* movements later emerged in Switzerland as political forces in the first decade of the twentieth century, almost simultaneously with the German Bund Heimatschutz, the French Société pour la protection des Paysages, and the British National Trust for Places of Historic Interest or Natural Beauty (id.: 115). The nature these movements defended was, however, still fundamentally different from contemporary concepts and also differed among countries, as each nation tried to define and differentiate its singular identity. History, archaeology, and geography became political sciences, often manipulated to build or confirm national myths and bind the populations around a

1968: straightening of the Isère River (*Dauphiné Libéré*, 02.02.1968)

reconstructed common past, generating a new interest for the history of the Frisian werd in Holland or for the Alpine lacustrine settlements.

In these turbulent times, the Swiss Heimatschutz fought to protect that first concept of nature from the zeal of the conquerors, since the picturesque Swiss landscape should be spared the trivialization imposed by new infrastructures (a criticism already made by Rousseau a century earlier) and commerce (with the development of road advertising), engaging in a definition of what should be considered as valuable. This process led naturally to the sacralizing of identifiable and spectacular landscape features, such as waterfalls, glaciers, peaks, and rocks. The first successful campaign of the Heimatschutz movement came with the preservation of an erratic block in Monthey in 1905, which was saved from being split and sold as construction material after a fierce campaign (Walter, 1990: 120). Eventually, authorities at all levels joined forces to buy the erratic rock: Federation, municipality, and the canton of Valais, the same canton that in 1907 'in the name of God Almighty!' engraved the funding of the Rhône and river corrections in its new constitution, still valid today (Etat du Valais, 8.03.1907). Erratic blocks and erratic rivers obviously belonged in two opposite categories. Not picturesque enough in its liquid form, and not suitable for navigation, the Rhône fell between the gaps of emerging cultural concepts, and long remained an insignificant element in the Valaisan landscape.

The interwar period constitutes a particular period in the cultural history of landscape. The sacralizing of national items was stronger than ever, but the objectification of nature was simultaneously being reinforced by economic depression. In 1914–18, European countries underwent the painful experience that, just as globalization could accelerate things, it could also rapidly crumble, returning countries to their former isolation,

Corrected river and railway running parallel through an ideal pastoral landscape (Hugo d'Alési, poster for the Chemins de Fer Jura-Simplon, 1895) • The Rhône valley after the correction (postcard, ca. 1919)

making food and energy autonomy a central item in national policies. In Germany and Switzerland, various initiatives called for an 'interior colonization', which included the planning, definition, and expansion of arable lands (Walter 1990: 155–159), and which in the Alps could mostly be made at the cost of remaining wetlands and floodplains.

In this political context, patriotism justified damming and reclamation, and opposing productivism came close to homeland treason, advocating dependency and surrender instead of securing national autonomy—an argument still heard today in the discourse of the movements opposing the widening of the Rhône. The ideals of environmentalists and agro-productivists occasionally joined forces to oppose hydroelectric dam projects, which were accused of destroying precious farmland and disfiguring the alpine pastoral landscape for the sake of the foreign energy market. However, this particular case aside, the transformation of almost all Swiss rivers and the subsequent quasi-disappearance of wetlands has long remained barely mentioned by conservationists, as well as the extraction of river materials that fuelled a century of urban sprawl. At a time of rapid urbanization, the Swiss Federation paradoxically defined its own identity around the alpine pastoral figure, through an image of cultivation and humanization of a beautiful but fundamentally inhospitable landscape. Two main categories of nature remained: on the one hand, the nature man should protect and venerate, the pristine and the overwhelming, only accessible by alpine heroes (and later cable cars); on the other, the nature humans were collectively entitled to tame and transform, ameliorate, and exploit for their own survival, displaying their genius and technical skills through the excavation of tunnels, the damming of rivers and the reclamation of wetlands—each dike breach or devastating avalanche being a lost battle in the long war against the wild.

Not all Western European countries obviously shared the Swiss representation of rivers, wetlands, and nature in general. The sacralizing of landscape features above landscape structures and the objectification of productive nature, however, was commonplace throughout the eighteenth, nineteenth and twentieth centuries. Everywhere, protection movements have not only concentrated on mounts, mountains, and cliffs but also on coasts and islands, for instance, with the case of the Wadden Islands in the Netherlands and Germany, or the coast of Normandy in France. Plains and valleys, in contrast, remained domestic space, where safety and functionality prevailed over cultural, ecological, and aesthetic value.

Paradoxically, the revaluation of rivers and floodplains at the end of the twentieth century seemed to be concomitant with urban sprawl and the progressive disappearance of large open spaces and natural fluctuations within metropolitan areas. Digging their way through frozen urban landscapes, insignificant rivers become particular and potential environments of ecological value, visual enjoyment, recreation, and identity-building again. The case of the Isar River in Munich is illustrative of this late evolution, as the river's status evolved from water machine to ecological corridor in the late 1970s and early 1980s, while environmental concerns also grew in the state's capital. New conflicts surfaced, as flood defence infrastructure was increasingly opposed as being destructive, and renaturation ambitions stumbled over the intrinsic variability of river landscapes and the risk this implies for vital urban infrastructures. It is thus in cities that environmental groups first advocated renaturation, often to be disillusioned by the difficulty in restoring free flow in a dense urban environment. Against the faith in absolute safety through technical know-how, '[t]he opposite snare of the green-minded is the belief in a once-and-for-all state of nature'. (Blackbourn, 2007: 74) Between contradictory demands of secure flood defence and 'free' streams, new ambitions of balanced and integrated flood policy could yet arise that changed again the perspective on riverine landscape and natural risks.

2

IN SEARCH OF COMBINATORY LANDSCAPES

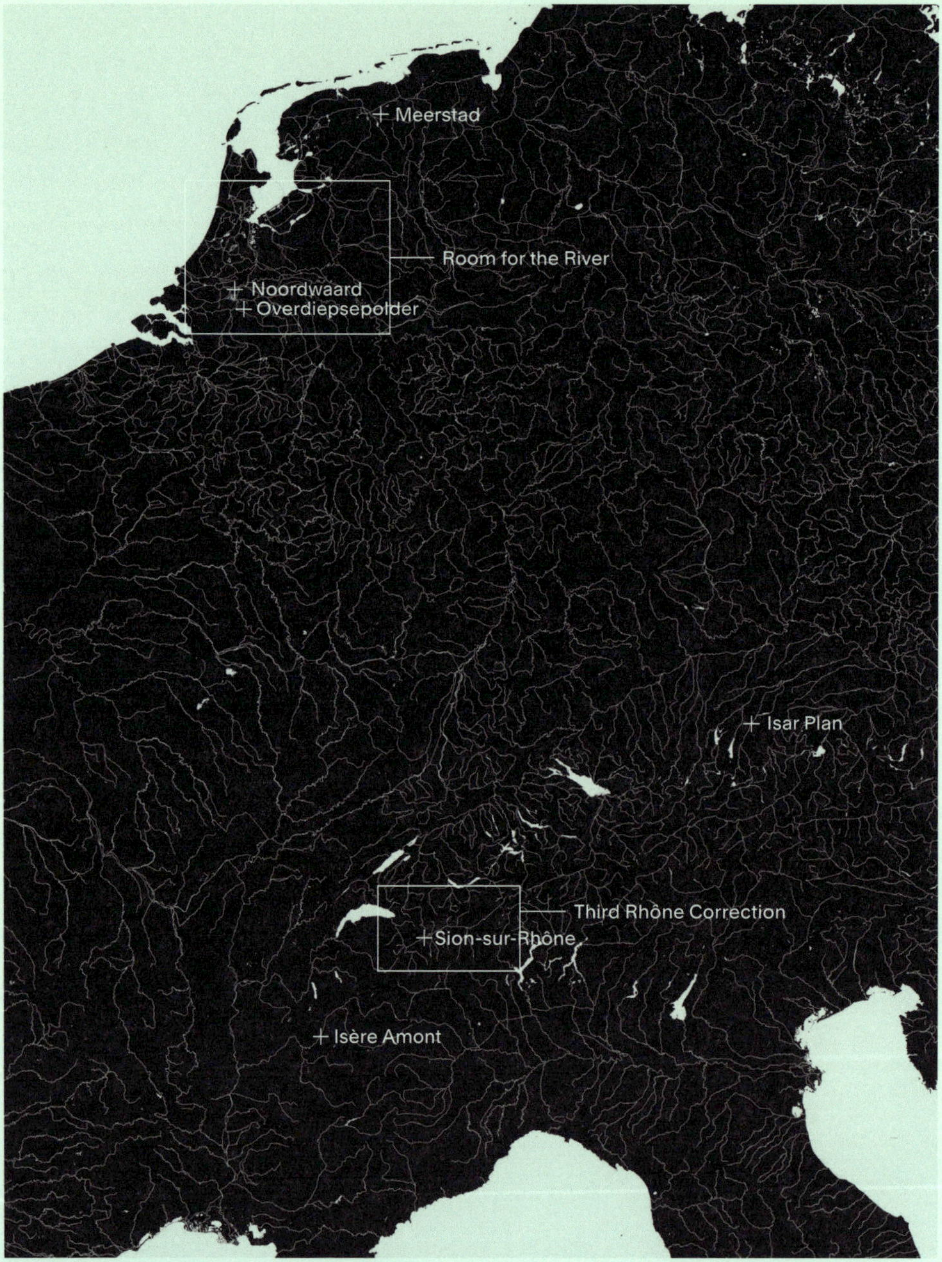

Location of the investigated projects in the Western Alps and the Dutch Delta

> 'The question to be posed about the current restoration and restructuring of river systems in towns and cities is: How can the multifarious functional demands on the design of urban water spaces be combined? How can these demands be reconciled with the natural internal dynamics within the water itself?'
> (Prominski et al., 2012: 15)

As seen in the first part, flood processes have long been present in landscape architecture, not only as a threat but also as agents of creation, as forces that could be shaped, diverted, captured, used, filtered, or spread. To a large extent, today's Western European landscapes still materialize the implementation of the 'modern' dream, modernity here being understood as the subordination and the organization of natural elements and forces into a permanently productive state, a state benefiting humans or at the very least harmless to them.

In the course of the eighteenth, nineteenth and twentieth centuries, most Western European streams and estuaries have been turned into sterile and petrified waterways, allowing for an unprecedented economic surge and a widespread ecological disaster. Yet as the Emperor Napoleon III noted in 1856, there is 'nothing easier then erecting civil works that temporarily protect cities like Lyon, Valence, Avignon, Tarascon, Orléans, Blois, and Tours from such floods. But as for the global system needed in the future to keep our rich valleys crossed by these great rivers safe from such curses, that is still missing and that we need to find promptly'. This open critique of traditional flood protection eventually led to an early 'room for the river' programme: the 1867 Comoy Plan, which today inspires a new approach to flood management. Contemporary river design tends more and more to facilitate and control inundation instead of excluding it, and, in this way, finally reintroduces flooding as an agent of territorial structuring and perception.

The following essays thus focus on three contemporary projects that have in common the reintroduction of flood within territorial process and spatial planning. An in-depth analysis of cultural motives and planning strategies reveal a set of combinatory tools applied in various geographical contexts, on a regional or metropolitan scale. Flood scape as ecological restoration, cultural statement, urban development, integral makeover, or discreet adaptation: this overview unravels the vast scope of possible approaches and combinations, providing tools and reflection for future landscape transformations.

In all the investigated projects, water dynamics have become an element of design instead of a problem to solve. The project Isère Amont in the French Alps creates a ribbon of controlled flood fields along the river, which gives it an extra 3000 hectares to expand into, in case of extreme high water. The Meerstad project merges urban extension and creation of a 600-hectare retention lake into one combined development. The canalized Isar River in Munich has been transformed by removing a large part of the straight embankment and restoring the accessibility of its floodplain to the population of the city, while improving its functionality in case of high water.

All projects no longer rely on vertical solutions only (dams, levees, pumps) to address flood risks but rather on horizontal solutions (widening, deviation, temporary storage). As seen in the previous chapters, this approach is not totally new: restoration of floodplains and integrated water management were already discussed before. Some of the principles applied today for this programme are indeed 'a rediscovery of time-honed practices that deserve the "traditional" label even more than dike building'. (Warner et al., 2013) However, the Loire long remained a rare example of such a strategy being implemented on a grand scale, and still inspires contemporary design, including the visionary 'Stork Plan' that prefigured the Dutch 'Room for the River' programme.

Landscape architecture today faces the challenge of reintegrating uncertainty into carefully designed living environments, in a perspective that includes possible short and dramatic events (floods, storms, landslide), as well as slow but potentially distressful processes (rise of sea level, salinization, increasing amplitude of river discharge, decrease of winter ice storage, drought). From this perspective, the physical, aesthetic, and philosophical definition of a 'healthy landscape' should today be reconsidered in a systematic way to allow for a flexible approach.

How to adapt the living environment so as to reach such a flexible and dynamic state, and how to make these qualities last and still remain adaptable to change? Creating spaces for temporary flood within urbanized regions once again creates the opportunity for the urban population to experience the reality of natural processes—provided these spaces are designed to actually be visible and accessible. Within an urbanization process still based on underground drainage and containment of water dynamics, these renewed answers to inundation threats reintroduce physicality and positivism, and open up perspectives for future living environments.

PROJECT	RIVER	CONSTRUCTION / PLANNING STUDIES	MAIN GOALS	MAIN MEASURES
Programmes				
Third Rhône Correction (CH)	Rhône	1995–2008 2006–2030 / 2050	Flood safety Restoration / consolidation of ecological, social and economic functions of the river	Widening of river segments and confluences Reinforcement levees Dredging of riverbed Creation of emergency flood corridor Creation of recreational routes
Room for the River (NL)	Rhine and Meuse branches	2000–2006 2007–2016	Making space for river Increase spatial quality	Dike relocation Flood bypasses Lowering floodplain green rivers Lowering of groynes Removal of obstacles Reinforcement levees Dredging of riverbed
Projects				
Sion-sur-Rhône (CH)	Rhône	2011–	Increase flood safety Urban renewal Public space development	Reinforcement of levees Dredging riverbed Creation of new park and promenade Building of urban riverfront
Isar Plan (DE)	Isar	1998–2006 2006–2010	Increase flood safety Development recreational uses Ecological restoration	Reinforcement of levees Softening of river embankments Restructuration of floodplain
Isère Amont (FR)	Isère	2004–2009 2012–2016	Lower river level at high discharge Development recreational uses Ecological restoration	Reinforcement of levees Local widening river channel Creation of emergency flood fields Creation of recreational routes
Ontpoldering Noordwaard (NL)	Waal and Merwede	2003–2011 2012–2016	Lower river level at high discharge Increase spatial quality Development recreational uses Ecological restoration	High-water diversion channel Lowering of levees for emergency flood mitigation Creation of wetland Creation of recreational routes Farms displaced onto elevated mounds
Overdiepse Polder (NL)	Bergsche Maas	2000–2010 2010–2015	Lower river level at high discharge Increase spatial quality	Lowering of levees for emergency flood mitigation Farms displaced onto elevated mounds
Meerstad Groningen (NL)	Eems-kanaal	2002–2011 2006–	Urban extension Nature development Creation of water storage capacity for flood situations	Creation of lake for recreation and water retention Housing development Creation of new nature areas Creation of recreational routes

Case studies overview

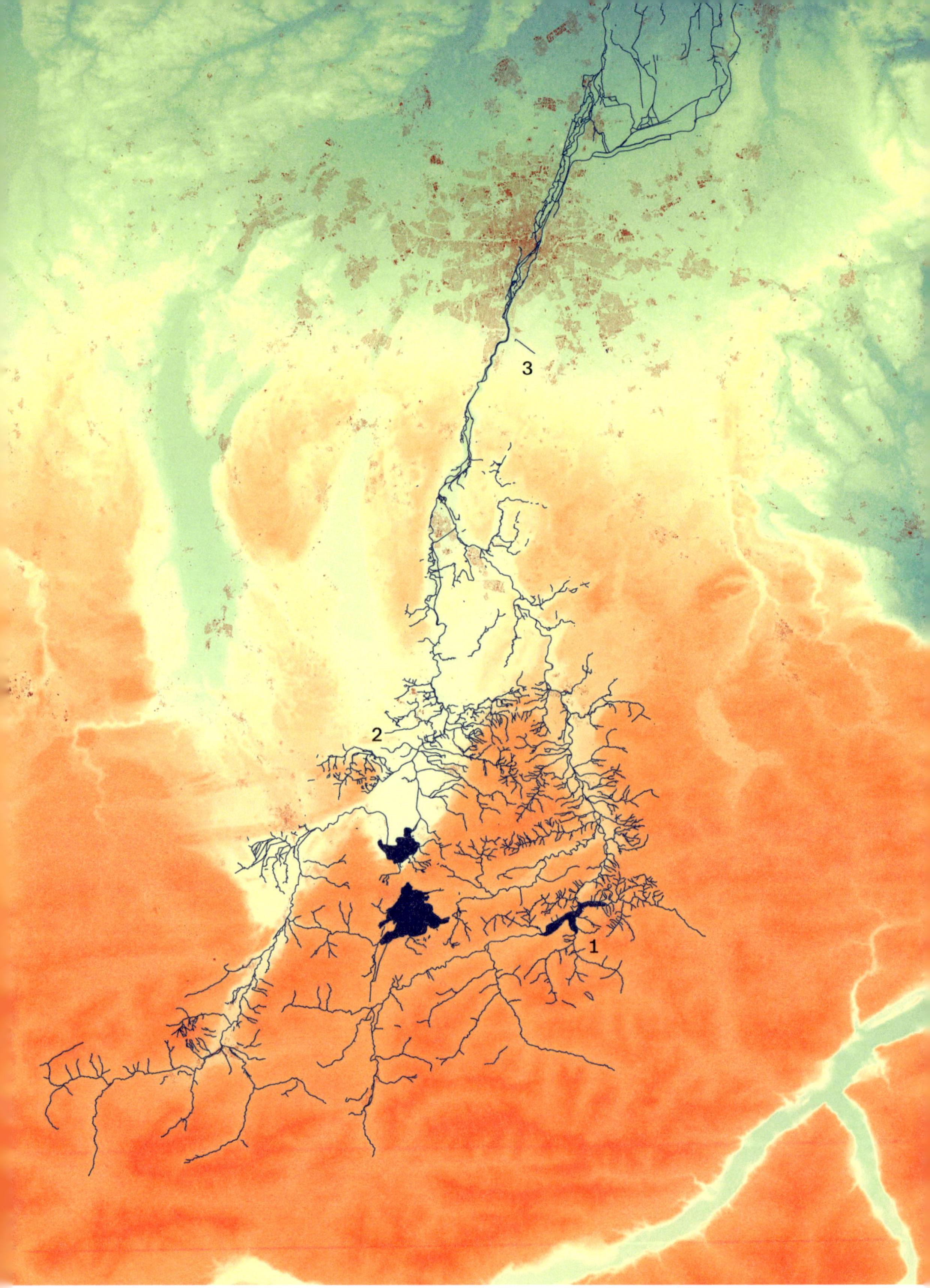

The Isar watershed upstream Munich: Sylvenstein retention lake (1), Loisach bogs and swamps (2), narrowed corridor through Munich urban area (3)

THE FLOODPLAIN AS URBAN PARK: ISAR PLAN, MUNICH

As seen in the first part of this book, the shift from a vertical containment approach to horizontal strategies was historically associated with the empowerment of ecology, particularly in Switzerland where flood-risk management has merged with the renaturation ambitions of environmentalist movements and federal authorities. This double agenda was found as well in neighbouring Bavaria, and brought a step further with the development of the Isar Plan in an urban context, right in the heart of the state capital Munich. While Swiss law strongly encouraged the softening of river banks and the widening of the streams in rural regions, the urbanized areas were mostly left out of this movement. In Munich, the Isar Plan began outside the city, but was consistently pulled all the way to its very centre, and is now planned to expand further downstream along the historical Englischer Garten. The Isar Plan could be called the star among river-restoration projects: widely published, internationally acclaimed, and, last but not least, much loved by the inhabitants of Munich, who on warm summer weekends come by the tens of thousands to enjoy the floodplain. The Isar can thus be considered as one of the most comprehensive floodplain restoration plans so far implemented in a Western European city, a plan that closely associates flood management with ecology and urbanity, addressing what may be the biggest challenge in the field. The Plan has so far involved only eight out of the 295 kilometres that make up the river, and a bit more than half of the river's length within Munich's municipal borders. Although limited in length, the new Isar has become an iconic space for the city, one that regularly advertises its quality of life through idyllic pictures of the summer crowds sunbathing along the river and athletic young people swimming and surfing over the river's streams and canals.

However, before being an iconic space, for a long time the river was an infrastructural element at best, and more often than not an enemy of the city, economically valuable for mining, transport, and power, but feared for its violent summer floods. At the time, it was hard to imagine that the floodplain south of the Museum Island could ever assume such a prominent place in the city's mental geography, comparable in popularity to the Englischer Garten created in the late eighteenth century, north of the city.

'How the tame Isar [river] becomes wild and beautiful' (Lieckfeld, photo Kapitza, GEO special, 2003)

The Isar is about the sole distinctive feature in the relatively monotonous relief of Munich's post-glacial plain, made of shallow slopes of limestone pebble with the Alps distant at about 80 kilometres further south. Yet the city never developed a representative riverfront or riverbank promenades, unlike most European capitals. The Isar was long prone to flooding: the Loisach, the Isar's main tributary, was partly naturally damped by the bogs surrounding the Kochelsee (Böhm and Wetzel, 2006: 787), but the upper Isar itself had no natural buffer to temper its violence before it reached the plains. Before modern dams curbed its fluctuations, the river showed the typical character of alpine rivers: strong erosion and massive material and sediment transport leading to frequent changes in the river's morphology; and high seasonal fluctuations, with increased discharge and velocity in summertime, with most of the major floods occurring in that season (id.: 791).

Besides being occasionally dangerous, the Isar has also long been unsatisfying to the eye, as Christiane Thalgott, Munich's former chief architect, says. The city 'has no urban waterfront because the Isar ... is not always a beautiful river', and, more precisely, it shows a unstable and chaotic image, with its bed sometimes almost empty and sometimes furious, with dead trees being carried along the stream through the city (Thalgott, 2013, interview). Munich thus long maintained a safe distance from the turbulent alpine river, as the location of Eastern Gate, the Isartor built in 1337, shows: it kept the Isar's main stream at a distance of about 400 metres. Four hundred years later, the plan drawn by Maximilian de Groth showed a still sparse urbanization between the Isartor and the stream, although the river had clearly been contained by a dam on its western side. On Von Riedl's 1802 map, the city still had no direct contact with the water. The Isar, however, played an important role in Munich's economic development: as a source of limestone collected in its wide bed, as a means of transport for alpine forestry and for passengers towards the Danube and Vienna, and later, with the development of power plants, as a source of energy. Several water mills placed on the side branches, as well as factories developing in the floodplain announced the coming industrialization of the city, and flood defence works slowly constrained the stream.

By the end of the eighteenth century, several dams and weirs had been constructed along the river, and segments were stabilized by light structures such as wooden faggots weighted down with stones and gravel (Rädlinger, 2012: 70). However, these various interventions were not yet following a general containment plan, which would come only at the end of the nineteenth century. In the 1840s the river banks were reinforced, and the floodplain between the Eisenbahn and the Reichenbach bridges was turned into a continuous stretch of meadow laid parallel to the river, flooded during spring/summer high waters, or else isolated from the stream by a steep stone bank. Through the Isar Vorstadt, the river was channelled and the floodplain was constrained by continuous levees, which brought

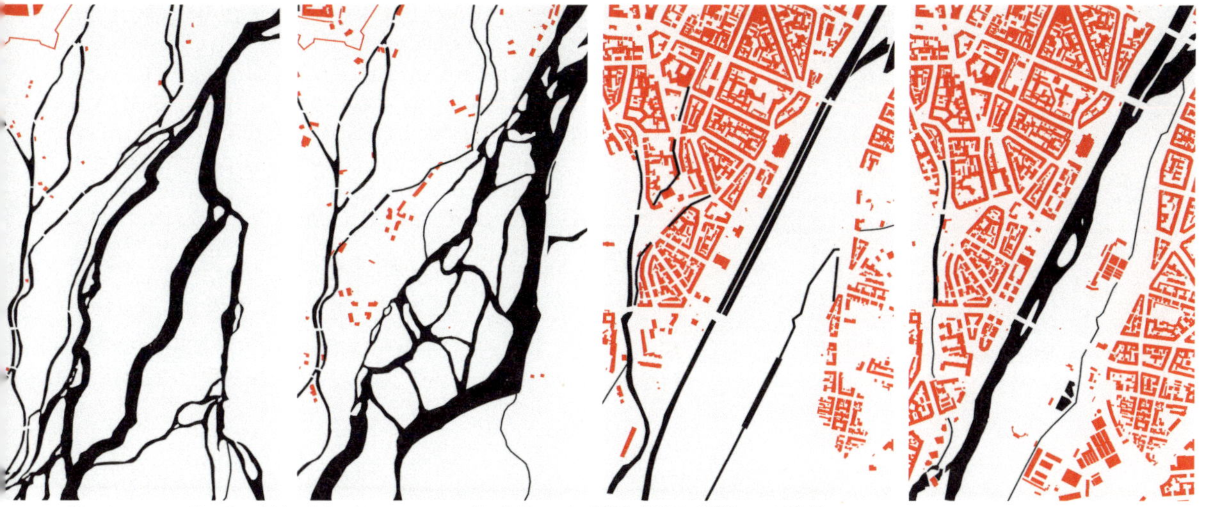

The *Isarraum* South of Munich: river arms and buildings in 1704, 1808, 1891, and 2011

it from about 500 metres wide down to less than 200 metres. A 'flood space' was defined, which gave clear borders and a fixed shape to the river and its floodplain, but a 'floodscape' remained absent: no particular attention was given to spatial quality, potential usage, or aesthetic value, nor to the relationship between river and city—this despite sporadic calls from local associations to allow more recreation in the area (Düchs, 2014: 17).

In the meantime, the lower Isar, downstream from the city centre, had given birth to one of the world's largest and most famous parks: the 'Englischer Garten'. Built into the river's floodplain, the English Garden diverted, and reshaped several branches of the river into a carefully calibrated water system, civilized, cleared from its alpine wilderness. South of the old city, on the other hand, the first meaningful urban intervention within the Isar landscape only came about after the devastating 1899 flood, with the construction of the Museuminsel in the middle of the stream. But there again, the water was not made accessible to the public; on the contrary, the Island was fortified against the river's fluctuations, and heightened to offer a view down upon the river from the new belvedere. The crowds passing the new Cornelius bridge could from that point on look down on the Isar like into a bear pit, with the river still roaring but kept at a safe distance. The former Coal Island, a place of industry and trade within the river's floodplain, dependent on its flow and exposed to its floods, was turned into a monumental ship of stone pointing its bow towards the Alps.

The various transformations clearly divided the Beautiful from the Functional. On the one hand, Nature's harmony was celebrated within the boundaries of the Englischer Garten, which integrated and reshaped secondary channels into an idealized wilderness. Benjamin Thompson's romantic layout recreated an alpine landscape there, made up of quiet streams, permanent waterfalls, and clear lakes; outside this microcosm, however, exploited river branches were dammed and turned into straight narrow canals, with the main riverbed being almost drained dry in order to feed the power plants, and bordered by steep levees reinforced with

Munich, collapse of the Prinz-Regente Bridge during the 1899 high waters

stones. The channelling of the Isar's southern section was so strict that it would later be referred to as a *Korset* by the promoters of 'renaturation' works.

The taming of the Isar became fully effective with the construction of the Sylvensteinspeicher dam, in 1959. The new retention lake since then curbs the variations of the river, and no summer peak comparable to the 1940 one has been seen since its construction. Dams have furthermore largely hindered the natural transport of materials and sediments from the Alps to the plains, while channelling and straightening have petrified the stream into a stable shape, that no longer reminds one of its alpine origins. This lost authenticity and hard-won permanence was precisely one of the reasons that would later motivate the nature-like transformation of the river.

FROM WATER MACHINE TO ECOLOGICAL CORRIDOR: FIRST RENATURATION PROPOSALS

Despite its lack of aesthetic quality, public amenities, and elaborated maintenance, the Isar floodplain was already used in the post-WW2 years by local young people, who actually appreciated the absence of control and the 'unregulated freedom' (Düchs, 2014: 15). In the late 1970s, environmental concerns grew in Bavaria's capital. The population had increased from about 832,000 in 1950 to 1,315,000 inhabitants in 1975 (Statistisches Amt München, 2012). Munich's rapid growth and high per capita land use

The Isar in its *Korset*, seen from the Deutsches Museum: A monotonous channel and meadow designed exclusively for flood management (1928 and 1939)

left little room for public green spaces, and 'parks are rather scattered islands, except the river Isar floodplain' (Oppermann and Pauleit, 2005: 79). Enriched by its flourishing car industry, the city also suffered from air pollution and was regularly covered in smog, with no natural winds evacuating the gas emissions. Furthermore, only two open corridors remained to bring a bit of fresh air into the city: the east-west oriented railway, and the northeast-southwest oriented Isar—making the latter an important factor in pollution dispersion, and a space for promenade, helping the inhabitants live with Bavaria's hot continental summers. Not yet a public space, the Isar floodplain south of the centre was acknowledged as a key space for the city's environmental and living quality, a knowledge that generated discussions around its potential redevelopment.

In the early eighties, environmental concerns became a topic in the local democratic institutions. In 1982, the Green Party made its notable entry into the Bavarian Parliament, winning almost five percent of the votes in its first appearance in the state's election (Bayerisches Landesamt für Statistik und Datenverarbeitung). It reached almost eight percent in the Munich municipal election held in March 1984. In those years, the city officially took the first steps that would eventually lead to the transformation of the Isar. The 1983 Urban Development Plan had already set the reshaping of the Isar into a 'natural way' as a goal, as well as the improvement of its hydrological regime (Referat für Stadtplanung und Bauordnung, 1983: IX. 3.2). In July 1984, through the motion 'Nature in the City', the Christian Social Union (CSU) stressed the fact that the Isar 'has lost its character of alpine wild river' within the city, and suggested eight measures to restore its natural character 'as far as possible' (CSU, 1984). The proposed measures carefully maintained the functionality of the river as a source of electrical power and as a tamed stream that needed a flood-proof riverbed, but suggested removing all obsolete infrastructure that hindered natural development. Although the Isar's potential as a public and recreational space was not yet explicit, the multi-purpose character of the project was already stated. The motion also formulated a perspective on nature that would remain a key feature in later development: a combination of voluntarism, protection, and restoration of an 'authentic' natural landscape through expanding the space available for erosion and sedimentation, and through the removal of planted alien vegetation. Although concise and precise, the motion was not free of romanticism as it suggested abandoning the winter bed's gravel banks to the 'free play of Nature's forces.'

In 1984, an alternative approach to cultural/ecological restoration was proposed by Günter Grzimek, designer of Munich's Olympia Park. Grzimek, who promoted user-oriented landscapes and a structure of 'connected de-concentration' (Jerney, 2003), began an initiative to promote 'useful technical landscape', with the Isar floodplain being potentially one of those urban multi-purpose spaces (Oppermann and Pauleit, 2005: 75). His proposal, however, raised a lot of protest 'by those experts who held up the qualities of a well-designed formal green space' (id.), and could not compete with the *Alpenfluss* restoration rhetoric.

The 'Natur in der Stadt' motion was followed by a second and more developed one in May 1985, this time proposed to the City Council by the Social Democrats from the SPD. The motion was entitled: 'Isar-Plan. Protection and restoration concept for the preservation and the development of the Isar area, meadows, and river banks.' The Isar was described as the main structural element in Munich's growing urban area—baptized 'Isarmetropole' for the occasion. The modifications brought to the river after 1900 were openly criticized for their purely technological stance: 'Not only did it impair the urban climatic function [of the river], it also impoverished the flora ... and strongly reduced the diversity of experience' (SPD, Antrag 636, 03.05.1985). It further specified the measures to be investigated and also raised the question of minimal discharge—since the Isar suffered not only from violent episodes of high waters, but also from a discharge deficit created by the multiple deviations that fuelled several hydro-electrical stations, leaving sometimes only a tiny stream in the main bed. To the goals previously identified the motion also noticeably added the improvement of the green network between the Isar and the city districts, as well as the creation of a cycle path along the river up to the town of Bad-Tölz, located 40 kilometres upstream. It also suggested the participation of external landscape architects in the design. While the first motion was clearly focusing on the environment, this second motion pinpointed the necessity to take public use into account, aiming at improving the accessibility and spatial quality of the river and its urban surroundings. This motion was retained by the City Council, which gave directions to the municipal Planning, Environmental Protection, and Building divisions to elaborate a concept for the Isar and propose a procedure to move the project further along.

The City Planning Department took the lead and eventually brought together a wide range of municipal instances for this first investigation: the departments of Construction, Real Estate, Public Utilities, Environmental Protection, City Treasury, as well as the Department of the local district health administration. This procedure led to a public decision made by the Committee for Planning and Building, validated by the City Council on the June 4, 1986 (Referat für Stadtplanung und Bauordnung, 04.06.1986). The text presented by the Planning Department related the river's history, made a balanced portrait of the Isar, and stated multiple ambitions for the future. The Isar area was depicted as the 'most distinctive element of Munich's urban landscape', a crucial recreational space to Munich's citizens, a regional green corridor, a protected natural environment, and as a source of electrical power and (still in 1986) a part of the sewage water treatment system. A historical overview related the role of the river in the city's commercial and industrial development, as well as the dangers brought by the river's fluctuation, before the nineteenth-century regulation constrained the average flow into a 45-metre-wide bed and constrained high waters within a 145-metre wide floodplain. The qualities mentioned in the previous motions were summarized: positive effects on urban climate, environmental value, recreational role, and identification with the city. New problems were identified, however, and added to the brief: the floodplain had been narrowed at several points by the construction of bridges, which could lead the river to flood over the side dams.

The restorative aspect was to take on a new and leading role in the proposed planning procedure, as 'the concept for renovation of the floodplain appeared to be the most important building block for the Isar Plan' (Referat für Stadtplanung und Bauordnung, 1986: 5), and 'all planning discussion should grant safety of high water discharge absolute priority'. This orientation, however, should not diminish the importance of spatial and environmental quality: 'Landscape planners and hydraulic engineers must work together' to elaborate the 'necessary in-depth planning and substantial calculations' (id.: 7). For the first time, the Bavarian Water Board, a State department, was invited to bring its expertise to the work group, although it was not legally responsible for the river within Munich's municipal borders, ever since the 1907 act that transferred this prerogative to the City.

The studies were further divided into four aspects: the design of the river space, the Isar secondary branches, the restoration measures and green connections, and the walks and cycling paths. The work group's elaborated list of concrete recommendations for the design:

- The straight winter bed concrete edges could possibly be removed;
- Part of the flat foreshore could be lowered and sloped towards the main riverbed;
- Deposits that diminish the riverbed capacity could be removed to prevent indirect flooding (with reference in that regard to the Isar status of 'river of first order', that could motivate a co-funding by the State of Bavaria);
- Replantation of (soft wood) groves could be considered after soil removal;
- Ecologically valuable areas should preferably be developed in insular situation to protect them from intensive recreational uses;
- Meadows could be laid as species-rich, meagre meadows adapted to heavy use on sunny days, rather than grass areas meant for intensive use;
- The reinforcement of the edges and bottom of the stream bed was seen as inevitable to prevent it being deepened by erosion, but should whenever possible be covered with pebble;
- Existing groups of trees would be in all cases preserved, and when located on islands, these should be protected from erosion by stone embankments. Secondary river branches would be located around existing or new groups of trees to enhance their presence;
- Valuable trees that grew on the dams would remain, with this linear horizon being possibly varied through planted front protections.

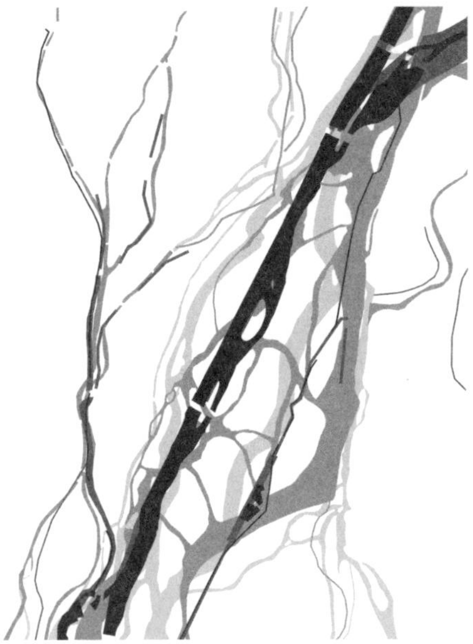

From braided river to single channel, the Isar in 1704, 1808, 1891 and 2011

Besides minor amelioration brought to this first draft (the existing concrete weirs would, for instance, later be removed and replaced by an open line of stones to facilitate fish migrations), this set of measures would be further detailed and implemented over the ensuing 25 years. From 1986 to 1989, the City of Munich organized several rounds of investigation and consultation, as well as an interdisciplinary colloquium in 1987. In 1988, the City Council decided to take various steps to move the plan further along: public information, a joint planning agreement to be sealed with the State of Bavaria, development of the landscape, and hydraulic solutions in collaboration with the state's water board. In 1989, a discussion podium was organized by the Münchner Forum, an association that was established in 1968 as a citizen platform and has since been playing an important role in Munich's planning policy. The outcome of the Forum pointed out the need to steer recreational uses and improve the river's accessibility and natural qualities; it showed the symbolic value of the Isar for the inhabitants, stating as well the need to integrate the 'alpine nature' of the river in its layout and give space to nature where the riverbed was not suitable for intensive recreation, increasing water inlet at low discharge and improving water quality to swimming level. Finally, it pleaded for collaboration between City and the State of Bavaria (Blasy+Mader, 1998).

In 1991, the water board proposed the first joint meeting with the city to investigate the possible combination of flood protection measures and landscape improvement. During the first half of the decade, various committees were set up to elaborate detailed studies on landscape ecology, fishery, forestry, and hydrology, finally leading to commission a preliminary design. The office of Blasy+Mader, specialized in environmental design and based close to Munich, was eventually chosen to propose a plan for the Isar from the Grosshesselohe bridge to the Eisenbahn bridge.

As the Isar Plan developed, it also became clear that recreational use was hindered by the poor water quality. Munich then had an efficient sewage water treatment system, but this was not the case for all the municipalities upstream. Furthermore, the bacteriological quality of the river was below the recommended level for bathing. As beautiful and attractive as the Isar could become, increased public use would therefore create new sanitation and safety problems, and create new conflicts 'if you [had] to tell people: do not go into the water, do not drink a drop or you get coli bacteria' (Thalgott, 2013). From 1998 onwards, the Bavarian capital thus put pressure on its upstream neighbours to improve their sanitation infrastructure and initiated roundtable discussions to address the problem simultaneously all along the river. As a result, several disinfection stations were built, and the Isar's waters reached bathing quality in 2005, six years before completion of the Isar Plan. Besides water quality, the renaturation project triggered another paradoxical question: how much water should be let through the main river channel in order to restore its wild character—and was it compatible with the needs of hydro-power plants? These two preliminary conditions for renaturation—water quality and quantity—would run parallel to the morphological transformation of the river, and give the whole project a vital base for its implementation.

The ten-year-long planning process organized around the Isar Plan made it possible to define and validate a rather detailed set of goals and the means to achieve them—leaving, as it appeared later, relatively few spaces for formal digressions. Blasy+Mader used the methodologies of *Landschaftsplanung* rather than *Landschaftsarchitektur* to elaborate their preliminary design for the stretch located south of the city. The firm elaborated a comprehensive atlas of fauna and flora, public uses and spatial qualities, merged into a conflict and value analysis of all aspects in relation to one another. This analytical approach was used to propose a scheme that was based on the faithful translation of the brief into practical interventions at local level. Following this preliminary study, the southern part of the Isar within the municipal borders was transformed by the city and the Bavarian Water Board, without design competition or broad public discussion. In the spirit of the popular Flaucher area, the last wide and braided river segment in the Munich region, the channelled river was transformed and softened: the straight embankments were removed, restoring the accessibility of the stream. Hard edges that formed a uniform boundary between main bed and floodplain were softened, slowing down the stream along the shore, and a part of the floodplain was left open for the frequent river's fluctuations. 'The Isar will flow again', wrote the *Süddeutsche Zeitung* in 1998 about the first segment being 'renaturated' (*Süddeutsche Zeitung*, 24.02.98).

THE DESIGN COMPETITION: THE WILD, THE URBAN, AND THE 'TRUE ALPINE RIVER'

Willing to pursue the transformation of the Isar but sensitive to the special situation of the central part of the river, in 2003 the Municipality of Munich and the Bavarian Water Board organized a landscape design competition for this 1.6-kilometre segment stretching from the railway bridge down to the Museum Island. The designs should improve the biological quality of the riverbed and its shores, diversify the stream structure, and facilitate the transport of boulder by the stream. They should also facilitate recreational uses in an open 'nature-like' (*naturnah*) landscape. Finally, the project should be 'aesthetically attractive' and enhance the cultural value of the Isar Space (*Isarraum*), combining an open character with prominent urban features (Baumeister, 2003: 10–12). Implicitly, the project should also integrate the constraints created by strong discharge fluctuations, the presence of hydro-power plant downstream, and sensitive infrastructures around and under the riverbed.

The first prize was awarded to the team led by landscape architect Irene Burkhardt, who proposed separating the main stream and a new recreational space through a sculptural levee. While the main flow would keep its linear character, the former floodplain was to be reopened to the stream, offering shallow waters, soft embankments, and a wide, open meadow for recreation. The levee itself was drawn as a hybrid structure, usable as a beach and as a pier for small boats. The second prize was awarded to the team led by landscape architect Winfried Jerney for an almost opposite proposal: a curvy informal design, suggesting a free flow forming its own meanders, islands, and edges through spontaneous deposits and erosion processes.

Although Jerney's project seemed romantic to the city's engineers in regard to what could actually be left to natural processes, it let the public believe in a possible restoration of natural fluctuations within the existing high-water channel. A heated discussion started around the two projects, which grew into a public quarrel placing supporters of an urban, functional, and 'honest' design in opposition to supporters of a full renaturation who came to doubt the reality of the technical constraints set by the city planning department. Public meetings were held in the neighbouring areas, where an active part of the population expressed its frustration and rejected the winning design for its built character and for the fact that the shorelines remained clearly defined instead of letting the flow trace its shore. The design was depicted as 'banal' and some baptized the central pier/levee the 'Concrete Monster' (*betonmonster*). Aggressive leaflets spread the slogan, 'renaturing instead of concreting' (*Renaturieren statt Betonieren*), and opponents called for the implementation of the second prize, against the choice of the jury. The city, through the Baureferat, took a neutral stance in the controversy, stressing that both first and second entries were feasible, and leaving the designers on the front line. Without clear support from the commissioners, the construction of Irene Burckhardt's proposal seemed compromised. After a period of public consultation and emotional discussions involving commissioners, designers, district representatives, and citizens, Irene Burckhardt eventually agreed to compromise and promised 'that the team [would] deal with the design in a critical way' (*Süddeutsche Zeitung*, 05.06.2003).

The professional press remained silent on the controversy. In an article published in the national *Garten+Landschaft* journal, Tilmann Zinsser from the Bavarian Water Board merely mentioned the difficulty of the assignment: 'To find a good compromise that, on the one hand, copes in its form with the contradictory urban and landscape aspects but also with the spirit of a natural restructuring, and on the other, integrates the various leisure opportunities present in the central recreation area' (Zinsser, 2003: 14). Zinsser wrote that 'the designs of first two winning entries will be further deepened and discussed', and that a final design should be presented in 2004—not mentioning the fact that the jury had actually awarded the first prize to the design of the Burckhardt/SKI team. The article was illustrated with photographs of the completed segments along with the 'option 1' plan from the 1998 preliminary design study. No illustrations actually showed the result of the design competition for the central segment of the Isar.

Only in January 2005 would the *Süddeutsche Zeitung* announce that a compromise had been found: 'The two prize-winners share the planning section' (*Süddeutsche Zeitung*, 12.01.05). This compromise presented by the Baureferat was finally approved two months later by the Munich City Council, opening the design phase with a delay of 18 months. It was

Schematic section of the Isar upstream the Reichenbach Bridge (Rossano/Orfanopoulos)

then too late for the plan to be implemented for the 2005 'Federal Garden Exhibition' ('Bundesgartenschau'), due to open in April of that same year. Ironically, the sharp geometries imagined by Gilles Vexlard for the exhibition main ground would generate similar negative reactions from Munich people; One 'really needs to get used to it', politely conceded Munich's First Mayor Christian Ude in a public Q&A session (Ude, 2008).

The alternative project elaborated by the two teams eventually ensured the safety of urban infrastructures while suggesting a relative freedom, with curvy shores and a pebble island. The design was published in *Garten+Landschaft* in 2010, after completion. Thomas Armonat—author of the article and former employee of Irene Burckhardt—described the 'near-natural urban landscape' and its controversial elaboration (Armonat, 2010). The pictures showed the striking contrast between the former straight banks and the new shallow curvy shores and islands. The article, however, did not shy away from detailing the hidden fortifications built to stabilize the 'near-nature' look: a 60-centimetre thick stone bed covered by grass to resist flood erosion, an up to two-metre section trench filled with stones along the water's edge, stone pavement to fasten the river banks and floor around the city infrastructures, a stone belt to stabilize the Weideinsel, and a metres-deep concrete wall poured into the levees to avoid cutting the existing trees. And although the paradox inherent in the 'renaturation' was not discussed further, a closer look into the design plan and sections left no illusions about the artificial character of the 'near-nature' curves.

'SOMETHING TO DISCUSS': DESIGN AS A STARTING POINT

How wild is the Isar after 'renaturation'? In terms of morphology, the riverbed is still very far from its original width and meandering nature, as a simple comparison with the river's past states shows: the boundary between regular riverbed and the floodplain has changed from wall to shallow slope and stairs, but the riverbanks remain as fixed as they were a century earlier; the biodiversity seems to have increased, but this is not thanks to more curvy shores, as these are the most intensively used by the public; finally, the variability of the river course is as limited as before, as the stream tends to follow the deepest channel on the left bank side (Engelmayer, 2013) and the river practically has not recovered a real meandering space. Near-nature or rather nature pastiche, the design for the central segment tries to combine two opposites: the impermanence of an alpine stream and the safe stability of an urban space, eventually simulating the first to achieve the second. Only the old binary variation—flood or dry meadow—has evolved towards a more gradual transformation of the river by rising waters, thanks to the new stepped banks and shallow pebble beaches.

Today, however, all the actors interviewed by the author separately stress that debate generated by the competition was necessary, despite the emotional and sometimes hostile discussions that took place in public meetings. As Daniela Schaufuß, today Head of Munich's Department of Water and Civil Infrastructure, explains, external designers and engineers brought in new ambitions, but city and water board were one step ahead: 'We had already constructed [eight kilometres of river], and we had gathered so much experience over the years that it was not easy for new companies, new engineering offices, to work with us because we knew what we wanted, we knew the problems. They started with great ideas, and we just said, no, it's not possible' (Schaufuß, 2013). However, for Schaufuß, who dedicated several years to the transformation of the Isar, a design competition was also necessary to reveal unrealistic expectations and expose the different constraints to a public enthralled by the Flaucher areas further out from the city centre. For Christiane Thalgott, former head of the city planning department, 'it was important to show, on the one hand, how little freedom there is when the river is so important, but, on the other, to speak with people about this limited freedom we have, because people perhaps expected something much more impressive.... These competitions were more an educational project. It was necessary to communicate'.

A point of view partly shared by one of the landscape architects and authors of the project, who regretted the irrational character of the debate but retrospectively supports holding a design competition in this context: 'People needed to see something to talk about it.... You need design as a sort of a benchmark to have any discussion at all. If the idea is that you can start public participation without some core idea to begin with, just the way it was before, then that would be just completely nuts and in vain' (Engelmayer, 2013). The intense public debate indeed was initiated thanks to concrete proposals, which clearly created a discussion platform for authorities, citizens, and designers. The designers, however, did not always feel welcome and supported, as the competition procedure followed a period were the Isar upstream had been transformed without too much spatial ambition and public discussion. 'Then there was this design competition that was initiated by the city council and which in a way was not something that the people who did the previous work ever wanted, they just wanted to continue in the same way' (idem). The debate further lacked the preliminary information needed to evaluate the proposals: in particular, the impossibility of leaving the stream totally free within the constrained environment of Munich's city centre 'was not clear and that was never made clear' (idem), and the discussion was rapidly caught in a caricature opposition of design-versus-nature that neither the commissioners nor the designers had anticipated, and ultimately resulted in an ambiguous compromise rather than a clarification.

CONTROLLED FLOODING AS VACCINE AGAINST DISASTROUS FLOODS

One could qualify the design made for the Isar River space as a non-design, since it lacks the formal or spectacular character one would expect in an urban public space, or call it transvestite nature that has the appearance but not the characteristics of the wild stream it pretends to be. Or, as Oliver Engelmayer clearly puts it: people 'thought [our first design] was wrong because [water] management was made visible. Now we still have the same issues, hidden beneath the greenery, and every time there is a flood you have to put it back in again'. Environmentalists may regret that the Isar still has not regained full freedom of flow. Architects and engineers may lament the absurdity of trying to mimic erosion, sedimentation, and

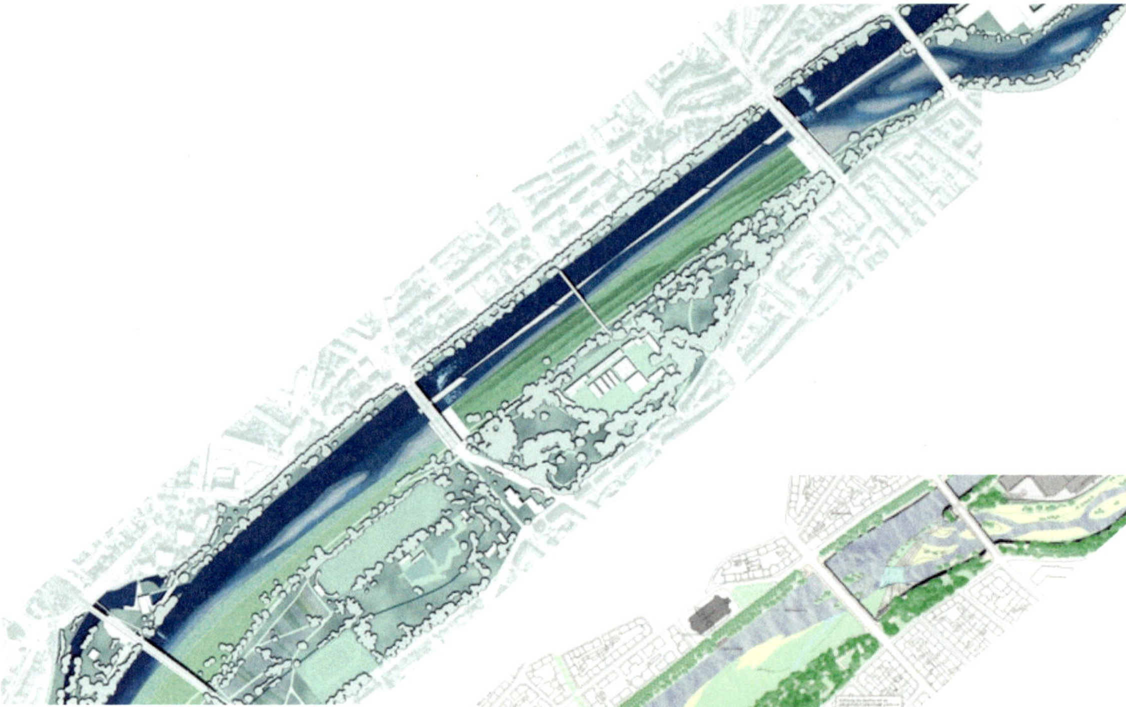

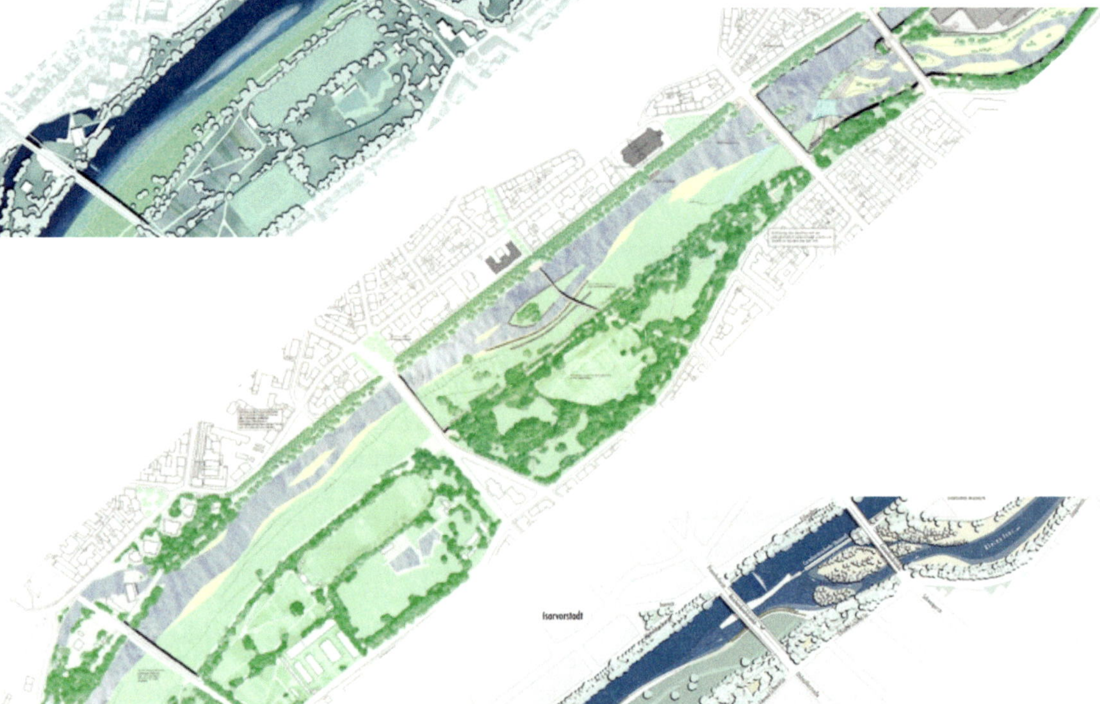

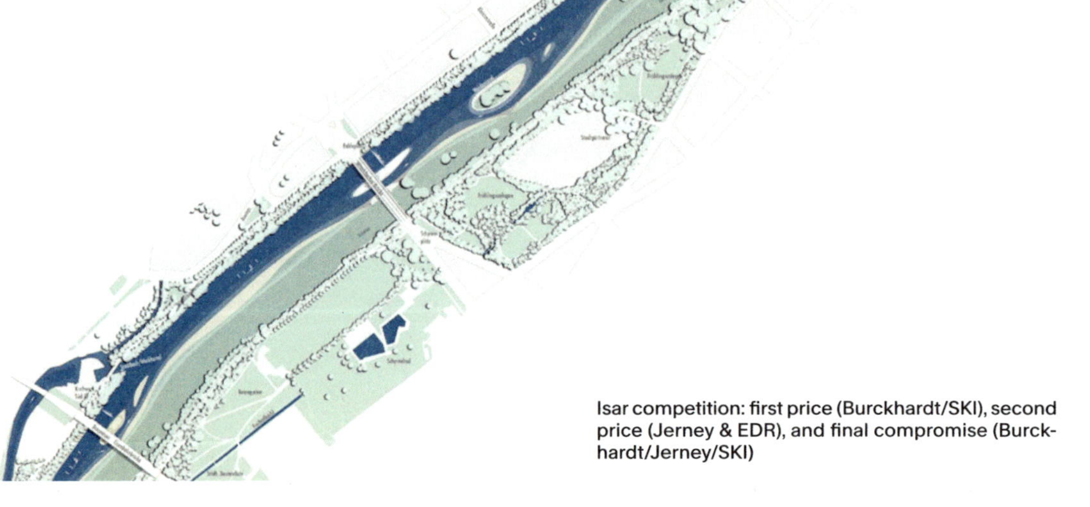

Isar competition: first price (Burckhardt/SKI), second price (Jerney & EDR), and final compromise (Burckhardt/Jerney/SKI)

From floodplain to urban beach: Terraced promenade along the water close to the Reichenbachbrücke; Meadow opposite the St. Maximilian Church (June 24, 140 m^3/s river discharge at station München and September 7, 2013, 55 m^3/s discharge. Rossano/Künzel)

anastomosis, even partially, when so much effort has been put into controlling these fluctuations to protect the city.

Whether beautifully informal or hypocritically nature-like, the 'Isarraum' is today enjoyed by the population of Munich. Appreciated for its low-key design and freedom of use, the Isar is officially not a park, and still is much more than a floodplain. The simple and robust design lets the river 'run the show', offering the possibility to see the shores and meadows disappear under a furious stream during summer high waters, as they did in June 2013. They reappear after a few days or weeks, and are rapidly restored by the city, at a cost that is still far below the regular maintenance cost of formal urban parks. Beyond an ambiguous design, the Isar' urban design reveals its intrinsic value every time water covers and uncovers paths and meadows: Public space functions here as a reality check. By removing the hard edge that used to isolate the stream from its monotonous flood meadow and by dealing with the regular but minor damage caused by flooding, the design has provided public access to a space of indecision, underlining fluctuation instead of neutralizing its effects.

While the original nineteenth-century design strongly accelerated and separated the main stream from the floodplain, making it even more dangerous and inaccessible, the Isar's new soft edges do not differentiate risk areas from safe zones anymore. Outside major flood events, the floodplain remains accessible, leaving the responsibility to evaluate the degree of risk they are willing to take when entering the 'Isar Space' up to citizens—in striking contrast with most European floodplains bordered with fences and warning signs. This openness alone constitutes a radical break with a centuries-long schism between collective management and individual adaptation to natural dynamics. It also provides a permanent and free education in living with these dynamics, free from deluge angst and beyond the myth of 'dry feet' (full safety, from the Dutch expression 'to keep dry feet'). In the quest for a new paradigm in terms of the perception of flood risk, and a stronger awareness and resilience on the part of individuals as well as societies, the experience of the flood is able to bring to the fore notions of natural processes and acceptable risk, and can serve to train citizens to appreciate the pleasures and dangers involved in an adequate way, much more efficiently than communication campaigns or dramatized news or apocalyptic fictions do.

ISAR PLAN: KEY FACTS

Clarifying expectations of nature and authenticity is a prerequisite for interventions in a human-made landscape.
Design competitions, even if controversial, trigger useful public discussions.
Simple, robust, and low maintenance design resists seasonal flooding and overcrowding.
Through adequate design, a floodplain can function as a park open to the public.
Reconnecting with a fluctuating river allows for a greener city and more conscious citizens.
From lost hopes of 'renaturation' a successful compromise between nature restoration, flood management, and urban life can emerge.

Groningen and the Woldmeer/Meerstad area east of the city

FLOOD AS INVESTMENT: MEERSTAD GRONINGEN

The new neighbourhood Meerstad—literally Lake City—was born out of a rare combination of flood risks and real estate ambitions, which drew the attention of local authorities, developers, and the regional water board to the same area east of Groningen (the Netherlands)—albeit for different reasons.

The region of Groningen generally has two main geological profiles: higher sandy grounds, flood-safe and naturally drained, and clay or peat grounds in the lower parts. The city itself was conveniently founded on the northern tip of a long sand ridge, between five and ten metres above sea level. The old city is thus settled on safe ground and is still protected against floods by its elevation, just like the many ancient villages of the province built on natural and artificial mounts (see chapter 'Early Adaptive Landscapes'). The Meerstad area on the contrary was located in one of the deepest polders of the region (its deepest point laying more than two metres below sea level, two-and-a-half below the nearby Eems canal), about four kilometres east of the city centre.

From the nineteenth century on, the construction of large drainage canals such as the Eemskanaal between the city of Groningen and the Dollard estuary, and the progress of drainage techniques has allowed urban growth to spread further away from the city and deeper into the lowlands. Lacking space to extend onto high ground, and following a logic of proximity rather than flood safety, the city grew first in its immediate surroundings, later in more distant settlements, and eventually, at the end of the twentieth century, the whole region became a destination for urbanites in search for affordable 'green living'.

In the 1990s, in the midst of a real estate boom, most land around the centre was in the hands of local developers and speculators, happy to benefit from the strong demand for new housing, but reluctant to participate in public infrastructure investments induced by their projects, such as civil engineering works or public transport. The municipality was successful in handling small-scale projects and relating them to the surrounding neighbourhoods, but had trouble coordinating its own urban growth, with a housing demand still higher than the private market could satisfy. Another source of concern was the rapid growth of neighbouring municipalities, which provided opportunities for free-standing housing a short distance from Groningen, thus depriving the city of a wealthy portion of its population—that still made used of the city's public facilities but was not contributing anymore to its revenues. While surrounding rural villages welcomed the new middle-class dwellers, Groningen struggled to maintain its tax income.

This extensive growth into the open countryside was also contradicting the national policy as defined in the *Vierde Nota Ruimtelijke ordening*. This Fourth Memorandum Spatial Planning and its follower, the 1990 Fourth Memorandum Spatial Planning Extra (or VINEX) had fixed national ambitions and conditions for spatial planning. It stressed the necessity of keeping cities compact, while answering a strong demand for low-density housing, fuelled by a healthy economy (Min. van VROM, 1988 and 1990). In the document '*De stad van straks*' (the City of Tomorrow) issued by the municipality in 1994, the area that would later host the Meerstad project was designated as a suitable extension for the city, since it answered the national ambitions for coherent growth and the tendencies of the real estate market (Gemeente Groningen et al., 1994). During the ensuing years, several investigations took place, involving civil servants and designers, which tended to confirm the site's potentialities. Informal contacts were made with National and Provincial authorities, which brought political support to the initiative.

In addition to the urban ambitions set forth for the area, other regional and national housing projects suggested promising synergies with nature development and water management authorities. The Main Ecological Structure (EHS, Ecologische Hoofdstructuur), elaborated in 1990 by the Ministry of Agriculture, Environment and Food Quality, envisioned a national network of existing nature areas, due to be completed through the creation of new ecological connections. One of the figured connections that was meant to link the coastal Wadden area, northeast of Groningen, to the southern nature areas of Drenthe Province, crossed the municipality of Slochteren, a few kilometres away from the city. The later Meerstad project seemed perfectly located halfway between city centre and the future nature reserve, reinforcing its suitability for an urban extension close to the city that would provide an attractive green environment.

An unexpected flood alert gave the project a new turn, by the end of the nineties. Although the northern provinces had escaped the massive evacuations following river overload in 1993 and 1995, heavy precipitation in 1998 dramatically reminded them that water safety was far from guaranteed. In October of that year, massive amounts of rainwater flowing from Drenthe towards the north saturated the canals of Groningen province. The Winschoterdiep, one of its main drainage canals, reached its maximum capacity, and its dikes came close to breach point. Many houses were flooded, a residential neighbourhood close to the canal had to be evacuated, while dikes were temporarily reinforced with sandbags. The commotion led to a compassionate visit by Queen Beatrix on the 29th of October, drawing the attention of the country to the exposure of the northern provinces. Besides the fear and destruction caused in the countryside and small towns, the province capital city was hit at its heart: the new Groninger Museum, completed just four years earlier, had to be evacuated, and its lower floor was flooded, damaging part of the museum's collections. Although the risk was identified by the water board before its construction—and was due to a design mistake rather than to high water—the event served, as a side effect, to empower public and political consciousness to the fact that flood exposure at the urban and regional level was calling for action and that raising dikes higher would not be enough this time.

Analyzing the event, the water board soon came to the conclusion that the flood was caused by two combined factors: widely spread and heavy rainfall, on the one hand, and a north-westerly wind that had pushed the sea water up into the Eems and Dollard estuaries, on the other, blocking the region's main outlets for several days. The origin of the problem was thus not only the discharge capacity of the regional water network but also the lack of temporary storage capacity whenever the outlets were blocked by natural phenomena—in this case an abnormally high water level in the estuary, all combined with a systematic failure to include flood risks in architectural and urban development. Consequently, after 1998, the Hunze-en-Aa's Water Board worked out a new combinatory approach, which could simultaneously strengthen the existing water structure (its traditional task), increase temporary water storage capacity on its territory, and raise awareness among decision-makers and the general public. Local urban ambition, environmental policy, and a major flood warning were paving the way for Meerstad.

FROM WOLDSTREEK TO WOLDMEER

After a decade that saw the inner city becoming a successful laboratory for architectural and urban renewal led by city councillor Ypke Gietema, the 1990s witnessed a new interest in the periphery and countryside. Designers were regularly involved in plans and visions developed for the open Groninger landscape. In 1993, the by now former councillor Ypke Gietema, constructor H. Koop, and investor H. Venema, joined in a private initiative. Together they called for an urban development in the Woldstreek area that should integrate environmental and water issues through the combination of housing construction, a vast water landscape, and the creation of a new lake, 'the most beautiful water reservoir in the Netherlands' (Archined, 2000). Soon after, they asked landscape architect Adriaan Geuze and architects Karelse & Van der Meer for a spatial proposal for the area. The resulting project envisioned a large-scale nature and urban development from Groningen to Delfzijl, following the Eemskanaal built in the nineteenth century. A 20-kilometre stretch of new forest and wetlands was proposed to link the two cities, putting the Harksteder polder and Rijpmapolder under water to form the new 'Woldmeer'. Together with existing lakes and future wetlands, they were meant to form the 'Loire River of Groningen', offering a wide and open 'square' connecting existing villages with a new 'village' planned on its northern shore (Archined, 1999). Comparing these future neighbourhoods to the recently redeveloped Amsterdam docks (partly reconfigured by Adriaan Geuze), the designers stressed the need to address uncontrolled sprawl in the rural area, the decay of cultural landscapes, and the decline of agrarian activities due to the land's mediocrity.

In September 1996, the plan 'Groningen above water' was publicly presented at the Keuningscongres, where designers and politicians were invited to discuss and envision the future of Groningen's eastern lowlands. In addition to Adriaan Geuze's proposal, Dirk Sijmons and Christophe Girot were asked for a general proposal, Girot focusing his contribution on the existing landscape, praising the characteristic structure of the province' open polders structure, Sijmons suggesting to further develop the existing ribbon developments (Wildeman, 1997: 40).

The Geuze and Karlse van der Meer design was then updated and received support from the local magazine for landscape and planning, *Noorderbreedte*, which praised the project as a necessary reconsideration of national planning policy that split urban and environmental planning: 'Main ecological structure often means juxtaposing science-conforming ecological typologies. This leads to a landscape that has nothing to do with the Dutch cultural landscape and that is only liveable for animals that often were never found in the area before. The Woldmeer [project] offers more than just nature: not only the grass snake but also man is at home in and around the lake. From the city, the Woldmeer can be reached by bicycle within 15 minutes. This is not just ecology, but a design that fits in the landscape structure, along with all recreational possibilities. The design for the Woldmeer does not turn its back on the existing landscape; it is not "nature for nature's sake".' (Abrahamse, 1999). Abrahamse also co-authored the website 'Groningen boven water' dedicated to the promotion of the Woldmeer project, and hosted by Archined, then the largest online architectural platform.

This integrative approach reflected a discussion that had been going on in the Netherlands since the eighties about the relation between ecology and landscape. The experiment initiated by the Stork Plan, which introduced the concept of two-speed landscapes and the bundling of various interests into an integrative approach of nature, had been narrowed down to a technocratic approach during the elaboration of national environmental policies, favouring the development of isolated and specific ecosystems rather than a reflective practice on a project level (De Jonge, 2009: 9). The Woldmeer project was thus not only a spatial proposal but also a pamphlet against landscape fragmentation and against a dominant technocratic form of ecology, which tended to isolate Nature from man's living and working environment, and reduce 'Nature' to 'natural habitats' considered only for specific targets, plants, or animals.

In 1998, the study *Voorbeeldplan Maximum laadvermogen*—commissioned by the city of Groningen, the neighbouring village of Slochteren and the national Department of rural areas—stated that intensive land use and new landscape development could provide excellent basic conditions for housing development in the region (Gemeente Groningen et al., 1998). The proposal generated an enthusiastic reaction from city, province, and ministry, and later became fully part of the brief set up for the design team. The water board, however, soon appeared less univocal about the project. Although the creation of buffering capacity close to the Eemskanaal, one of the province's main outlets, perfectly fit in the water board's new strategy following the 1998 flood, planning residential developments in the lowest and most exposed area of the province seemed

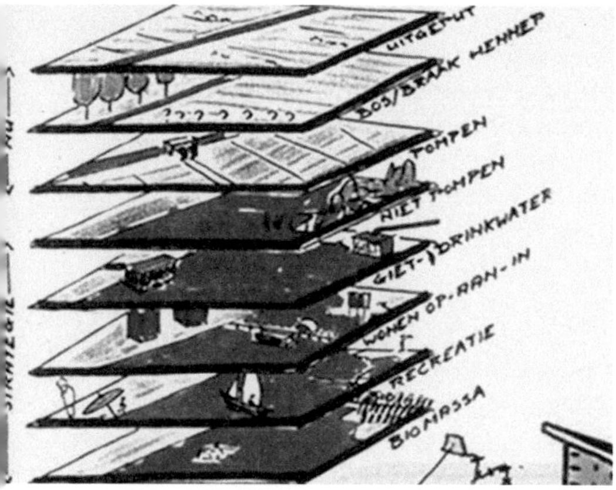

The Woldstreek landscape's 'maximum load' strategy: from an 'exhausted landscape' to a combination of water, housing, leisure, biomassa production.... (Gemeente Groningen, 1998)

questionable to its Chairman (Van Hall and Kuipers, 2013). The very reasons that made the area suitable for water buffering provided identical arguments against urban development: the land lay several metres below the average level of the Eemskanaal flowing along the area. The canal could thus easily be offloaded by gravity into a new mitigation lake, but it could just as well flood the future housing area in a very short time, if a dike was to weaken, slide, or break open. But as before with the construction of the Groninger Museum and its (too) low basement windows, the warning of the water board remained unheard in the context of general enthusiasm for a visionary project that reconciled nature, landscape, housing, and flood control in one coherent plan.

Following the contacts made in the previous years, in 2000 several public parties officially decided to collaborate in the development of the project. An *Intentieovereenkomst* (agreement of intention) was signed in March 2001 by the Province of Groningen; the Ministry of Agriculture, Nature, and Food Quality; the Department of Rural Areas at the Ministry of Economy; the Hunze-en-Aa's Water Board; and the municipalities of Groningen and Slochteren (most of the project's area being located in this small rural municipality). The agreement stated a new development aimed at creating 8,000 dwellings around a 650-hectare lake, representing an investment of one billion Dutch guilders to be earned back by the sale of land and buildings. A few months before, the same players had selected a team of designers to work out a development strategy and turn the programme into a spatial framework. Alle Hosper Landscape, a renowned landscape architecture office from Haarlem, associated with Kees Christiaanse, Rotterdam, was given the commission to lead the design process from 2001 on (Provincie Groningen, 2001).

LANDSCAPE ECONOMICS

Private developers were not left out, since successive Dutch governments from the early 1990s actively promoted public-private collaborations through the so-called P.P.S. programmes (*Privaat-Publieke Samenwerking*), especially on big infrastructural projects and urban developments. The Fourth Memorandum on Spatial Planning urged extra effort to develop this type of joined planning processes, and involve private companies in the investment, design, construction, and marketing, and in 1999 the Ministry of Finance created a dedicated 'Knowledge centre PPS' to foster new initiatives (Keuter, 2007).

Some local developers already owned part of the area, and big national real estate companies were attracted by the perspective of securing positions in a massive housing development, comparable in scale

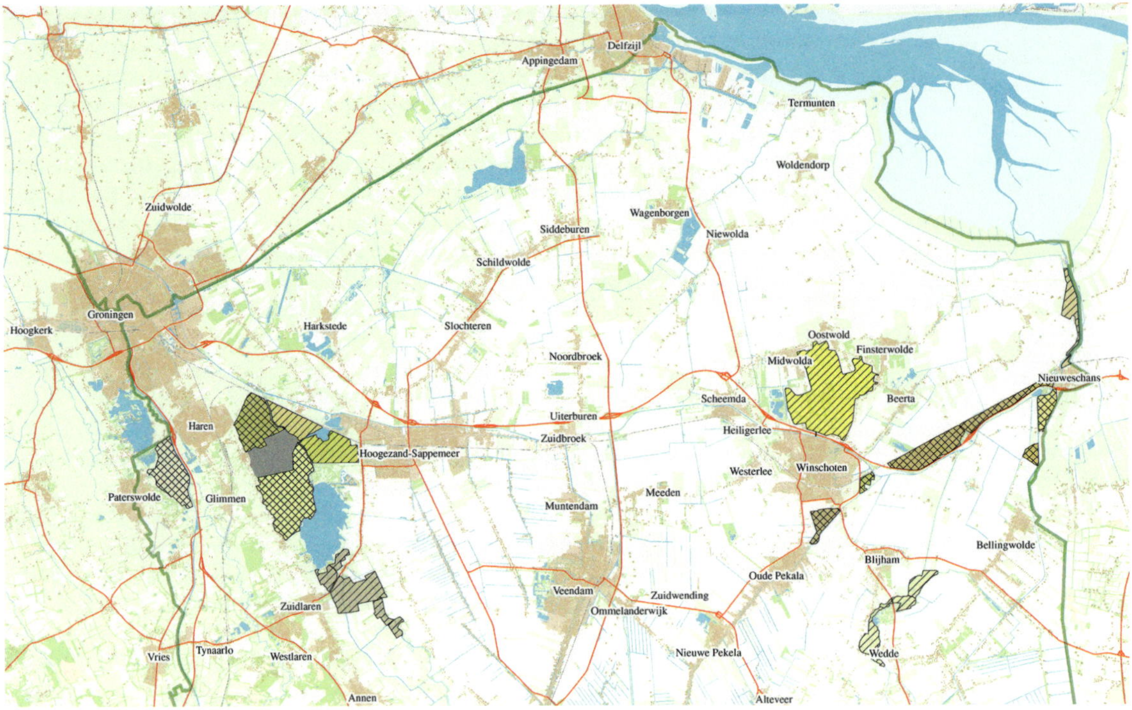

Areas available for emergency storage, Simple hatch: average usage frequency 1/100 years or less; double hatch: 1/100 years or more (Water board Hunze-en-Aa's, 2013)

to the VINEX projects under construction around most Dutch cities. In January 2002, the public parties involved signed a Collaboration Agreement ('samenwerkingsovereenkomst') with two consortia of private investors: the Grondbank Meerstad Groningen CV, regrouping national players AM Grondbedrijf BV, Koop Holding BV, Heijmans IBC Vastgoedontwikkeling BV, and BPF Bouwinvest BV; and the Twentse Combinatie formed by regional developers Rotij Projecten BV, Timmerhuis Projecten BV, Groothuis Woningbouw BV, and Plegt-Vos Bouwgroep. Meerstad was identified by the Ministry of Finance as one of six PPS pilot projects. Private stakeholders, however, were not involved in the selection of the design team—'luckily', according to landscape architect and project-leader Remco Rolvink, who later found that developers were soon inclined to lower the city's ambition and question the cost of the design process (Rolvink, 2012).

Although neither programmatic nor juridical aspects were yet quite settled, the planning organization was set up to start an ambitious project that would from then on assume the official name of Meerstad—literally Lake City. The new denomination, however, already generated discussions among commissioners and designers. The study area first led to difficult negotiations among investors, authorities, and designers: while investors wished to restrict it to the area necessary for the water and housing programme, the designers insisted that it should include existing and perennial farmland, space for nature areas, and future developments. A study area of nearly 4000 hectares was eventually validated, a size equivalent to the conurbation of Groningen. The name Meerstad, besides being semantically ambiguous (*meerstad* means lake city and *meer stad* means more city—a difference hardly perceptible in oral language), the new name was also disputable in terms of programme. Was Meerstad a new city, a city extension, or an inhabited landscape? Was it one lake, or a diversified water landscape? According to Remco Rolvink, private developers finally opted for the name Meerstad for its supposed persuasive value, but in doing so contradicted their own reluctance to include elements of urban typologies and density in the plan.

At the start of the actual design process, most of the plan area was still cultivated by farmers but already for the most part owned by the project's shareholders. A part of the area closer to the city had been bought earlier by the company Mega-Vastgoed, which declined the invitation to join the private-public agreement, and consequently lost influence in the planning. The Ministry of Agriculture, Nature, and Food Quality had purchased a large area east of Harkstede in 1998 to secure a future ecological area, within the framework of the national E.H.S. nature corridor (Ecologische Hoofdstructuur). About 1500 hectares of farmland had been bought by private investors, seemingly without much difficulty, since the price offered for what was then seen as building land (about 20 euros per square meter) comfortably topped the regular price paid for farmland in the region (around 2.50 euros at the time)—but still 'an honest price in terms of the expectations' (Leverman, 2013). Although sceptical about a city extension onto what then still seemed very far from any urban pressure, the farmers were aware of the poor quality of their land. Many, getting close to retirement age, were also uncertain about the future of their exploitation, and accepted selling their land, while still remaining active for the time-being via ground-lease agreements. Extra land was purchased close-by to make it possible for those who wished to remain in the region to do so—with five farmers ultimately making use of this possibility.

During this active period of trade and negotiations, the presence of private investors proved to be decisive in getting control of land ownership, giving the stakeholders all the freedom to develop a coherent plan, reorganize the current and planned land use, and match land, infrastructure, and construction investments with sales benefits prospects in a balanced financial plan—a capacity that the city of Groningen had lacked in its previous urban developments.

FROM PROGRAMME TO PLAN

Before private developers joined the team, the public authorities that had signed the Collaboration Agreement had already set several targets for the project, based on regional prospective studies: the construction of 8000 dwellings, the creation of a wide lake dedicated to water buffering, the creation of an 'attractive recreational element' (the lake itself), the development of a durable ecological connection to reinforce the city and region's ecological structure, and the development of a new high-quality landscape of substantial dimensions. Ambitious but then still vague, these targets would later be made specific with the help of public departments involved in the planning process, the input of development companies, and external advisors commissioned to investigate specific aspects—water system, landscape and ecology, recreation, transportation, housing, and economy.

The design phase started at the beginning of the year 2002, led by a Management Board composed of the public authorities and private developers, under the leadership of Marc Calon, Provincial Deputy. Besides Calon's regional roots and his understanding of territorial projects, the choice of a provincial deputy was also politically motivated. Although Meerstad was initiated by the city of Groningen and mostly driven by the city's development ambitions, most of the land lay on the territory of Slochteren, a vast municipality hosting a modest population of less than 15,000 in 2000 (Gemeente Slochteren, 2012), compared to Groningen's 173,000 inhabitants (Gemeente Groningen, Statistische jaaroverzichten). The role of Marc Calon was thus not only to chair the Management Board but also to prevent possible tensions among the two municipalities.

The city of Groningen provided the project coordinator in charge of running the new dedicated Bureau Meerstad and organizing a complex process involving six public authorities and their sub-departments, with two consortia regrouping eight private companies, along with a population that expected to be informed and involved. Together with the energetic Remco Rolvink from Alle Hosper Landscape Architects, they formed the driving task force, hosting multiple workshops with the groups of experts and shareholders that were formed around each specific aspect: masterplan, water, sustainability, economy, agriculture, leisure, housing programme, and mobility. More work groups were created to work on specific procedures related to the project: the elaboration of the *milieueffectrapportage* (or M.E.R., Environmental Impact Survey) and the *streekplan* (Provincial Planning Document). A specific work group dealt with communication. Last but not less influential, a work group focused on the financial balance, with the support of external advisor PAS. More than a simple work group, this assembly hosted crucial discussions about programme, density, and housing typologies regarding what private developers considered as suitable for the housing market, and what designers and public authorities considered as suitable for Meerstad and Groningen in the long run. The quantitative aspect played a key role in these discussions, since the expected income had to match the expenses—which explains the raising of the initial 8000 dwellings programme up to a 'possible' 10,000 in the final masterplan and back to about 6500 planned in 2018.

Specific to Meerstad was also the fact that the M.E.R. environmental survey, generally following the design process, ran parallel to the design process here. The company Grontmij was hired by the municipalities of Groningen and Slochteren to work out the survey simultaneously with the elaboration of the master plan. This rather unorthodox procedure promoted by Remco Rolvink was, however, accepted by the national M.E.R. Commission and allowed designers and surveyors to interact in a continuous evaluation and adaptation of the different scenarios, based on the pros and cons each of them raised regarding environmental aspects. This led to a 'somehow chaotic but productive project development' (Rolvink, 2012), but facilitated the elaboration of a balanced and consensual plan. During the year 2002, landscape architects and urban planners

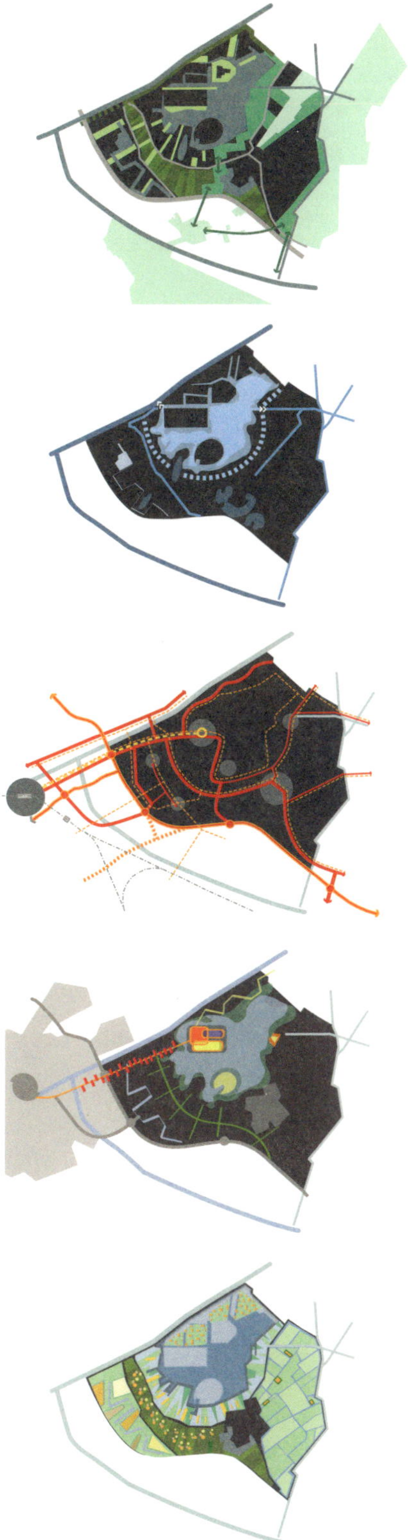

Meerstad development principles: landscape, water, traffic, key urban features and housing areas (Rossano for KCAP, 2004)

(including the author) elaborated dozens of variations and hypotheses for the distribution of programmatic elements, water, nature, housing, industry—occasionally inviting participants of the workshops to play with coloured paper sheets cut in proportion to each function. This method—based on multiple scenarios, followed by a rational evaluation of each one and a combinatory development—was at the time widespread in Dutch planning (Salewski, 2012) and practiced by Kees Christiaanse's office, which had made scenario-based planning one of its specialties. This methodology also supported the chosen 'open plan process' that involved inhabitants and stakeholders in the elaboration of the design.

In 2002, a preliminary plan (*discussiemodel*) was presented, that fulfilled most of the programme's ambitions. The housing and industry programme fitted into the framework, distributed into various landscape types. A 600-hectare lake stretched from Groningen to the East, allowing 50 centimetres of fluctuation in water level, sufficient for a three million cubic metre buffering capacity expected by the water board, and formed a spectacular, attractive, and accessible landscape, suitable for water sports and leisure. A dedicated nature area was located east of the planning area, away from the city but contributing to Meerstad's overall framework. And, above all, the plan provided a legible landscape structure based on natural topography and elevation, which integrated cultural elements such as the ribbon-developments along ancient roads (*linten*) and provided aesthetic and functional qualities for future and existing housing. Pursuing the goals already stated in the earliest design studies, the masterplan elaborated by Alle Hosper and KCAP favoured the integration of building and nature rather than functional segregation, playing on the association of various landscape types and urban typologies to create an optimal diversity, while still keeping legible landscape entities—lake, shores, ribbon-roads, wetlands. Moreover, in the small concentrated 'urban' node located halfway between Groningen and Harkstede, roads and buildings played a modest role compared to its situation at the lakeshore and its future public beach. The structuring of the landscape, mostly defined by water, wetlands, and tree plantations, carried out the initial ambitions of public authorities and landscape architects to pre-invest in a robust landscape structure that could attract, shape, and withstand long-term developments.

In the ensuing years the masterplan was praised for its ambition, its integration of urban development and water management, and its collaborative planning process. Niek Verdonk, former director of the Groningen

Urban Development department and promoter of the 'compact city' policy, praised the indirect effects of the project, which 'spares valuable landscapes and cultural areas' by concentrating multiple ambitions into one area (Abrahamse, 2002). Meerstad was proposed by the province to be one of the national 'model projects for development planning'. The same year, the magazine *De Water*, published by the national water agency Rijkswaterstaat, also presented Meerstad as an 'example for development planning', praising the combination of flood protection and regional development (Bijnsdorp et al., 2002). In 2005, Meerstad was nominated for the 'Real Estate Award' during the Rotterdam Biennial of Architecture (Architectuurbiënnale Vastgoed Award).

The 2006 governmental planning memorandum 'Nota Ruimte' (Min. van VROM, 2006: 28) would later again praise the integrated character of the project: 'For Meerstad in the region Mid-Groningen, red [housing], blue and green functions are implemented within one integral regional development. Agreements among public and private parties take concrete shape in the masterplan. Cost recovery and settlements are managed by the municipal exploitation company, allowing part of the profits made on farmland to be used to finance green areas in the exploitation area and to empower the Main Ecological Structure in the area' (id.: 63). The bill also referred to Meerstad as a model for the redistribution of investments and benefits among private and public parties, thanks to the new planning law regarding land exploitation that 'makes municipalities not only stronger but also protects them against "free-riders" [in English]: parties that collaborate insufficiently but fully benefit from improvements'. Combined goals and shared investments were presented as successful 'instruments for taking advantage of opportunities' (*Instrumenten om kansen waar te maken*).

All in all, the project stood out for its voluntarist and collaborative nature, materialized into integrated landscape development, water safety, and market-driven construction intended to meet the strong demand for green living that threatened the Dutch countryside.

RED FOR BLUE AND BLUE FOR RED

The integration of landscape and buildings was, however, also a source of tension. The Environmental Impact Survey issued in 2003 echoed the criticism by the environmentalists from the Milieufederatie environmental organization but also relativized their relevance in regard to the global benefits the project could generate. Following the usual evaluation procedure, the Survey compared the 'most environment-friendly alternative' (*Meest Milieuvriendelijke Alternatief*, MMA) with the 'preferred alternative' (*Voorkeursalternatief*, VKA). Where the 'environment-friendly' alternative isolated constructions from nature areas, the 'preferred alternative' dissolved the building programme into less dense and more diversified ensembles. This mix of housing and nature weakened the integrity and continuity of dedicated nature, but also increased its total area. Involuntarily highlighting the core dilemma of Dutch environmentalism—a segregative approach versus an integrative one, protection versus accessibility to nature—the general conclusion of the report stressed the positive aspects of the plan supported by designers and commissioners: 'There is less diversity in living environments in the [environment-friendly] MMA than in the [preferred] VKA, this because the MMA focuses on the ambitioned physical ecological aspects. However, partly in relation to its long-term character and the uncertainty of future housing markets, a plan with a wider distribution of typologies over the different areas provides the flexibility needed to provide housing that corresponds to the demand. This is a strong positive effect' (Grontmij, 2003: 140–141).

This conclusion was remarkable on two levels. It acknowledged the need to provide an attractive living environment for the development of the whole plan and therefore stated, in this context, a reflexive dependency between urban expansion and nature development. And, by putting the plan in a timescale perspective, it also identified flexibility as an essential factor in the success of future development—while environmentalists aimed at a stable framework to provide a steady habitat for specific species to develop. The authors of the Environmental Impact Survey chose, above their sectoral concerns, to validate a holistic perspective that allowed human activities to interfere with nature, and the other way around, trading the risks of disturbances for the benefits of greater joint achievements.

Taking control over the land proved to be relatively easy and fast, but joining assets into one common structure and defining the respective roles appeared to be more difficult. Although the masterplan was globally known in 2003, it was only validated in March 2005, since it took two years of negotiations for the participants to engage in one single juridical structure, the Grondexploitatie Maatschappij Meerstad (GEMM), created on September 28th, 2005. The Twentse Combinatie, initially part of the Private-Public-Agreement, pulled out of the planning process at the end of 2004. Another disappointment for the project's initiators was the reluctance of the water board to participate financially in the new structure. Despite political pressures, the water board stubbornly refused to invest in the joint operation, raising frustration on the city's side, which considered the water board as being 'mildly enterprising, not collaborative and reserved', and suspected them of passively making a profit from the fact that the project would be implemented anyway by other parties (Van de Bospoort, 2013). For the water board, it was not only a strategic issue but also a question of juridical status. Albert van Hall, then *dijkgraaf* (dike-warden), recalls: 'The province asked us for both [Blauwe Stad and Meerstad] projects to carry a share of the risks. This led to a quarrel, and we made clear that we would not risk taxpayers' money. I wrote in a letter that I would not join and would not share in the profit either—they expected high benefits back then. We were consciously [passive and reserved], because we are a public authority that is only allowed to raise taxes to fulfil water-related tasks, and not to participate in a big business, or in province or city [investments]' (Van Hall, 2013). Participating financially in the implementation of a retention lake thus remained within its prerogative, but joining the GEMM structure also meant investing taxpayers' money in real estate speculation. The water board thus considered that only the extra buffer capacity added to the lake could legitimize a financial participation from its side, thus running contrary to the hope that all parties would happily join the great enterprise.

In the meantime, the project followed the usual procedure. Following the critical but positive environmental impact survey and the creation of the GEMM after long negotiations, the masterplan elaborated in 2001–2002 was finalized and summed up in a comprehensive document presented in September 2005. In 2006, a permanent accommodation for the Bureau Meerstad was opened in the plan area. A municipal zoning plan (*bestemmingsplan*) was prepared for the central part of the project, Meerstad Midden, which covered a large portion of the lake (350 out of the planned 600 hectares) and a large part of the housing programme. A sub-area named Meeroevers (Lakeshore) was chosen to be the first part to be constructed, which already symbolically included a piece of the lake relatively close to Groningen and addressed the principle demand for low density housing in a blue-green setting with 217 row-houses, attached and freestanding villas. The architecture office De Zwarte Hond was commissioned to work out the urban design, delivered in November 2009. De Zwarte Hond happened to be the designer of a similar project, though less important in programme, the Blauwe Stad ('Blue City'), located 25 kilometres east of Meerstad. After 25 years of dreams, visions, and projects, Meerstad was about to come true.

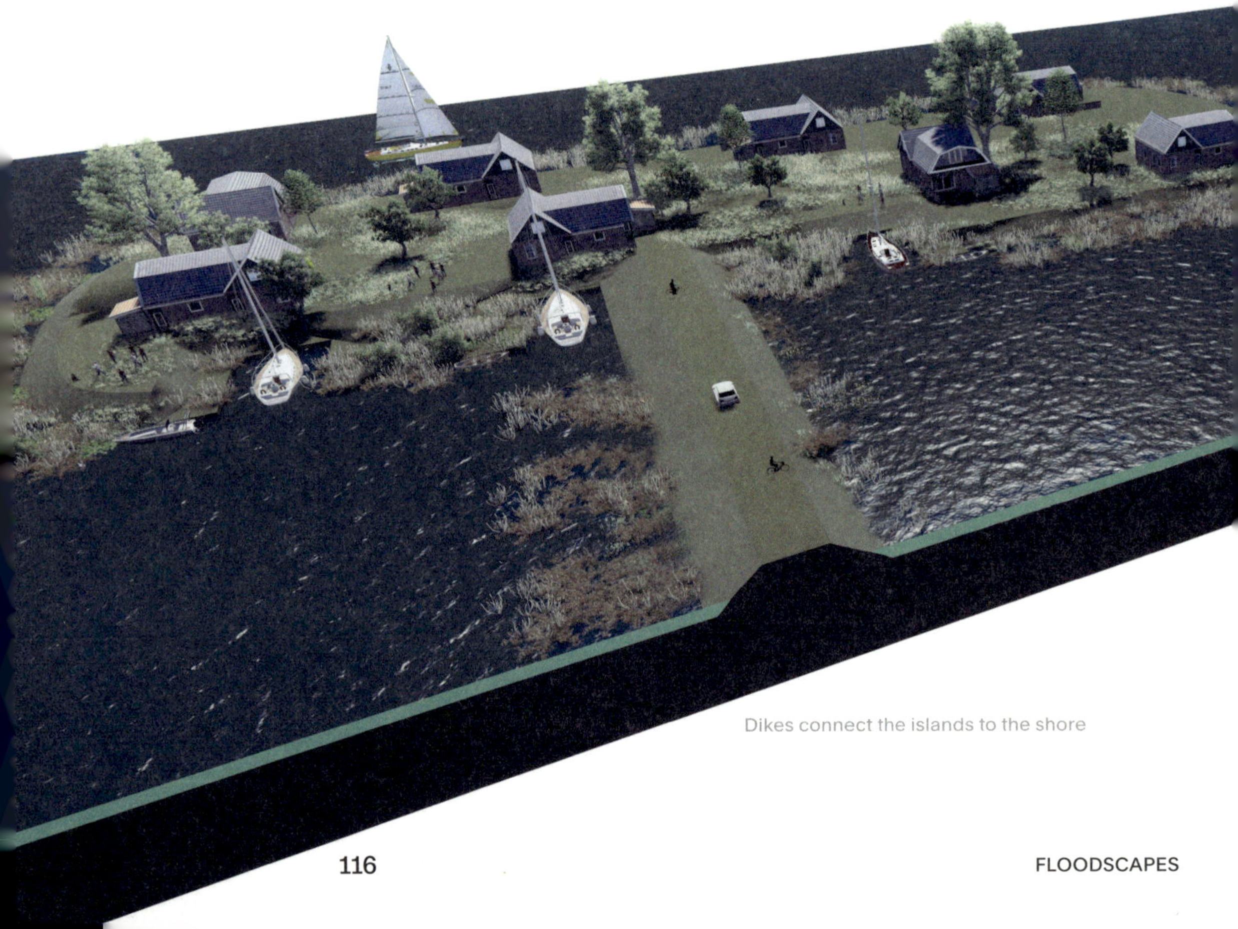

BEYOND THE HOUSING CRISIS: MEERSTAD WITHOUT STAD

The financial crisis hit the Netherlands pretty soon after it started in the United States. A long period of economic growth and accessible mortgages, largely subsidized by the Dutch government, had generated an overheated housing market and an over-exposed banking sector. Dutch private finances were particularly exposed to the real estate market, since almost half of Dutch households financed their own house with a mortgage, and had the highest long-term debt in Europe—almost 120 percent of Dutch GDP (Notten, 2012: 208). The financial crisis had rapid repercussions for the financing capacities of potential buyers, and the housing market saw prices and the number of transactions decrease from 2009 onwards. While the Dutch Office for Statistics recorded 2200 households moving to houses they bought in 2006–07 in the East Groningen region, only 1600 did so in 2009–10 (Das and Van Daalen: 11). After a decade of 5 to 15 percent yearly inflation in the provincial housing market, prices stagnated from the end of 2008 and then declined dramatically in the ensuing years. It quickly appeared that the development of Meerstad would take longer, if ever completed. The land that was purchased for about 300 million euros and was accounted for 356 million euros in 2010, was clearly not worth its estimated value, which was consequently downgraded to

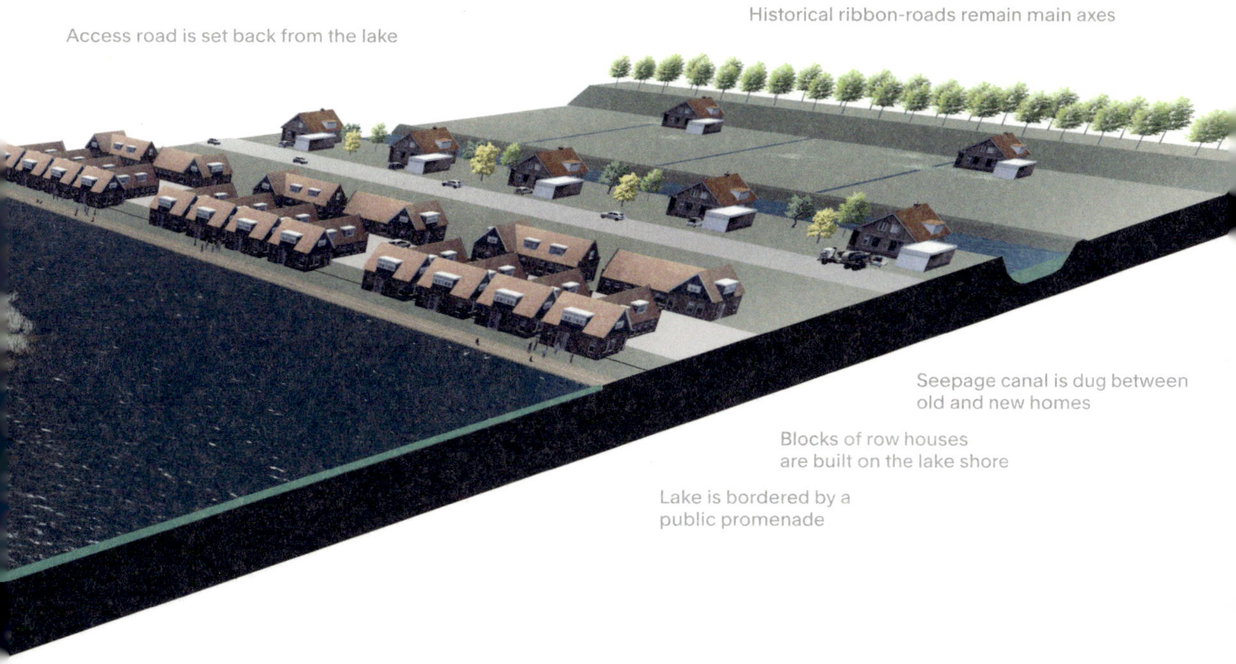

Schematic transect of Meerstad new inhabited landscape, from existing ribbon-villages to new lake (Rossano/Orfanopoulos)

2005 Masterplan (Alle Hosper and KCAP)

317 million in 2011, and again lost 25 million in 2012, lower than its initial purchase value. The limited building and sale prospects drove the four private partners to withdraw from the joint venture on February 29, 2012. Associated public authorities withdrew as well at this point, leaving the municipality of Groningen as sole shareholder: the joint venture had dissolved after only seven years, turning a page on public-private collaboration in Meerstad.

Meerstad however survived the crisis, though not without difficulties. In 2013, 125 hectares were under construction or completed, accounting for about a hundred dwellings—far from the objective set in the initial masterplan to 'deliver 700 dwellings before 2010' (Alle Hosper, KCAP, 2005: bijlagen, 113–114). Along with the financial adjustments and the management reorganization, a new development strategy was presented by the end of 2011 by Bureau Meerstad and the urban designer earlier commissioned for the design of the first phase, De Zwarte Hond. The new strategy did not fundamentally question the ambitions nor the spatial structure set in the masterplan, but reversed its approach: 'we did not drop the masterplan ... but we are thinking from small to big, and not from big to small anymore' (Leverman, 2013). The strategic document subsequently identified the local landscape qualities, existing or planned, that could be exploited to qualify and develop each sub-area, stressing the programmatic flexibility of the plan. The core area, 'Heart of Meerstad', which was initially planned in combination with a public beach, a marina, and a fast public transport connection, remained only as one of many 'possible developments'.

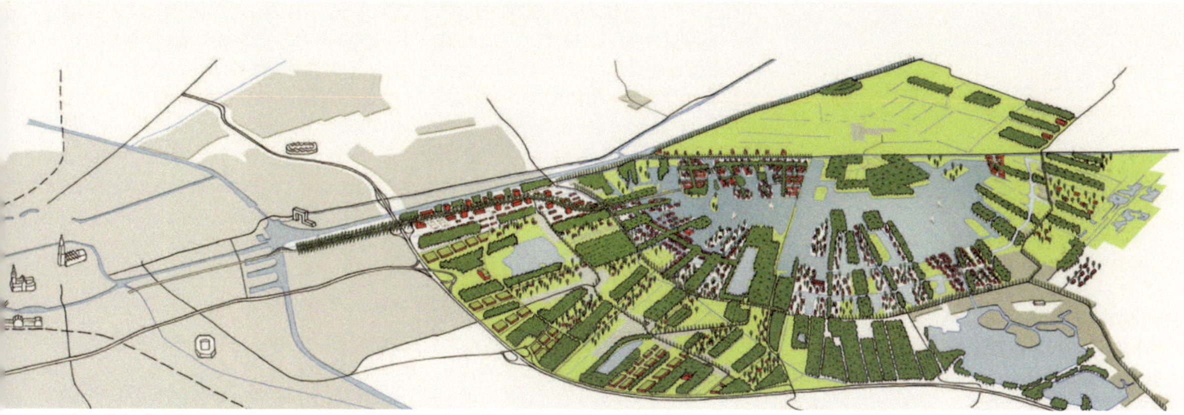

Possible final development stage of Meerstad, De Zwarte Hond, 2011

Jan van de Bospoort, landscape architect and one of the early participants for the municipality, is today critical about the rigidity of the 2005 masterplan: 'We knew already back then [during the design phase] that it would not happen the way we were planning it. To me it is a pity that it became such a clotted plan. It could have defined flexible orientations better.' To him, the housing programme was simply too big for Groningen, with its countryside already providing many opportunities to live outside but close to the city. 'We had to much competition from other living environments around Groningen, and there was too little regional coordination to make this happen.' Not only was it impossible to concentrate the regional housing demand into one big project, but for Jan van de Bospoort it was also retrospectively the wrong orientation, made under pressures of time and socioeconomic circumstances: 'We should have taken a time-out when we realized that we were not able to build in small areas, and take a break before we made the—forced—big jump.' Acknowledging the problems induced by sprawl for the regional capital, he retrospectively doubts that Meerstad came up with the right answer: 'We should not have surrendered to the political and social pressure. Is it that bad for Groningen if people go look for a house in Roden or Zuidhorn [two small towns close to Groningen]? It is of course a concern when people outside of Groningen benefit from its public facilities without paying [taxes], but this should be solved through financial agreements. Groningen had to provide everything, urban and green environments; retrospectively it was a bad idea to develop a big residential area there.'

For landscape Architect Remco Rolvink, looking back at a ten-year process, 'the pre-investment principle is a reality, but implemented phase after phase, with each step expanding the landscape elements and strengthening the area as a whole' (Rolvink, 2012). It is thus not so much the spatial structure that is questioned but the far too optimistic construction plan, its static building programme and overambitious financial scheme that led to a financial disaster for its stakeholders. 'The early 2000s were characterized by growth and enthusiasm, which to a great extent explains the massive scale of the project. Retrospectively, the project could have anticipated a more progressive development.'

Meerstad, island and 'sail-in' villas under construction • First villas buit at the Wolmeer lake (Rossano, 2013)

Today, however, from the windows of the newly finished villas, the lake already seems to reach to the horizon, although it is bordered by a temporary dike awaiting future extension. The subsequent phases are in the planning stage, with the development of 'Lakeshore 2', along with a distant eye on the completion of the Meerstad Midden area anticipated in municipal zoning plans. The completion of the project within the framework ambitions of 2001–2005 seems far away today; Meerstad counted 550 dwellings in 2017, with a moderate ambition of about 2500 houses in 2030. However, the original goals of the commissioners and designers—designing a green and blue framework that could facilitate diverse and unpredicted urban developments—have emerged reinforced from this self-critical phase.

Paradoxically, the careful and integrative elaboration of the project, from the first ideas developed in the eighties to the master plan validated in 2005, can be held responsible for the financial failure as much as it can be praised for its capacity to bundle various interests and motivate all actors through an open process. Multiple goals had a stimulating effect, since 'all these functions still had to come about—nature, water, water buffering—and this brings constraints, but constraints bring about creativity. [Interdisciplinary work] was not a limitation, it just led to a different way of working' (Van de Bospoort, 2013). Furthermore, 'The combination of multiple actors and ambitions, together with the involvement of the population, was a strong factor of integral quality' (Rolvink, 2012), but it was also a strong factor in the delay of the development of the plan, which eventually came too late to benefit from twenty years of booming real estate.

What can this experiment teach us about integrated regional development and water management? Meerstad might have been seen as 'a little pond for real estate developers' (Van de Bospoort, 2013). From today's post-financial crisis perspective, the project was based on far too optimistic an economy—but this could be said about most development projects that were developed at the turn of the century, at the height of the housing market, and fully hit by the financial crisis in 2008. This however reveals a more essential discrepancy between the fluctuating nature of the housing sector, short-term expectancies of private investors, the relative slowness of regional development, the urgent response needed to address flood risks after the 1998 events and the necessary long-term vision of water management.

The price paid for poor farmland was obviously based on a reflexive inflation, generated by programmatic ambitions based on short-term perspectives but uncertain in the long run. It did, however, give local authorities lasting control of a still promising area, securing potential for valuable long-term growth. The water board had to look elsewhere to satisfy its short-term need for reliable water storage, but will doubtless make use of the capacity offered by Meerstad in the future, since climate change projections remain pessimistic. New nature areas have already been implemented, offering new ecological and recreational quality to the region and the country. Meerstad thus still retains its original potential and, provided that authorities remain faithful to their initial ambition, it should eventually become what it was meant to be: a vast and generous mix of water, nature, and housing, organically grown within a unique landscape, addressing simultaneously urban growth and flood safety.

MEERSTAD GRONINGEN: KEY FACTS

Emergency water retention is a constraint that can be transformed into an opportunity.
Flood prevention can be combined with recreational uses, and support landscape enhancement and urban growth.
Private investment can co-finance and energize floodplain redevelopment on the short term, but different expectations need synchronization.
Water authorities should be consulted about and associated to adaptation plans, but must always maintain their independence and autonomy in order to ensure efficient management in the long term.
Over-ambitious programme might need downscaling, but a robust landscape resists political and economic setback.

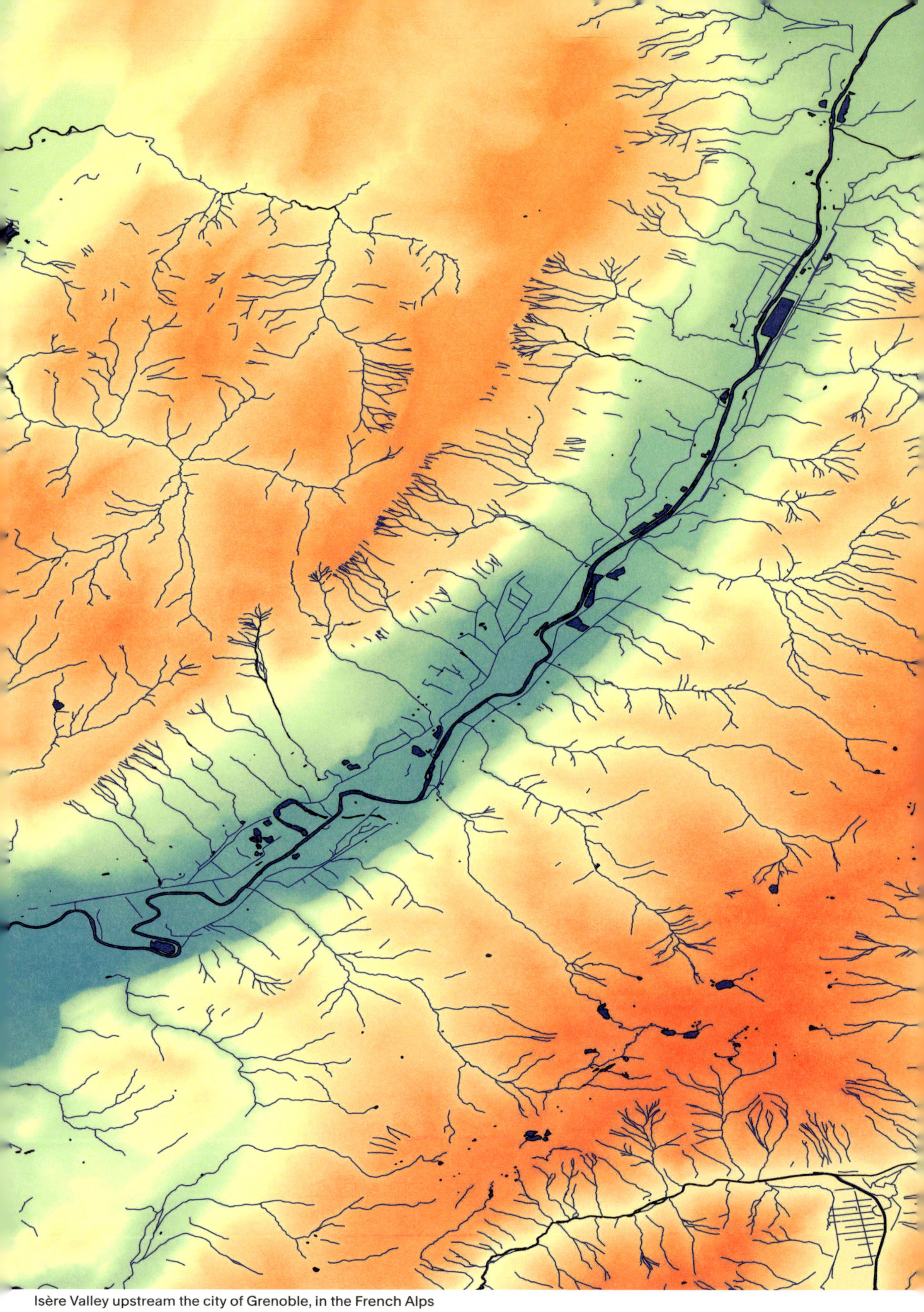

Isère Valley upstream the city of Grenoble, in the French Alps

FARMERS AS RISK-CATCHERS: FLOOD FIELDS IN THE ISÈRE VALLEY

The French Isère Amont project stretches over a 50-kilometre segment of the River Isère, from the border of the Savoie Département upstream to the city of Grenoble downstream, with the name of the project ('Isère-Upstream') referring to the situation of the city. A discrete project in terms of landscape impact, Isère Amont is nevertheless radically changing the pre-existing paradigm regarding flood risks in the valley. The project was initiated at the end of the 1990s, following upon a risk survey that had shown a critical state of the river's levees at several segments, and insufficient capacity to face even a thirty-year occurring flood without damage, which would put more than 100,000 people at risk. In order to protect the capital city of the Département, local authorities took the decision to facilitate flooding rather than resist it, and to seek to control and mitigate its effects. Sixteen 'controlled flood fields' planned in combination with other structural measures should be fully operational in 2021.

The River Isère has its source in the Western Alps, draining an area of 11,800 km^2 from the Italian border down to its confluence with the Rhône River. Fed by alpine torrents and rivers, the Isère within the project area has a torrential regime but usually meanders slowly, following the valley's flat topography and the confluence with the River Drac downstream from Grenoble. Prior to dam and dike building, which occurred mostly in the nineteenth and twentieth centuries, the valley had been through a series of metamorphoses following climatic evolution and the first anthropic influences. Throughout human history, the valley showed a relatively stable hydrology until the sixteenth century, allowing for a cultivated landscape to develop already from the ninth century onwards. During the 'Little Ice Age', however, the Isère evolved from a single bed to an unstable braided morphology, as the river bottom rose, carrying large amounts of coarse sediments brought by mountain torrents, a phenomenon enhanced by the extensive deforestation of the Alps. In 1651, the level in the Isère rose 6.5 metres above the low-water mark, reaching the second floor of the city's buildings, and frequent floods followed during the seventeenth and eighteenth centuries (Bravard, 1989). The river was then still a slow stream, with a pre-photography engraving showing boats and people bathing in the centre of Grenoble, but its surges were frequent and devastating. The 1733 inundation was particularly costly and revealed

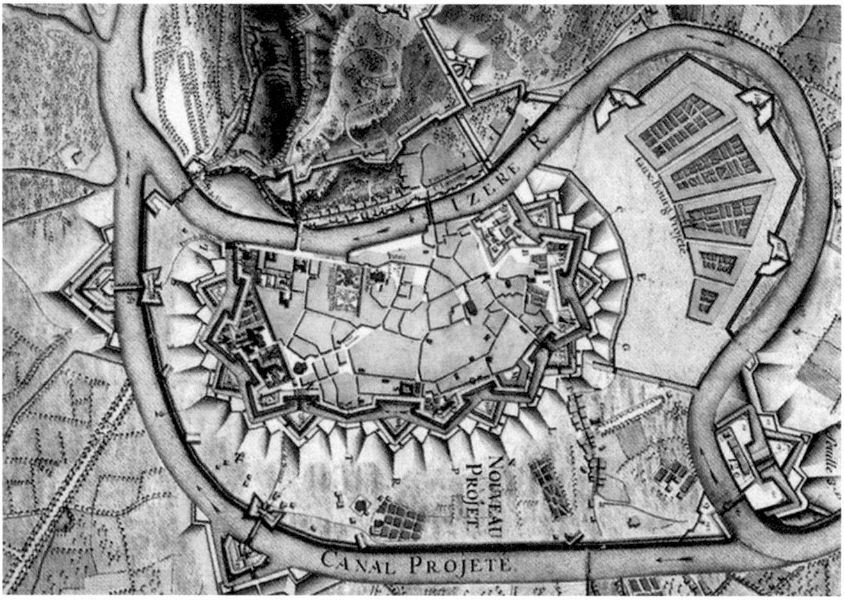

Design for the fortification of Grenoble and a diversion channel south of the city, made by the Ingénieurs du Génie in 1747 (Service Historique de la Défense, Vincennes)

the city's exposure. It would remain famous thanks to the verses written by the poet François Blanc, beautifully illustrated by Rahoult and published in 1860 under the title 'Grenoblo Malherou' (see chapter 'From Myth to Phenomenon'). In this agitated century, fear of invasions and of floods inspired an ambitious project conceived in 1747 by the 'Génie', the French royal military corps of engineers. Since the Isère was squeezed within the old city, threatening to destroy its bridges and flood its heart, the engineers proposed splitting the river into two branches, by creating a new channel encompassing the city on its southern border. Not only would this new channel double the river's discharge capacity and deviate a major part of the stream but it could also create a second ring of military defence, and allow for the construction of a new neighbourhood sheltered against future attacks from the river or foreign armies. All-in-one—urban design, military defence project, and flood protection—the proposal showed a holistic vision of urban planning, pioneering present combinatory projects. Although visionary in its integrative approach, the project, like many of those that followed, was not implemented due to a lack of financial support by the state—with the local economy being far too modest to fund such a *Grand Projet*.

Following the 1778 flooding—known as the Saint-Crespin Déluge—the national administration acknowledged and again asserted the need for a global project along the Isère, which would combine levees, flow straightening, and warping (colmatage). Engineer Rolland described the situation in 1787: 'Floods pour over the valley, where streams develop due to the height of the overflows, obstruct old passages with gravel deposits, and open new ones through the best parcels ... part of the best land is already covered with sterile gravel ... the river, when destroying its sides, widens its bed, divides its waters, and does not have enough strength to carry the materials brought in by the torrents, so that its bed rises up to the point that the smallest raise [in water level] will lead to overflow' (Rolland, 1787, quoted in Coeur, 2008). A new project was drawn up, which proposed straightening 51 kilometres of the river and redrawing no less than 24 kilometres of its original meandering course. The outbreak of the French Revolution, and the financial and political difficulties that followed, however, also postponed this plan (Pinhas, 2013).

1870–1914: THE ISÈRE CHANNELLED 'FOR GOOD'

In the first half of the nineteenth century, Isère was still bordered and managed by a large number of individual landowners, locally grouped into numerous *Syndicats de riverains* or landowners' associations (Blanic, 1975). The levees visible on Raoul Blanchard's map of 1845 formed a discontinuous protective shield, managed by 36 associations from Grenoble to the border, without any coordination (Pinhas, 2013). Upstream in the neighbouring Savoy Département (then part of the Kingdom of Sardinia), the Isère River was systematically channelled and its floodplain artificially raised (Pinhas, 2009 and Girel, 2010). The channelling of the Savoy section had unfortunate consequences in the French part, increasing sedimentation where the river slowed down due to the valley's natural topography.

Further generalized in the second half of the century, levees were continually raised by the associations, supported by public subsidies (Blanic, 1975, Castex et Belleudy, 2005), while parallel drainage ditches, the 'chantournes', were dug to drain adjacent farmlands—this, however, without any coherent plan nor any single authority to coordinate the various interventions. These protection works led to a prosperous period of agricultural development until the early twentieth century, when their disruptive side effects started to have a negative impact on land safety and productivity again. Without cutting the meanders, the correction had, however, considerably reduced the former floodplain (up to 1500 metres at its widest) down to a 100-metre-wide single channel. By putting an end to the natural spreading of sediments over the valley, the channelling thus induced the development of stone and gravel islands within the channel. These islands were further consolidated by silt deposits and pioneer vegetation, provoking a gradual and irregular raise of the riverbed. This consequently reduced the river's discharge capacity and provoked increased damage during flood occurrences, which remained relatively frequent. The rising riverbed further led to overflowing, infiltration and a rise in the water table, which, in turn, transformed part of the valley into marshlands. In 1926, it was estimated that about a quarter of the valley was suffering from an excess of water (Blanic, 1975).

THE 1929 SCHNEIDER PLAN: A MANUAL FOR FLOOD MITIGATION

In 1927, a competition was organized to elaborate and implement a global river correction plan. The Schneider Company won the competition with a 260-million-franc proposal that combined the deepening the river, raising of the existing levees, and the restoration of the canals and ditches draining the valley. More specific in the proposal was the creation of areas for temporary retention in order to lower the water levels at peak discharge and protect the city downstream (Pinhas, 2009). Spillways were planned between Pontcharra and Saint-Gervais to let the flow spread over farmland and return to the main riverbed after the flood.

The idea, already suggested by the Gentil Report in 1860 (Castex and Belleudy, 2005: 18), was further developed and generalized in the Schneider proposal, which defined and described the principles of a dual system, adding to the correction of the riverbed itself the precise adaptation of the floodplain's topography to facilitate and control flood propagation: 'Where the riverbed is established for good, the valley on each side will be divided into distinct compartments, protected upstream by a perpendicular flood-proof [insubmersible] levee. These compartments, defined by existing dikes along the river, would be delimited downstream by the transversal edge of the next compartment.... The dikes would be set at the height of the ordinary flood level (3.3 metres above low-water level) and built as an intake so that it can be topped off without being overturned. The length of this intake will be calibrated so that the compartment to be submerged can be filled up to the level of the intake, before it flows over the upstream levee. Each compartment will have aqueducts with a valve to let waters flow out when the flood is over. This would guarantee the system's insubmersibility for all ordinary floods. For large floods, waters would, on the contrary, enter the compartments by overflowing the lower part of the dike. The role of these dikes would be a storage in the valley, similar to what would happen by dike breach, but with very limited damage, as it would prevent the formation of rapid streams' (idem).

This first integral mitigation project, meant to be fully financed with public funding, was unfortunately abandoned for financial reasons but was soon followed by a major reorganization of the water administration: the various landowners' associations became members in 1936 of one single management authority for the river's levees, the 'Association Départementale des collectivités intéressées à l'Aménagement des plaines de l'Isère, du Drac, et de la Romanche', which included not only private landowners but also the Municipality of Grenoble and the Isère Département authorities.

In 1940, a large flood proved a reminder of the urgency of a defence project. It followed upon an exceptional discharge of 1500 m^3/s measured in La Gâche and provoked a general dike breach upstream at Brignoud. The Isère covered 14 km^2 of farmland, causing widespread damage. The event also confirmed the potential of 'preventive flooding'—although unsolicited—for the protection of Grenoble: The river discharge was almost divided by two (900 m^3/s) thanks to the involuntary spreading of the flow a few kilometres upstream from the city. The flood also secured the financial basis for the correction project in the finance bill passed in 1941. The Schneider project was further investigated and developed by the Neyrpic Company (whose engineering branch, the Laboratoire Dauphinois d'Hydraulique, became Sogreah in 1972, and continued working on the Isère correction until the preliminary studies for Isère Amont in 2000). Neyrpic retained the main principles from the Schneider plan, but added a new proposal: cut three of the river's meanders.

Between 1950 and 1969, the river correction was partially completed and was indirectly supported by urban developments: Grenoble's population soared from 102,000 inhabitants in 1946 to 161,000 in 1968, the year of the Winter Olympic Games. Out of the four million cubic metres of mostly gravel extracted out of the Isère before 1965, 2.5 million were sold for the construction of buildings and infrastructure for the city and its booming agglomeration (Blanic, 1975). In total, 15 million cubic metres were extracted from the river upstream from Grenoble (Pinhas, 2009): The river correction was then not only an enterprise of flood protection but also, and maybe given priority, a mining enterprise—with the extraction authorizations being actually issued by the governmental mining authorities.

On February 2, 1968, the 2.5 km long meander called Bois Français was cut off by an 850-metre-long and 110-metre-wide channel. 'At 11:17 a.m., Grenoble time, the Isère River has obeyed the engineers of the Ponts et Chaussée and changed riverbed', reported the *Progrès* newspaper (*Progrès*, 2.02.1968, quoted in Pinhas, 2009–2011). The local paper *Dauphiné Libéré* celebrated lyrically and with a slightly misogynous tone, the superiority of man's will upon nature's unpredictably, the 'victory of "Man" over her, the river' (the French word 'rivière' being a feminine noun): 'Today the hand of Man dictates the way, and she surrenders.... At 11.24 ... the unfaithful Isère has adopted the bed offered by Man, who dug for her a 900-metres-long channel' (Dauphiné Libéré, 2.02.1968). With this new bypass, inhabitants upstream hoped to reduce sedimentation by accelerating the stream, following the strategy promoted by Tulla in the early nineteenth century; the City of Grenoble downstream believed the extraction had supplied an answer sufficient enough to prevent a further rise in the riverbed (Blanic 1975), and thus had agreed with the measure. Despite studies to fill and exploit the river branch cut off from the stream, it remained in place and would later be divided into five ponds dedicated to sports and leisure.

CORRECTING THE SECOND CORRECTION

Just like the first channelling, this long-planned correction (1929–1969) again generated new problems and new risks. The riverbed dropped rapidly, under the combined influence of intensive extraction, increased water velocity, and a decrease in sediment input held by hydroelectric power dams built upstream. In the years following the river's transformation, the riverbed dropped, on average, thirty centimetres per year, up to three metres in some sections (Blanic, 1975). Furthermore, the urban section of the Isère was not adapted for large amounts of water, and the transformed upstream section was not designed to retain or delay the river discharge, worsening the risk of flood and damage within the city.

A new plan was elaborated in 1971–72 to mitigate the effects of the previous project and again to protect the agglomeration of Grenoble that had by then largely invaded the river's natural floodplain. A stop was put to private extraction from the riverbed. Walls and dikes were raised locally within the city to resist a 1500 m^3/s discharge. Although this figure was still below the discharge level of the 1859 flood (then estimated at

1800 m³/s, the highest known), it became the reference discharge, since the studies based on a 1/100 model built by the Sogreah Company showed that the city bridges and embankments could not cope with a higher discharge. Raising the discharge capacity in the valley further thus seemed counterproductive as long as the inner city remained a bottleneck, a situation that could only be changed by gigantic works and unrealistic investment. The project included (again) the deepening of the riverbed, meant to allow another 300m³/s discharge, through digging and through the creation of a reinforced 50-metre-wide and four-metre-deep channel in its centre. A sunken dam downstream would maintain the minimum water level and prevent a drop in the water table, along with the damage this could provoke in the wooden foundations of ancient buildings.

The Sogreah studies went further than the sole aim of increasing the capacity of the riverbed. Flood propagation over the valley upstream Grenoble was also investigated, as well as possible mitigation effects benefiting the city's safety. It reintroduced the idea of creating 'flood fields' to protect the most valuable areas, an approach earlier proposed in the Schneider plan; a mathematical model made it possible to consider and study the floodplain as a whole system of communicating basins. An attempt was made to apply this principle in Le Touvet, with the construction of a levee in a tunnel crossing the A41 motorway. The levee, meant to facilitate and regulate the flood propagation between two areas of farmland, was soon removed, most probably by farmers considering it to be too big a disruption in their daily activity and movements (Jardin, 2012). Although engineers no longer considered absolute safety a realistic ambition and already began work on flood deviation and mitigation principles to protect urban area and spread the risks over the valley, the principle, however, still appeared to be unacceptable to the valley's main users.

The following decades saw no major floods, and the levees progressively degraded as maintenance was neglected. Meanwhile, the number of people and properties exposed to flood risk soared, as the valley's urbanization continued at its strong pace. A study conducted in 1995 showed that a 200-year flood would impact 10,000 dwellings, 120 farms, 6000 companies as well as many public buildings such as schools and hospitals. 34,000 jobs were involved, and the damage could cost 500 million euros, according to an updated survey made in 2005 (SYMBHI, 2009a: 7). Although no major flood occurred in this period, concerns and awareness grew around the doubtful safety of the Isère levees and the growing exposure of the agglomeration.

A STRING OF FUSES FOR THE ISÈRE RIVER

The Association des Digues, responsible for maintenance but having no mandate to invest and conduct large-scale structural works, in 1991–92 initiated a Project of Public Interest (*Projet d'Intérêt Général*) for the Isère Amont area. The procedure reinforced the prerogatives of public authorities that prepared a new Flood Risk Prevention Plan (PPRI), which became compulsory for all French municipalities in 1995. A diagnostic study was made of the river levees, which revealed great disparities in their composition and stability. The SOGRÉAH Company was again commissioned to

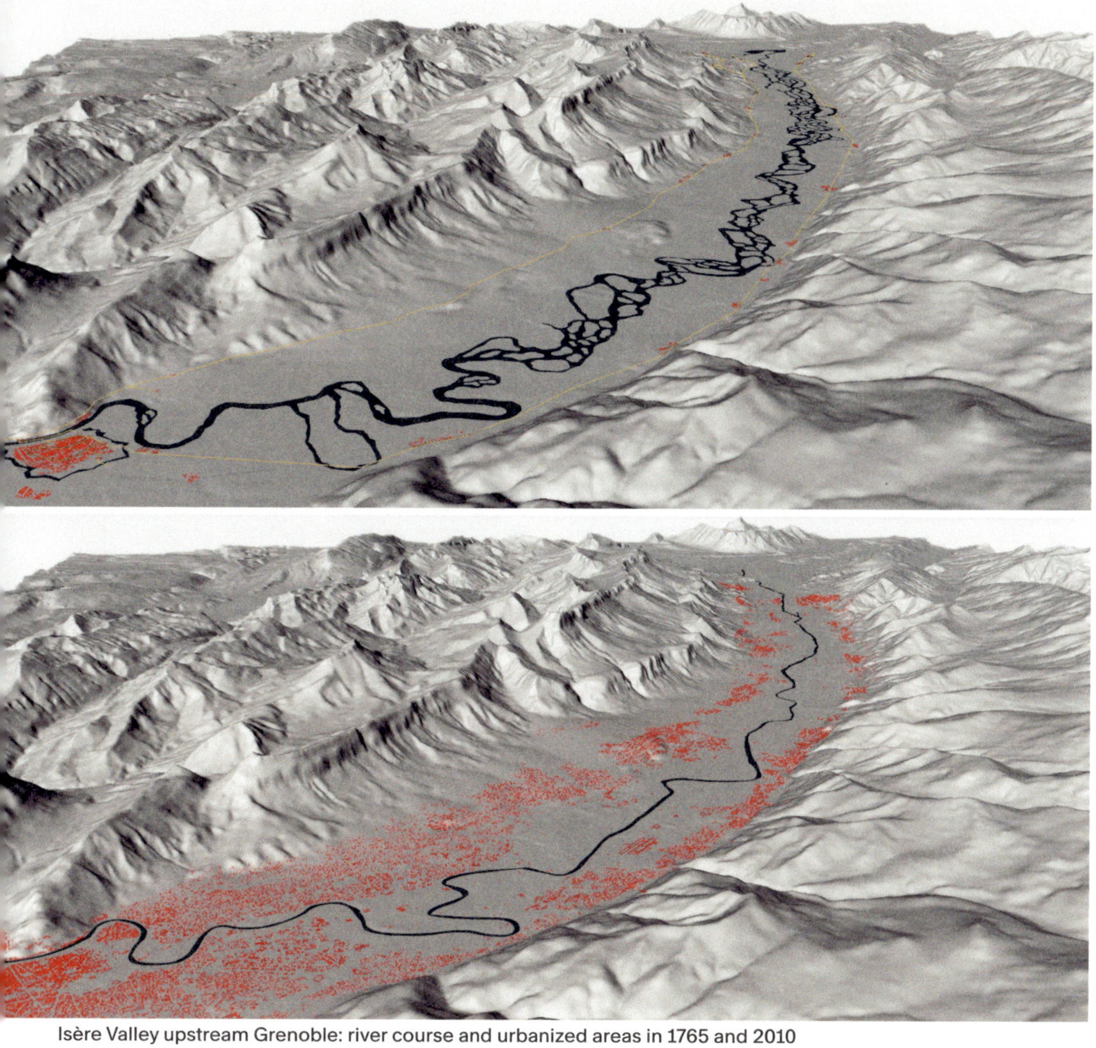

Isère Valley upstream Grenoble: river course and urbanized areas in 1765 and 2010

assess flood risks within the valley's now densely urbanized context, over a territory involving 29 municipalities and belonging to two administrative communities: the Communauté d'Agglomération Grenoble-Alpes Métropole and the Communauté de Communes du Pays du Grésivaudan. In parallel to this risk assessment, SOGRÉAH was also commissioned to work on a global project to address flood risks from the northern border of the Département, down to the city of Grenoble, a stretch of 25 kilometres, where so far no authority had succeeded in implementing a coherent flood defence system or strategy.

Conforming to the national PPRI regulations, SOGRÉAH again took the highest known water discharge, the 1859 flood, as a reference in order to identify risk areas and potential water retention areas. The 1859 high water was, however, prior to reliable discharge measurements and the estimates were based on available historical information, such as local water levels and flood propagation. These data, already used for the 1972 plan, were thus interpreted and developed through the use of statistical models that had great territorial consequences. So did some conclusions of the report, which identified new risk areas—contradicting previous planning documents—and frustrated the expansion possibilities of several municipalities, while locally boosting land prices (Jardin, 2012).

This generated a new controversy among local authorities. Doubts were raised over the reliability of flood risk projections—allegedly over-estimated according to local representatives. In a letter sent by the municipality of Champ-près-Froges to the Conseil Général (department council), a virulent critique was addressed to the authors of the new PPRI, in particular with regard to contradictory studies made on flood risks and on the impact on infrastructure, as well as with regard to the preference given to the worst case scenario for the tributaries' discharge, exaggerating flood risks along the Isère River. The letter also expressed a more general frustration regarding the area that should be reserved for temporary water storage—and thus subtracted from potential urban growth. This was perceived as a sacrifice on the part of small municipalities caught between the effect of river channelling upstream, and the urban growth and increasing exposure of Grenoble: 'Squeezed between Grenoble downstream and the Savoie Département upstream, the municipalities of the Grésivaudan Valley have to stand alone against the Isère big floods; upstream in Savoie, the dikes built in the nineteenth century have closed off the former floodplain, and the high floods are not absorbed anymore; downstream, close to Grenoble, between Isère and the railway, the last flood fields have been eliminated' (Letter from the Municipality of Champ-près-Froges, 2006).

For this small municipality, the new flood regulation was seen as an illegitimate limitation of its development for the sole benefit and safety of the big city of Grenoble, echoing a conflict seen in most mitigation projects. Flood risks were allegedly overestimated, as a form of political blackmail to force small municipalities to cooperate with the big plan: 'We cannot accept our construction zone being frozen in accordance with unreasonable principles that exceed the regulatory framework' (idem). 'To get politicians to act, one needs to be maximalist, irrational in a way' (Jardin, 2012). Whether accurate or exaggerated, the maximalist view ultimately succeeded in mobilizing all the municipalities concerned: the project proposed by SOGRÉAH could proceed with the design and building phases, led by SYMBHI, a public agency specifically created in 2006 to elaborate and implement the project Isère Amont.

FROM SCENARIOS TO CONSENSUS

In 2004, three scenarios were defined by the engineers from SOGRÉAH and SYMBHI, based on the goals set for flood protection, environmental restoration, and development of recreation along the Isère. Although the word 'scenario' suggests an open investigation based on various choices, the planning process instead followed a dialectical pattern: thesis, antithesis, synthesis. The first scenario favoured dike reinforcement and riverbed dredging upstream, with negative consequences for the river's fauna and flora. The second scenario proposed moving the levees further away from the river and incorporating the remaining riparian woods within the riverbed, a positive option for environmental development but negative for 180 hectares of farmland that would be flooded every five years on average—an option that was incompatible with vegetable farming. The third scenario was elaborated later to synthesize the remarks made by elected representatives, environmental associations, and farmers during this first round of public participation. Without adding to the existing nature reservations, it proposed associating and balancing the various approaches: allow natural water fluctuation into unexploited areas, in particular existing riparian woods; reinforce 50 kilometres of levees; and create 'controlled flood fields' (*champs d'inondation contrôlée* or 'CIC') to mitigate the effects of future floods.

In 2005, SYMBHI communicated the scenarios at about 45 meetings, eight of which were public. Despite the risk of delays and conflicts —such as the above-mentioned conflict opposing the Département to Champ-près-Froges—the planning process involved extensive communication, thus creating room for discussion and improvement. Furthermore, the project was promoted by deputy Robert Veyret, vice-president of the Conseil Général in charge of water policy and president of SYMBHI, but also the long-time mayor of a small village and as such very keen on involving the local population and elected representatives in the planning process.

This third scenario was validated on 2 February 2006, by the SYMBHI Conseil Syndical. The project was further elaborated segment by segment—with seven areas being defined for the development of the draft. For each one of them, two meetings were organized with the local representatives, and one public meeting and two workshops involving local inhabitants. Comments were then tested by the engineers from SOGRÉAH and SYMBHI, with three-quarters of them being integrated into the draft, which was again presented to the public in October 2006. According to SYMBHI's final report, most of the comments concerned the future water system of irrigation and drainage, the recreational infrastructures, the landscape integration, and ecological measures, as well as the protection of buildings located in the flood fields. The project was finalized in 2007, and again presented to the elected representatives and the public between January and March. It was also presented to the authorities in charge of water, agriculture, environment, public and private infrastructures (railways, highways, electricity networks).

In the meantime, a specific discussion began around the elaboration of a specific compensation system for farmers whose land was located in the future flood fields, and thus would be potentially flooded, starting from a thirty-year flood situation. The local Chamber of Agriculture represented their interests in the negotiation held by SYMBHI. These negotiations lasted for more than two years, leading to new discussions and conflicts with regard to not only the financial compensation system but also the localization of recreational facilities, parking spaces, and tree plantation—which was eventually reduced in order to maintain land productivity. The discussions saw some rather heated episodes and incidents, the most striking being the dumping of manure in front of the Conseil Général office in Grenoble. The agreement was finally signed by both parties in February 2010 and covered various types of damage that could concern farmlands designated as 'controlled flood fields', as well as land exposed to a higher risk after the project's implementation. The compensation not only covered damage to buildings or equipment but also the loss of cattle and harvests. Trucking activities, their irrigation system, drainage networks and greenhouses were in essence more exposed to permanent damage than regular crop fields. The agreement thus included, in these particular cases, compensation for potential loss in sales contracts, and opened up the possibility of implementing extra flood protection whenever the exploitation's sustainability was affected by a flood event (Conseil Général de l'Isère, 2010).

1860–2010: CONCLUDING A 160-YEAR PROJECT

The project implemented by SYMBHI can today be read as the achievement of almost 160 years of designs, great and modest, that successively stumbled upon controversies and financial backlashes. Although based on a new urban situation and updated hydrological models, many elements of the design can be traced back to the plans made by the engineers Cunit (1850–60), Gentil (1860), de la Brosse (1860), Schneider (1927),

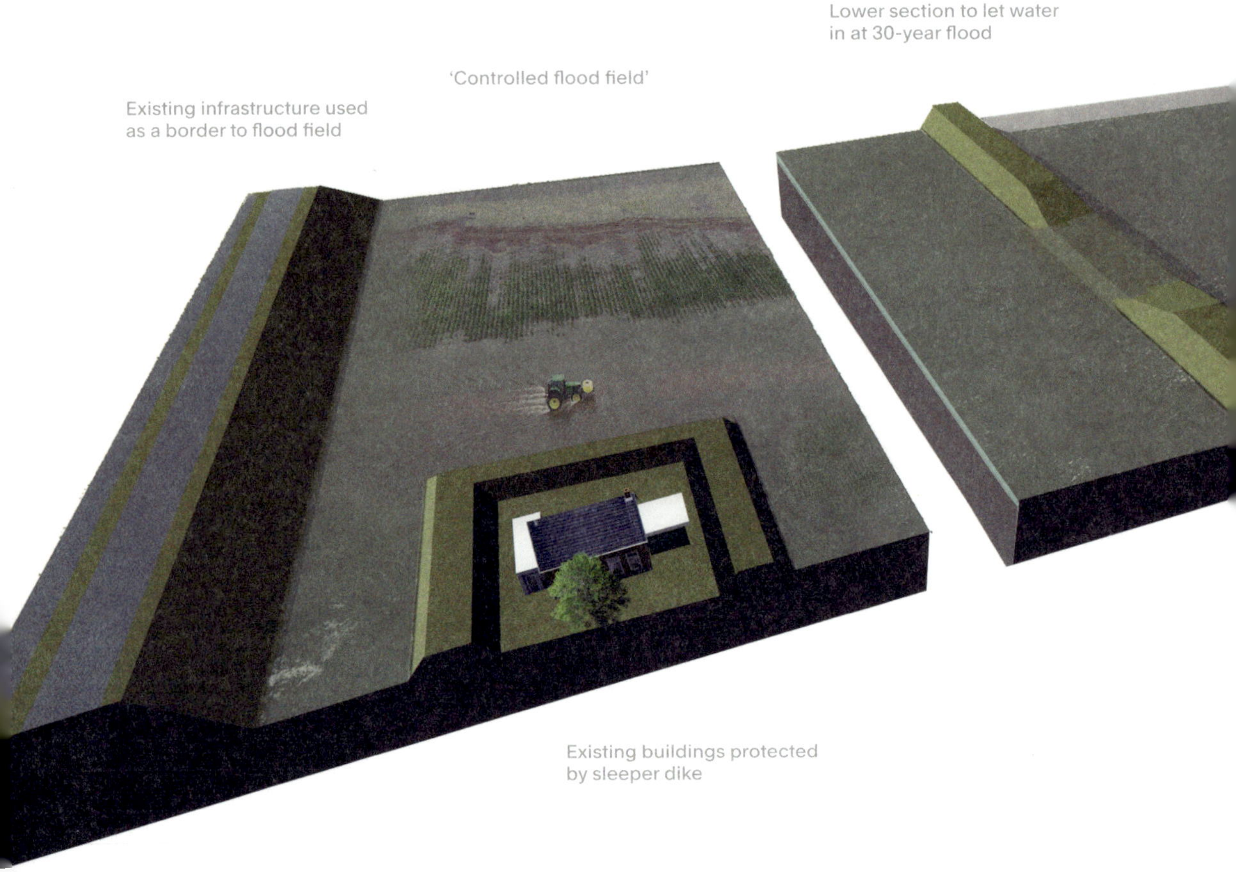

Neyrpic (1942–44), and SOGRÉAH (1972). The 2009 project synthesis claims three objectives: protection of urban areas against a 200-year flood and farmland against a 30-year flood; ecological improvement of the river and adjacent space; and development of nature-based recreation.

Within this combined programme, the hydrological aspect is, however, the most developed and the most significant in the context of contemporary adaptation strategies, measures in favour of ecology and leisure remaining modest in terms of studies and investments. The hydraulic project involves the strengthening of dikes and embankments, the removal of pebble islands within the main riverbed, the creation of two deposition area in order to hold excess materials from the stream and, most importantly, the creation of sixteen areas that can be inundated, based on a 30-year flood. Once completed in 2021, these 'champs d'inondation contrôlée' will allow for 25.5 million m³ water to spread temporarily over 3400 hectares (SYMBHI, 2009), eight times the retention capacity of Meerstad lake.

The flood mitigation system based on the controlled inundation of farmland is now finally to be applied on a large scale in a coherent and dynamic system. The distinction made between urban areas and farmland allowed for a differentiated approach to risk. The flood mitigation goal stated a new role for the valley's farmland, from main asset to risk catcher, a role only made acceptable by the compensation system for potential

Schematic transect of the flood fields implemented along the river Isère (Rossano/Orfanopoulos)

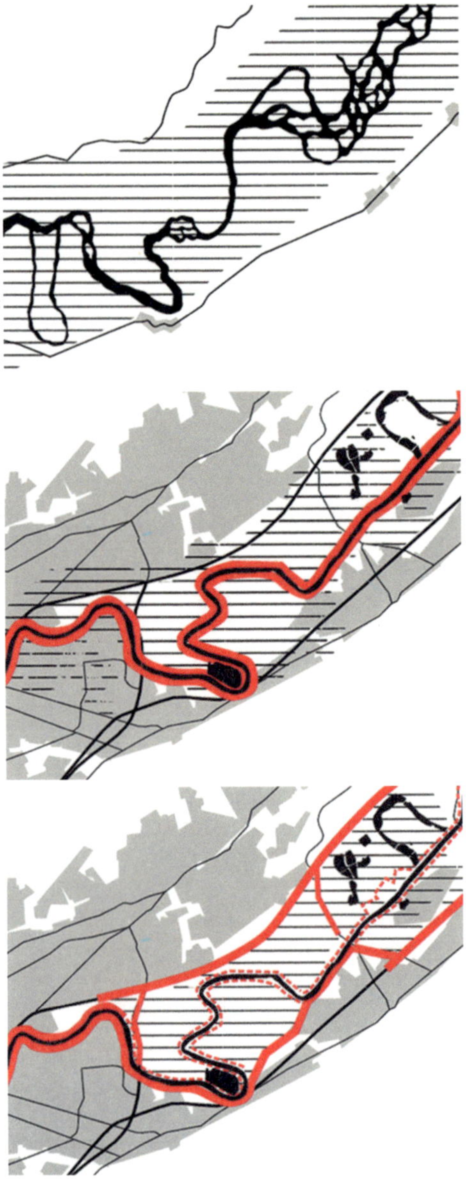

Isère Valley east of Grenoble: river, built areas and floodplain in 1765, 2010 and after implementation of the 'controlled flood fields'

loss. The two levels of protection—or risk tolerance—not only introduced a differentiation in value but also installed a functional relationship between urban areas and farmland. Furthermore, quantifying the protection level introduced a dual land status that implicitly defined the land's functional destination and value for decades at least.

The second stated objective of the project was the restoration and development of natural habitats. SYMBHI's final report presented a diagnosis of the ecological situation of the valley, pointing out the high value of the alluvial forest. The alluvial forest that would naturally cover most of the valley, thanks to the Isère frequent floods and divagations, was in danger of disappearing, however, by urban extension, by agricultural development and drainage, and by the influence of the previous river works, which had interrupted the regular floods and lowered the ground water table. While 50 percent of the valley's open space consisted of arable land, 20 percent was made of hardwood forest populated by species disseminated from the mountainsides, and 30 percent was occupied by a 'mosaic of poplars, hedges, groves, and soft wood' (SYMBHI, 2009). Original alluvial habitats were not even quantified in SYMBHI's final report, whose authors however pinpointed the rarefaction of marshes, of wet grasslands and of natural ponds, respectively turned into arable land and fishponds. The disappearance of dynamic alluvial habitats has indeed been an ongoing process in the Grésivaudan Valley, though stretching over a longer period than in neighbouring Savoy, where river channelling, valley drainage, and landfills were completed in a shorter and more coherent way in the early nineteenth century. In Grésivaudan, on the contrary, the lack of global and coherent maintenance of flood defence works has left a fragmented landscape and sporadic riparian vegetation, too often cut off from the river stream by previous channelling attempts. The vegetation developing exclusively under unstable conditions, in particular under the influence of water, sediments, and organic materials brought in by occasional floods, is hard to find along the river's steep and calibrated embankments. The challenge faced by restoration supporters is therefore to combine a stable and reliable flood defence and a regular and controllable water system needed for profitable crop farming, with a dynamic stream needed for alluvial habitats—along with several endangered species.

The first design sessions tried to develop a 'renaturation' concept that would allow the river to fluctuate again onto larger areas, within a profile compatible with the dense use of the valley. This goal soon proved to be elusive, however: 'We thought that by letting the river erode fifteen metres of embankments, we would give it back its

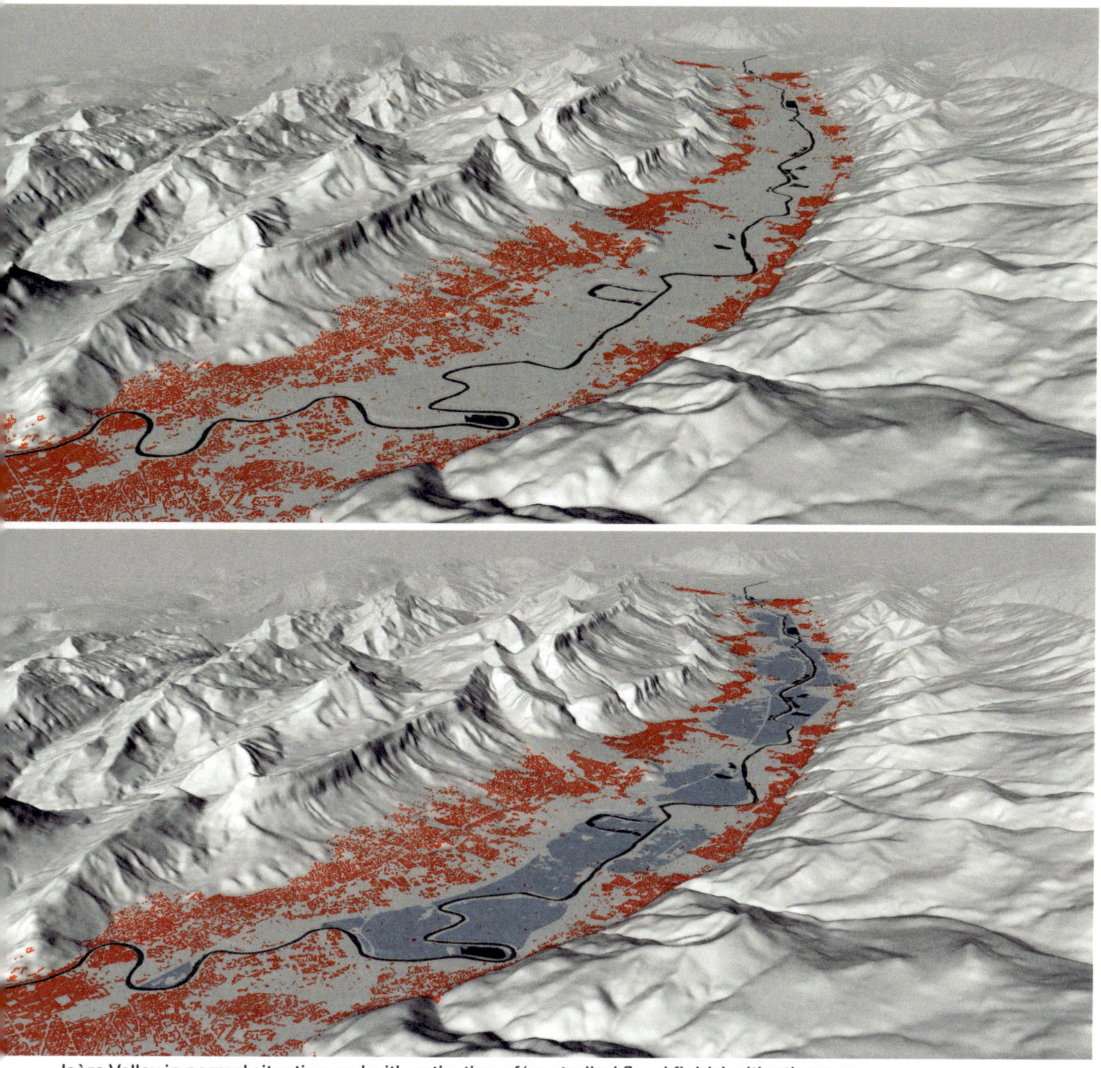

Isère Valley in normal situation and with activation of 'controlled flood fields' mitigation areas

freedom', recalls dike master Michel Pinhas, but 'freedom for a river should be total, or else you have only its downsides: damaged levees, dramatic floods, unstable profile' (Pinhas, 2013). The ribbon of trees on the levees and within the channel was valued as a strong landscape feature and a last ecological asset. However, its presence also appeared controversial during the preliminary investigations: trees could weaken the levees, cast shadow onto farmland, and eventually did not belong in the channel of a regulated river: 'A river flowing between sandbanks, with trees, woods and islands, looks very pleasant, but one does not realize that it is a scale model of the Isère [compressed] between the levees, and that this model will be wiped away by the next flood, at best. In the worst case, the channel will be obstructed and the river will flow over the levees' (idem). The attention thus rapidly shifted from hopes of partial renaturation towards more modest ambitions: reinvigorate the 260 hectares of former alluvial woods by putting them in relation with the river floods again, and reinforce the continuity of the green ribbons along the river by planting about twenty hectares—'generally selected for their limited agricultural value' (SYMBHI, 2009: 20).

The valley's existing landscape was investigated, but through an analytical lens rather than a design-oriented one. In the final report, it was schematically divided into three parallel stretches of 'residential hillside' on the north side, an 'industrial trench' on the south side, and 'woody river between highway and railway' (id.: 9). In a longitudinal direction, the authors of the report identify five segments from east to west: nature and leisure, industry, woods, suburbia, and city, again more a 'fait accompli' than a strategy. Wedged between two suburban ribbons, the Isère Amont project also tried to address the issue of connectivity, here again with restraint. Before the project started, the Isère levees were already used for recreation and slow traffic, particularly between the university campus and the city centre. Taking advantage of the civil works induced by levee reinforcement, promenades and cycling paths were enlarged, or created where lacking, on and along the levee; narrow paths along the river are kept or developed on shorter segments to provide intimate promenades closer to the water. Static elements, such as steps allowing visitors to get closer to the water, and picnic areas, were spread out along the river. New accesses to the water were planned for water sports lovers.

METROPOLITAN RIVER: A MISSED OPPORTUNITY?

Thanks to a threefold approach—flood risk, ecological restoration, and recreational quality—the Isère Amont programme could claim an integrated tactic. Still, unwilling to enter the scope of urban planning, the design avoided addressing spatial challenges and restricted landscape architecture to visual analysis, local recreational amenities, and plantation schemes. As Michel Pinhas, chairman of the Isère Water Board, retrospectively notes, 'At the beginning it was a very hydraulic project, and then later one thought "Let's try to do it right"'. Yet the enormous potential of the river as structuring element and agent of change was left unexploited. The Isère river flows along 27 municipalities in the project area alone, and links major urban functions along its way: four leisure and camping areas (Le Bois Français, la Terrasse, La grande Lône, La Taillat), several industrial and retail areas, the university campus, the Innovallée technological centre, the regional hospital, the headquarters Grenoble-Alpes Métropole, and the city hall of Grenoble, whose park had recently been extended down to the riverside. Since old centres are located on the hillsides, initially settled at a safe distance from the Isère floods, most of the post-war urban extensions were built towards the river, potentially a factor of risk, but also of living quality and good connectivity. Nine bridges alone cross the Isère between Grenoble and Pontcharra, one every four kilometres on average. None offer a cycling lane that could safely lead inhabitants to the recreational lakes, riversides, and promenades. Eventually forming

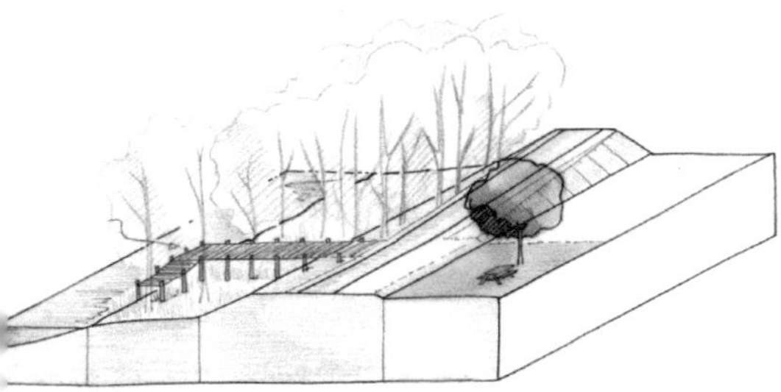

Former riparian wood being reintegrated within the main riverbed (Rossano, 2013) • Landscaping interventions planned along the Isère levees (SYMBHI, 2009)

a mosaic of mono-functional zones—diverse in their substance but uniform in their isolation—residential, retail, and industrial areas were juxtaposed according to municipal ambitions and real estate opportunities, long left uncoordinated by an administrative tax system that let local interests prevail over regional planning. The abovementioned conflict opposing the Municipality of Champs-près-Froges to the Conseil Général was symptomatic of urban development policies that to this day are still defined locally, despite the growing integration of the valley's economy, housing market, leisure practices, and transport system.

In this context, not using the project of securing the river to lend coherence and structure to the scattered urban fabric retrospectively appears to be a missed opportunity for the young agglomeration. The three scenarios put up for debate by SYMBHI in 2005 were evaluated according to the respective costs and benefits for farming and nature as competing components, excluding a symbiotic approach that would seek mutual benefits. Farming activity and ecology were taken into account, but separately. The limited mandate of SYMBHI made it difficult for planners to envision a positive symbiosis for farming, nature, and recreational uses. Accessibility for urbanites was limited everywhere, by lack of infrastructure or, voluntarily, in order to protect biodiversity and exploitations. Paradoxically, keeping the project within the scope of flood prevention has probably avoided long and complex debates, and ensured a swift development and implementation.

The growing number of citizens walking, running, cycling along its shore, shows that the river's potential and its role in the growing alpine metropolis still remains untapped in many aspects. The creation of the 'controlled flood fields', although bringing no added value to the agglomeration in terms of spatial quality, however, by default makes room for future visions, and might be a historical move towards the development of a vast metropolitan landscape, open by necessity, available for the alpine agglomeration to improve its living environment in the future.

ISÈRE AMONT: KEY FACTS

River channelling and widening strategies alternate through history, and can today better be combined than opposed.
River management is an ongoing transformation process with elements held only in temporary equilibrium due to their dynamic nature.
Flood defence can lead to higher ecological value, but 'free flow' restoration is an ideal that is incompatible with dense habitat and land use.
Controlled flooding sets necessary limits to urban sprawl and therefore can secure large open spaces in metropolitan areas; a spatial project is however necessary to exploit the potential benefits in terms of quality of life.

3

UPSCALING FLOODSCAPES DESIGN

New submersible path through the Biesbosch area in South Holland (Rossano)

> '... Each project of river channelization highlights the social dynamics and allows us to understand the political dimension of man's relationship with his environment. The eighteenth century proclaims without acting, the nineteenth century conquers but is weakened by its divisions, the twentieth century is arrogant and certain of its rightfulness. The twenty-first century produces a modest version in which the "Minister-Engineer" suggests to return to the river part of what it is due.'
> (Bender, 2004)

As seen in the previous chapters, flood management questions the boundaries of urban and landscape design: boundaries between city centres and peripheries, with new synergies developed in Groningen; boundaries between urban space and nature areas, with the necessary compromises made for the transformation of the Isar river in Munich; boundaries between river space and dry land, a boundary made porous in the Isère Valley in order to protect the city of Grenoble. In all these projects, interdisciplinary approaches were used, bringing with them new difficulties and new benefits in terms of safety and quality of life, a double goal that guided the design and the transformation of these landscapes.

Is there a maximum size to floodscape design? The Meerstad project aimed, at some stage, at building 10,000 dwellings, inundate or transform 4000 hectares of land, for a total budget of 800 million euros. Had it exceeded a feasibility limit when the whole enterprise almost went bankrupt, before going back to more modest ambitions? How to engage in flood management at the scale of a valley, a plain or a watershed, far beyond the city limits? This might seem out of reach in democratic countries with decentralized political systems, a common idea being that large-scale landscape design is the privilege of authoritarian regimes or the result of major crises, such as the post-war and reconstruction era that saw the Upper-Rhine river radically transformed for the sake of energy production and shipping. A large-scale perspective is however necessary from a hydrological point of view, in order to understand and influence flood processes, and integrate the seasonal changes and variations in climate and land use which altogether determine the frequency and intensity of flooding. How should one then build clear narratives, consensual goals and efficient governance systems at the regional scale, over a period of several decades? As seen with the nineteenth-century Comoy Plan made for the Loire river, large ambitions can stumble upon political, economic and climatic variations that are difficult to predict or integrate in time and space.

It is *a fortiori* interesting to trace the history of two contemporary European projects of an extraordinary size and approach: the *Troisième Correction du Rhône* (Third Rhône Correction) in Switzerland and the *Ruimte voor de Rivier* (literally 'room for the river') programme in the Netherlands. The decision to address river safety on a larger scale was taken almost simultaneously in Valais, where the investigations for the 'Troisième Correction du Rhône' were launched in 1995, and in the Netherlands, when the 1996 Ruimte voor de Rivier governmental guideline was approved (Min. van Verkeer en Waterstaat et al., 1996). Room for the River concerns the Rhine estuary and its various branches (Maas, Lek, IJssel, Waal), a total of about 450 kilometres. A budget of over two billion euros had been set aside for its implementation, and expenses had reached 2.3 billion in 2017 (Min. van Landbouw et al., 2006 and Min. van Infrastructuur en Waterstaat, 2018). The programme meant to improve the safety of four million people living in the Dutch river region, was validated in 2006 by the governmental Key Decision (*Kernbesluit*), and was completed in 2017 (although some projects along the IJssel, the northern arm of the Rhine, are not expected to be delivered before 2022).

The Third Rhône Correction encompasses 160 kilometres of stream. The Rhone river is being transformed and locally enlarged in order to protect a 14,000-hectare flood prone area (11,000 in the Canton of Valais and 3000 in Vaud), and about 100,000 people living in the valley (Projet Rhône, 2011: 2). A substantial investment reflects the project's ambition: with a budget initially estimated at 2.5 billion Swiss Francs for the full implementation of the project (Projet Rhône, 2008: 66), equivalent to 2.3 billion euros. In terms of absolute value, 'Rhône III' is equivalent to the Room for the River programme. Considering the regional context, the 'Rhône III' budget is largely superior: equivalent to 16 percent of the GDP of the Valais Canton in 2008, and corresponds to an investment of 25,000 Swiss Francs per inhabitant living in the risk area. Needless to say, the threat is not limited to the population physically residing within the Rhône floodplain: with 77 percent of its territory covered by steep forests or considered unworkable, the Valais sees securing the Rhône Valley as a vital issue for the entire canton.

From the start, both programmes marked a turn in the history of flood defence, by favouring spatial measures over channelling infrastructure, and by reintroducing flood mitigation into regional planning. For both programmes, also, authorities sought to generate understanding and consensus around the chosen strategy, although using very different narratives. The Dutch government chose to communicate with a general and simple motto, fully expressed in the name given to the programme: Room for the River. The government of the Valais Canton sent a message that was altogether more specific and more complex when introducing the Third Rhône Correction. The name itself suggested, in the Dutch case, a

Rivers addressed in the Room for the River program in the Netherlands and the Third Rhône Correction in Switzerland

historical shift towards horizontal measures, while the Swiss denomination implied a repetition or at least a continuation of the two previous river corrections implemented in 1863–1894 and 1930–60. Despite similar scales and strategies, the Swiss and Dutch authorities made different choices, the one emphasizing the innovative character of the project, the other stressing the historical continuity, each programme coping with the vastness of its planning area and the historical proportions of the enterprise in its own way.

On top of their extraordinary scale and cost, these two grand projects didn't 'benefit' from a strong authoritarian government, both countries being democratic and regionalized, or from an emergency situation (although local floods did accelerate their implementation). They have instead developed through complex processes of participation and consulting, aiming at building a large consensus, sometimes at the risk of overlooking divergences or betraying original spatial and ecological ambitions.

The two following stories show that simple principles don't lead to systematic solutions. For the Rhône as well as for the Rhine, the full and continuous widening of river channels, whether motivated by flood safety, environmental restoration, or spatial quality, seemed out of reach in urbanized floodplains: historical land use and structures resisted systematic measures and urban and infrastructural bottlenecks formed strong obstacles. Both in Switzerland and in the Netherlands, initial visions of global transformations were soon split up into combinatory and differentiated approaches, resulting in negotiated risk distributions and landscape new deals rather than grand gestures. Eventually, the most useful lessons are hidden in the detail of these two exemplary cases.

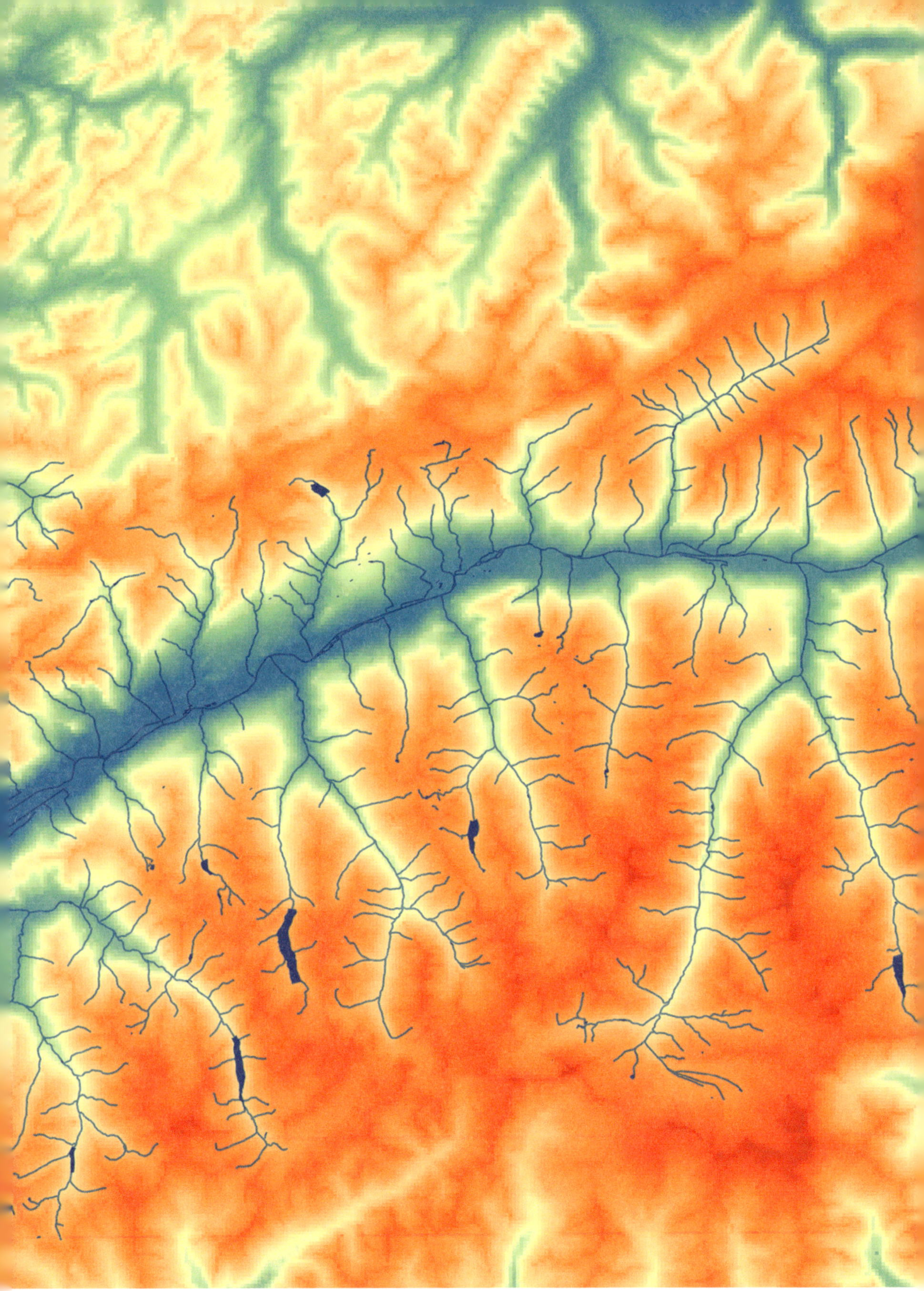

Upper-Rhône Valley in Switzerland

THE RIVER AS *GESAMTKUNSTWERK*? THE THIRD RHÔNE CORRECTION

With its length of 160 kilometres, thirty years of construction, and a budget of more than two billion Swiss Francs, the Third Rhône Correction is a project of extraordinary proportions. The project stems from a long history of human interventions on a limited space, long dominated by the river, but is also the result of relatively recent cultural and legal shifts that took place in Switzerland at the federal and local level. Born in the early 1990s and planned to last until 2030, the programme 'Rhône 3' is expected to deeply transform the landscape of the Valley through the widening of the river channel, breaking with a century of coercive engineering measures. Willing to integrate all components of the densely populated valley into a global approach, since its origins the Third Rhône Correction has been presented as a Gesamtkunstwerk, a comprehensive work of civil engineering capable of reconciling forces that were traditionally opposed, such as hydro-power, urban growth, intensive farming, recreation, and alluvial ecology. Although it is still too early to estimate the effects of the programme on the landscape of the Rhône Valley, it is already edifying to analyze the origins, ambiguities and successes of this grand enterprise set in an international perspective.

THE RHÔNE RIVER, FROM LANDSCAPE TO INFRASTRUCTURE

As seen in part 1, for millennia the Upper Rhône Valley has been a space of activity, transit and trade, object of a continuous negotiation between the river's fluctuations and human efforts made by inhabitants to adapt their environment to their needs, and adapt their activity to their environment. The systematic channelling of the Upper Rhône finds its origins in the first agreements between the states of Valais and Bern in the seventeenth century, by which both states agreed about the width the river should remain along their common border (DFI, 1964: 41). In 1756, the two states had a precise map drawn of the river between Martigny and the Geneva Lake, including all the piecemeal works constructed along this segment. In 1768, the first general plan was developed for the regulation of the same segment, in the hope of making the river's course fixed—and thus the border between the two states. Upstream Martigny, the Valais canton together with the municipalities around Martigny, agreed

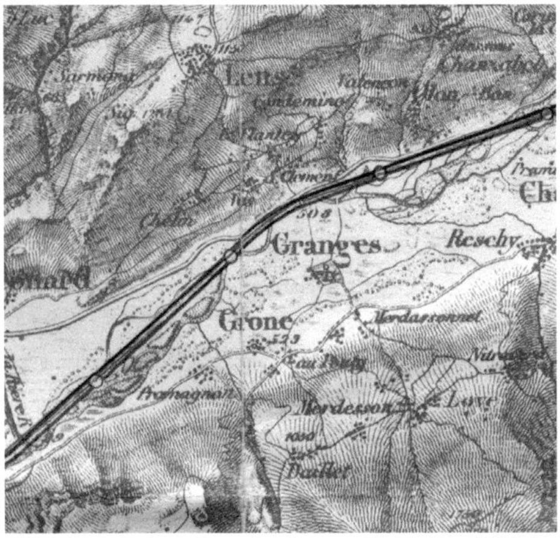

1863 Correction plan over-layered with the 1860 Dufour map

on stabilizing the banks of the river, which first fixed the course of the Rhône and led to the first coordinated correction at municipal and cantonal level.

At the turn of the nineteenth century, a full correction was actively promoted by the federal commissioner in Valais, Franz S. Wild. A second plan followed in 1825, drawn by Ignace Venetz and Adrien Pichard, respectively engineers for Valais and Vaud cantons (which had then both joined the Swiss Confederation). The plan was met with strong resistance from the municipalities, then still in charge of maintenance and civil works on the river (id.). However, the two cantons signed a convention in 1836, allowing for a global approach of the whole stream across cantonal borders.

The Valais canton was hit by a series of inundations in 1855, 1857, and the most catastrophic one of 1860. After this catastrophe, local authorities agreed on the necessity of a general correction. The canton asked for the support of the Confederation, which was first denied, as the Swiss constitution contained no provisions for temporary measures in case of natural hazards. However, the constitution allowed for the funding of lasting civil works of national interest, and the Rhône river did form a threat to national infrastructures. The Confederation thus offered to finance a large part the works provided the river was fully channelled, in a systematic way and on its full length (see chapter 'A European Self-colonization'). Local authorities engaged in a global project, on the basis of a preliminary plan established by Ignace Venetz before his death in 1859. As with many correction projects of the time, the first ambition was to regulate the erosion and sedimentation processes of the river, which in many places silted up and regularly flowed over cultivated areas. It was then a common idea that a well-dimensioned channel could bring a stream to an optimal velocity, and have it carry away the materials it had gained from the mountains upstream, without eroding its bed to a point that it would threaten the stability of the levees and bridges.

Discussions about the 'ideal width' for the Upper Rhône had started already in the seventeenth century. The 1602 convention signed by the states of Valais and Bern had proposed setting the river width at 185 metres. In 1760, engineers had estimated the ideal size at 92.5 metres (id.: 41–42). Hartmann and Blotnitzki proposed calibrating the Rhône within fixed, flood-resistant levees from 100 feet (32,5 metres) apart in Brig to 200 feet (65 metres) in Monthey, equipped with groins along the entire length of the river. They strongly recommended the correction of all tributaries, and the construction of 200 retention dams to regulate the discharge in the Rhône itself. Noticeably, they recommended the simultaneous combining of the project of a new railway with the Rhône

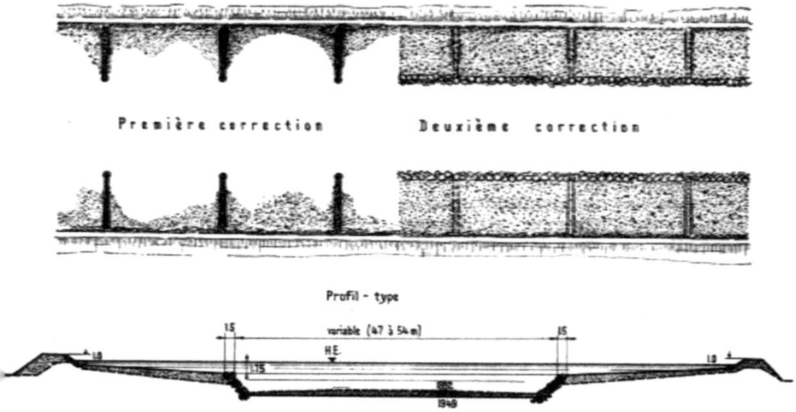

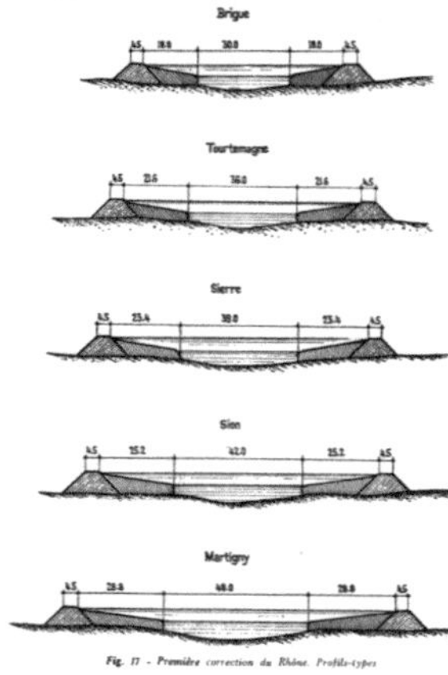

First and second correction, typical plan and section • Correction from Brig to Martigny (DFI, 1964)

correction: if the railway line could 'follow the rear edge of the Rhône, or even use this edge as backfill, it would then be located at the highest point of the valley bottom'. The railway would then 'not only be protected against floods, but costs would also be lower' (Blotnitzki and Hartmann, 1861: 303). The two engineers, who advocated a radical channelling of the meandering river, pointed at the careless way the railway had been built so far. The railway, they lamented, had damaged the curvy topography of the valley by pulling straight elevated railroad embankments, constructed by peeling hundreds of feet of land on both sides of the levee: a waste of land and a missed occasion to associate the two works of civil engineering. 'We believe that forcing [the future railroad] to come closer to the path of the Rhône correction, would, in our opinion, be only a slight atonement for the way it ravaged the Valley below Sion' (idem), in the interest of the railway company as well as in the interest of the Valais Canton. They concluded their first report by stressing that the Correction should be fully implemented 'with strong will and sufficient energy', for 'as long as the embankment of the Rhône and its tributaries only happens at important spots and along small portions … it will be difficult to achieve the necessary result. Construction too often interrupted will always be partly removed by high waters' (id.: 22).

FIRST EPISODE: THE CORRECTED RIVER

These calls for rapid and systematic intervention were answered by local and federal authorities: in the following years, a full survey was established, the project was detailed, and its costs more precisely estimated. An agreement was reached in 1862 with the railway Compagnie d'Italie to follow the Rhône channel, to integrate the design of future railways and river, and to fund part of the flood defence infrastructure. The Decree issued in 1862 by Valais' Grand Conseil specified the project organization in its grand lines: the canton would supervise the Correction; Federation and railway company would jointly fund the works, up to a still to be defined

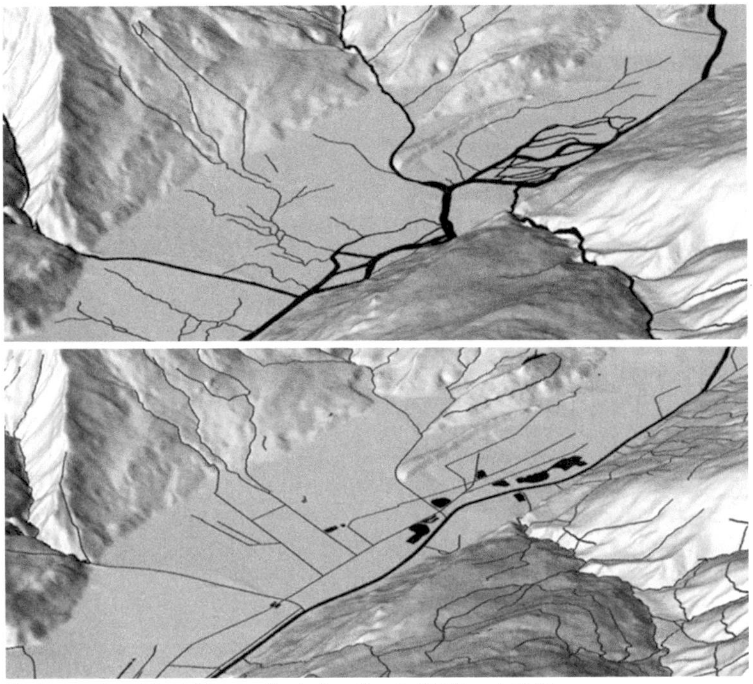

Hydrography of the Rhône valley in Conthey in 1860 and 2013

percentage for the railway company; what was not covered by the Federation and the company, would be paid for by municipalities in francs or in natura; municipalities and 'bourgeoisies' (local form of community organizations) would provide the land necessary for the embankments, and would afterward retain their maintenance duties; and the correction should be fully implemented in 12 years' time.

Furthermore, the decree specified that the forestry administration would be reformed in order to better prevent deforestation of the mountains, which was then largely held responsible for the great floods seen in the Rhône and Linth Valleys. 'We will make a point of having our experts check the application of this article', replied the Federal Council in the message of July 23, 1863, in which the Federation agreed to fund Valais' project, despite the doubtful participation of the railway company (Conseil fédéral, 19.01.1863 and 24.06.1963, quoted in DFI, 1964: 57). The federal support was decisive to start the herculean works that required, according to the Council, 'altogether an absolute, energetic and coercive vigour', and were 'the product of our Swiss constitutional institutions and, above all, a pure emanation of our feelings of confederal fraternity' (id.: 58–59).

The braided river was straightened and narrowed within a 72- to 96-meter-wide channel, erasing numerous islands and meanders; its curves were stretched and enlarged; some parts of streams were fully displaced, noticeably at the confluence of the Rhône and Vispa rivers, which was moved 800 metres upstream. The First Rhône Correction ended in 1884 for the main embankment works. However, frequent high waters and dike breaches occurred during this period, documented almost yearly between 1863 and 1898, leading to a continuous adjustment of the levees. The correction eventually took twice as much time as the 12 years originally planned.

In parallel, from 1870 to 1884 a secondary campaign took place with the reclamation and drainage of the low areas newly isolated from the river. The valley might be protected against flood and run-off, but rainwaters were caught in many places between the tributaries' alluvial fans and the new levees, leaving damp areas unsuitable for cultivating. In twenty years, eight areas along the Rhône representing 7160 hectares in all were drained by the Canton, financially supported by the Federation (DFI, 1964: 65). Drainage canals were dug and underground aqueducts were built to lead collected water in the tributaries further downstream into the Rhône. Warping enterprises were undertaken to raise the land and expand farming areas, with the help of special sluices invented by Engineer Ignace Venetz to collect rich waters from the river bottom. More wetlands disappeared, increasing Valais' scarce arable area to the satisfaction of the Canton and the Federation. Once completed, the First Correction was praised as a great success by the Swiss engineers, who saw the radical transformation of the river, and praised it both for the safety it seemed to ensure and the economic development expected from the new reclamations.

THE PLAIN 'IMPROVED' FOR THE BENEFIT OF FARMERS AND DEVELOPERS

In the first decades of the twentieth century, it appeared that the Correction had not fulfilled its goal of creating a self-regulating channel. The riverbed was rising more and more, especially in the central part of the valley, the most densely occupied and exploited part, increasing by 1.80 metre in the Canton's capital city between 1892 and 1935 (DFI, 1964: 80). Inside the riverbed, the space left open between the groins was carved by whirlpools and remained partly empty instead of filling up to narrow the central channel; the stream strayed and spread deposits instead of maintaining its permanent bed straight and clear. A correction of the correction seemed inevitable.

This second river correction was planned in 1928 to complete the first and remedy the continuing rising of the riverbed. The complementary works consisted mainly in dredging the central winter bed, connecting the groins by longitudinal dikes, and filling the spaces left between the groins with the extracted material. More segments of the Rhône, its tributaries and confluences, were reshaped and narrowed, sometimes with the financial participation of local industry (the Lonza factory built in the Rhône floodplain, close to the Rhône-Vispa confluence, paid 23 percent of the local works to increase its own flood safety level).

A major flood in 1935, caused by a dike breach in Conthey, forced the canton to take emergency measures again during the planning of the Second Correction. A hundred hectares were flooded as well as downstream Sion, and a breach had to be opened in the right bank in Chamoson to let the waters flow back into the channel (id: 93). Once again, a destructive flood significantly accelerated the river transformation: the second correction project and its budget were promptly validated six months after the catastrophe, in February 1936. Federation, Canton, and the railway company supplied most of the funds. Municipalities were accountable for 16 percent of the works, a total divided among them through a proportional system based on financial capacity (2/10) and arable area to be protected by the correction (8/10). This agreement allowed the programme to move forward and be implemented in three main phases, with the last being completed in 1961.

The drainage works outside the levees also appeared insufficient soon after completion of the first correction, and many areas remained too wet to be cultivated. Various drainage works were already undertaken in the early twentieth century. Drainage infrastructure was built from Saillon to Fully in 1910–1919 and 1920–27, from Riddes to Martigny between 1916 and 1934, from Sion to Riddes in 1920–1954.... The transformation of the Rhône Valley from a riverine landscape into fully drained intensive farmland lasted 90 years in all, from 1870 to 1960. It added a total of 10,000 hectares of arable land, drained by 215 kilometres of new canals, and absorbed more than a quarter of the total funds spent on river management in the Rhône Valley between 1863 and 1961 (DFI, 1964: 130, 131). In the meantime, as the Rhône thinned and the valley was drained dry,

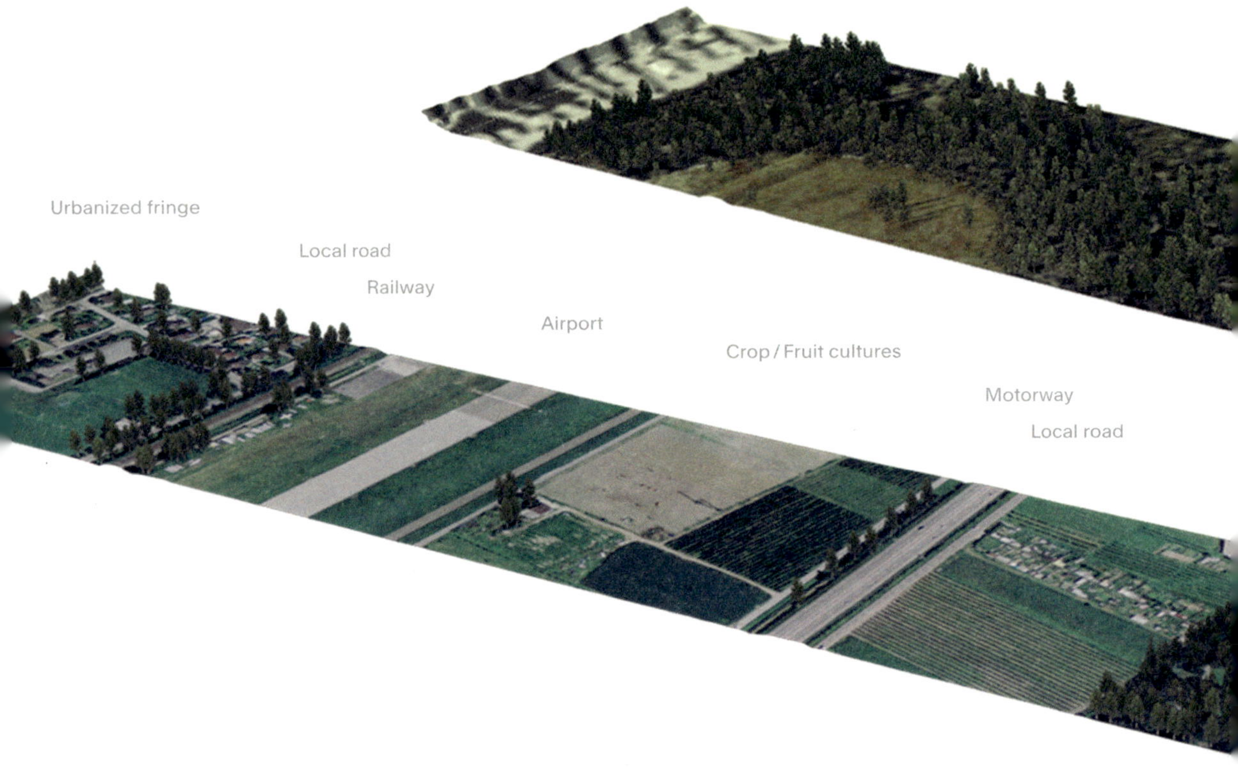

Evolution of the Rhône valley between Sion and Conthey: a densely exploited area gained from the river's fluctuation space

infrastructure works could start and expand. In spite of the various vicissitudes, the railway extended further up and reached Oberwald and the Rhône glacier in 1915. Between the two world wars, several airports were built that benefited from the flat lands made available by the corrections. The airport of Sion opened in 1935, occupying about 100 hectares of drained floodplain, later followed by the airports of Turtmann (1939), Raron (1942), Ulrichen (1942), and Münster (1943).

In September 1948, following a rainy week, two breaches occurred in the levees in Fully, one of them over a length of 160 metres. About 15 million m^3 spread over 1200 hectares between Fully and Martigny (Hydronat, 2000). The *Confédéré* liberal newspaper underlined the fact that waters had spread into two municipalities where 'the river [was] higher than the valley' (*Confédéré*, 06.09.1948), laying bare the risks inherent in the channelling of the Rhône above the valley floor, and its disconnection from the rest of the plain—making each dike breach disastrous where water spread fast and couldn't find its way back to the channel. Yet at the time, this dramatic event did not lead to a change in the coercive policy conducted by the Canton's engineers, nor did it prevent the growing occupation of the former floodplain. Day-to-day opportunism or religious fatalism seemingly delayed coordinated vision and action. Wasn't the Rhône beyond human influence, as stated in the *Nouvelliste* Christian

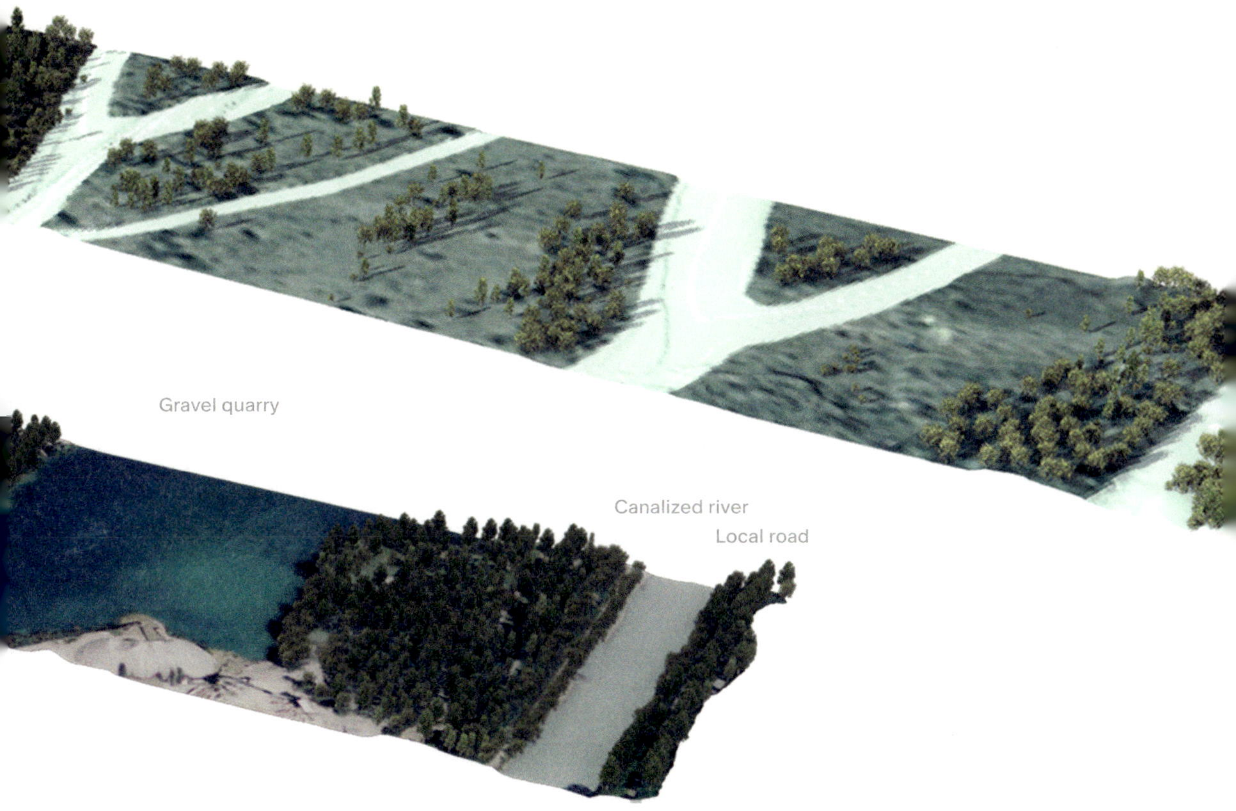

paper: 'The most precise science cannot predict the deluges of rain that make rivers overflow.... God knows when the time comes to clear the black clouds' (*Nouvelliste*, 07.09.1948).

Beyond doubts, Ruckli, Director of the Federal Department of Roads and Levees, could yet proudly write that 'the Rhône Correction differs from other similar enterprises in the fact that it has transformed a full, large valley, once covered with reed and swampy fields, into a vast prolific and quasi Mediterranean orchard', stressing the economic side of the project in addition to its safety aspects. This perspective was far from unique, certainly in the European context, where most river channelling works were for centuries motivated by the need for farmland as much as by the fear of flooding (See Part 1). Ruckli's claim rather confirms that in the first half of the twentieth century, as in the nineteenth, land improvement and production increase were still given absolute priority.

During these times of political and economic turmoil, the valley bottom was seen foremost as a space of conquest and opportunities, and rarely ever as a natural habitat or a living environment. Hardly any trace of concerns over these two aspects can be found. At the most, the federation did appear concerned about upstream deforestation—for the torrential floods generated more than ecological damage. Civil engineers barely thought of the spatial quality that could be created together with the hundreds of kilometres of canals and levees, although some were concerned about the lack of coherence among the various infrastructures built in the valley. For the Valais people as well as for the growing number of tourists, nature and beauty were to be found uphill, and the brutal conversion of a diverse riparian landscape into a utilitarian and mono-cultural area seemed to all a legitimate enterprise, if not a duty, in modern minds. As a result, only 20 of the 160 kilometres of river still showed any natural morphology at the end of the twentieth century, about 12 percent of the Upper Rhône's total length (Projet Rhône, 2008: 80).

The Autoroute du Rhône motorway was built gradually from the seventies onwards, mostly in the floodplain. Since the Rhône remained quiet for four decades, apparent safety and permissive regulations encouraged an unprecedented construction fever in its former territory. Farmlands, infrastructure, and built areas conquered the floodplain thanks to river containment, but with no physical connection to the river that used to shape and rule over the valley. By 1980, the space gained on the former floodplain was largely built on, with the development of transport infrastructures boosting the construction of low-density housing. Paradoxically, soaring construction industry helped resolve the shortcomings of the two Corrections. Urban expansion benefited from the apparent safety brought by the levees and created a strong demand for river sand and gravel. Massive extraction—30 million m^3 of gravel extracted between 1936 and 1999 (Rhône.vs, sept. 2010)—countered the river's spontaneous silting, and made it unnecessary to raise the levees again (Projet Rhône, 2006: 14), just as the construction boom in Grenoble did for the French Isère River in the same period.

Hydropower installations built from the 1950s on also held growing amounts of materials behind the dams—with the Dixence alone collecting waters from 7 percent of the 5220 km^2 Upper Rhône catchment

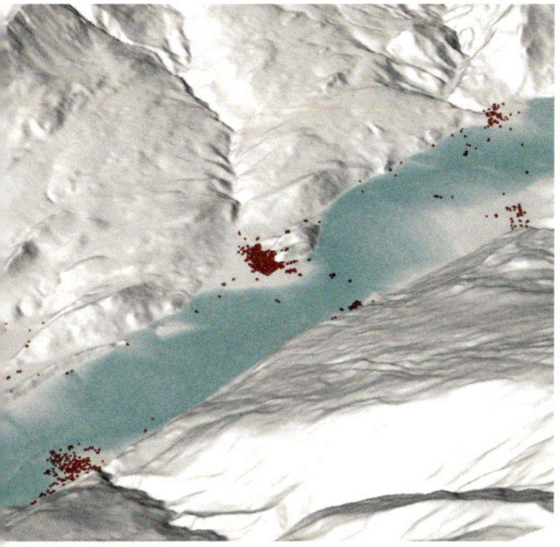

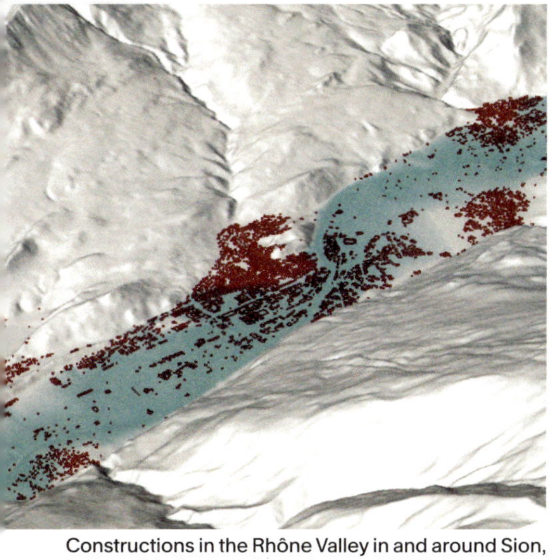

Constructions in the Rhône Valley in and around Sion, in the mid-nineteenth century and 2010

area. Slowly, the Rhône appeared to lose more material than it received from the mountains, showing signs of erosion already in the 1950s. There again, the permanently balanced and self-maintaining river hoped for by the engineers of the correction appeared out of reach. The new dynamic balance combining channelling and dredging for the building industry showed major flaws: apparent safety encouraged construction in the floodplain and increased exposure in case of flood; urban sprawl and induced infrastructures consumed scarce arable land; and riverine ecosystems were considerably deteriorated by the channelling and extraction of material. Eventually, all three factors combined into a new critical perspective in Valais and Switzerland in the 1990s, which led to the current Rhône Correction, the third of that name but intrinsically based on a different cosmogony.

AFTER THE CORRECTION, THE RENATURATION: A U-TURN IN FEDERAL POLICY

Starting from the early nineties, Swiss flood policies shifted from a model of vertical containment to a model of spatial prevention, with the 1987 floods being considered to mark the beginning of this turn from coercive to adaptive strategies (Reynard et al., 2006). The federal report on the 1987 inundations stressed the need to introduce passive strategies and to prevent damage through a more restrictive planning policy, based on better risk assessment (OFEE, SHGN, 1991). The various laws and ordinances on water protection and on hydraulic engineering passed in the following decades brought horizontal solutions more and more to the foreground.

Regarding flood defence measures, the Swiss Hydraulic Engineering Law of June 21, 1991 confirmed the attribution of flood defence executive powers to the Cantons, in line with the 1874 Federal Constitution. But as was already the case in the nineteenth century, the Federation remained the main sponsor of large hydraulic works, greatly influencing the decisions made by local authorities, and with this law it took a new stance on flood defence. The most striking change was explicated in article three on 'measures to be taken', which gave priority to 'maintenance and planning measures' and only when these measures failed to provide protection did it allow other measures to be taken such as 'corrections, damming, sediment collectors, and retention basins' (Assemblée fédérale, 21.06.1991). This change of perspective marked the end of unlimited river engineering in Switzerland and publicly stated the impossibility of achieving full protection through defence works, with priority given

Thur river riverbed and riparian wood, ten years after reconfiguration (Rossano)

to spatial strategies and environmental quality. This new stance constituted a U-turn from nineteenth-century federal flood management policy, which considered full river diking and straightening as the only valid strategy, regardless of environmental damage and long-term manageability (see chapter 'A European Self-colonization').

The Federal Agency for Water and Forestry (OFEG) developed and promoted this new approach: 'These last few years, specialists developed the concept of sustainable flood protection that does not seek to tame nature, but to give nature space through creating floodable areas.... According to this new concept, spatial planning plays an important role' (OFEG, 2002: 5). The same publication noted that 'Defensive civil works built for extreme situations often cost much more than the damages they are meant to prevent'. This shift was thus not only motivated by the belief that vertical solutions had reached their capacity limits in terms of safety, but also by the rising costs of defence infrastructure combined with soaring risk exposure, therefore stating again that state support was conditional on the expected outcome in terms of financial gain or savings on potential damages—and was not granted for a limited short-term response.

The 1991 law also reflected growing environmental concerns regarding the state of Swiss streams as well as the empowering of political ecology that soon occupied the centre of this new arena. The Confederation introduced environmental goals into flood prevention policies, supporting the *revitalisation* (in French) of rivers in order to heal the damage of full damming. The law further introduced quantitative and qualitative criteria for hydraulic engineering, as interventions on riverbeds should 'as far as possible respect the natural course' of the stream, in other cases restore it, and create the conditions for riparian flora and fauna to develop on its banks. The Confederation still would not fund maintenance works, a prerogative of local authorities, but would allow the funding of restoration works meant to 'bring waters affected by civil works back to a state close to natural' (Loi fédérale sur l'aménagement des cours d'eau, 21.06.1991, art. 4, 7).

Besides public institutions, the Swiss environmental organization Pro Natura also actively supported the 'renaturation' motto by stressing the combinatory virtues of this approach as being 'beneficial to humans and nature'. Environmentalists stressed the destructive effects of channelling and the possible gains of river widening for flood protection. In 1997 Pro Natura commissioned a study on the subject, which concluded that streams would need about ten times their regular width to allow natural meandering, and that the big rivers such as the Rhône and Rhine would need at least 200 metres to fluctuate again (Pro Natura, 1998: 4). Extrapolating from these results, the ensuing report calculated that approximately 0.5 percent of the Swiss territory should be reserved for river widening, tripling the area dedicated to rivers.

The law on water protection was supplemented in 2011 by a new article setting goals for the determination and restoration of a 'minimum space' needed for riparian ecology and flood protection (Waters Protection Act of January 24, 1991, version 2011 and later, art. 36a: Space provided for water). Eventually, the 1998 law on water protection was amended to integrate this normative and linear approach, defining a set of standard widths to be kept free along streams and rivers: eleven metres for watercourses with a channel bed of less than two meter natural width; seven metres plus 2.5 times the width of the channel bed for watercourses with a channel bed of two to fifteen metres natural width (Waters Protection Ordinance 28 October 1998, Status as of 1 June 2011, art.41.a: Space provided for watercourses). The width should even be increased wherever required in order to guarantee protection against flooding (idem), and the reserved area should not be used for agricultural activities.

The last two decades have seen numerous state-funded 'renaturation/Renaturierung/rinaturazione' projects flourish along the Swiss waterways, thanks to the engagement of nature protection organisations and federal authorities: the Thur river in Thurgau canton, the Wiese and Birs in Basel canton, the Aire river in Geneva canton, the Reuss and Kleine Emme in Lucern canton, the Töss river in Zürich canton, the Cassarate river in Ticino.... The early transformation of the Thur river, however, made it clear that the shift from systematic diking to river widening was going to be anything but consensual. Long conflicts arose between environmentalists and farmers about the level of renaturation and the surface of arable land to sustain within the floodplain areas. These conflicts revealed the gap between federal and local authorities, with the latter being more influenced by local communities and landowners, more sensitive to land productivity and more inclined to maintaining status-quo (Zaugg, 2002), as opposed to the pro-ecology federal authorities. These conflicts delayed the project for several years and soon exposed the controversial character of the 'renaturation' approach.

In terms of legal obligations, the 2001 laws also insulated environmental measures from urban policies, specifying that built areas could be exempted from river restoration. However, the most degraded streams were precisely located in inhabited valleys, were urban sprawl had long taken advantage of permissive building regulations. In that regard, the federal legislation failed to address the spatial quality and the transformative potential of river restoration in inhabited areas, and contradicted itself, as it promoted an 'integrated water management' considered 'in interaction with other domains'. Environmental requirements were instead kept outside of the living environment, where, one could argue, they were most needed.

FLOOD AS AGENT OF LANDSCAPE ARCHITECTURE: FROM RIVER CHANNEL TO RIVER SPACE

In 2003 four federal agencies—the Office for Forest, Environment and Landscape, the Office for Water and Geology, the Office for Agriculture and the Office for territorial development—jointly published a brochure entitled 'Guiding concepts. Swiss waterways. For a sustainable management of our waters' (OFEFP et al., 2003). This short publication summed

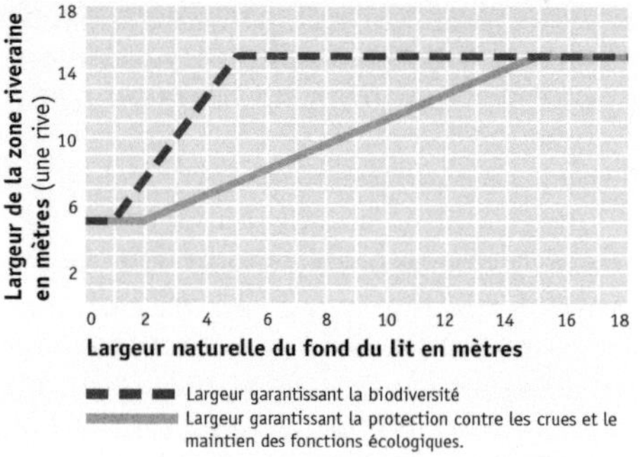

Calculation chart for the estimation of the width needed for biodiversity (black), for flood protection, and for ecological functions (grey) in relation to the 'natural width of the river bottom' (from OFEFP, et al., 2003)

up the new 'horizontal' federal policy, and detailed the goals and the measurements required to achieve them. Each river should be allowed to fluctuate according to seasonal changes, stated the brochure. A good profile should accommodate high water discharge and allow the river to perform its drainage function for the surrounding areas; it should provide enough space to guarantee structural biodiversity, both terrestrial and aquatic; it should provide enough space for typical species to develop and form a network of natural habitats; it should create an attractive site for the population to recreate and 'integrate this landscape into its cultural environment'; finally, it should leave enough distance between exploited land and streams to prevent water pollution. Space was thus needed to perform the river's three main goals: drainage, ecology, and recreation, with the addition of a new cultural dimension that was not present in the federal laws.

Although the new set of goals was to a great extent locally defined and subjective—especially regarding recreational and cultural value—the federal agencies collaborating in the abovementioned brochure agreed on dimensional standards to help local authorities implement these goals. A small chart made it possible to calculate the necessary width of the 'riverine zone' needed for a stream in relation to the 'natural width' of the bottom.

The official Swiss approach to river management was still linear rather than transversal, and normative rather than contextual. However, beyond the question of river width, the successive laws on water protection and hydraulic engineering used the notion of 'space' also in a wider sense but in cautious terms, as spatial planning remained in theory the exclusive domain of local authorities (an agreement, however, regularly challenged by federal laws on transport or housing, such as the controversial 'Lex Weber' passed in 2012 to regulate the proportion of second residences).

The 1994 ordinance, complemented in 1999, tried to put an end to construction on floodplains and flood-risk areas, first of all by compelling cantons to generalize flood risk assessment and actually define flood-risk areas (Conseil fédéral suisse, Ordonnance sur l'aménagement des cours d'eau, 2.11.1994: art. 21, 27). The law further excluded from federal funding any measures meant to protect new construction in threatened areas (id.: art. 2.6). More generally, the danger zones would have to be taken into account in all planning documents and 'other activities having an effect on spatial planning' (id. art. 21.3). Beyond the sole issue of flood risk, the Federal Law on water protection has, since 2011, compelled all cantons to define 'the space necessary for surface waters' (Loi fédérale sur la protection des eaux, 36.a). With this new obligation placed on local authorities, the notion of river space took on a larger dimension, integrating various states of the riverbed: the permanent bed, the high water bed, the space needed for meandering, and the space of maximum dilatation in a flood situation. It also gave this spatial aspect a real impact on river management projects, not least for the Third Rhône Correction, where the definition of 'necessary space' remains to this day the most difficult to resolve in any consensual way.

LOOSENING THE UPPER RHÔNE, FROM POLICY TO PROJECT

In the Upper Rhône Valley, four relatively quiet decades followed the 1948 flood, before threatening high waters reappeared in 1987. In 1992, the Sionne inundated Valais' capital city and paralyzed the region's railway network. In 1993, the streets of Brig were inundated by the waters of the Saltina stream and partially covered by several metres of alluvium. This series of local but dramatic events generated a rapid surge in emergency measures planned between 1994 and 1997, which mainly consisted of local reinforcement of the levees in Visp, Niedergesteln, Steg, Chippis, and Fully. Besides eliciting this short-term classical response, these events shook the trust placed in the existing flood defence structure (Hydronat, 2000).

In 1995, the Canton Council commissioned in-depth studies to analyze the situation and sketch future global response to flood threats, focused on the segment that appeared to be the most exposed. The first ConSeCru research programme, financed for 70 percent by the Federation, analyzed the restoration of the recently flooded areas, explored a larger scope of variables that could orientate future measures: meteorology, hydrology, sediment and driftwood transport, flood-risk areas, along with studies for a flood protection concept that would integrate spatial planning policy and disaster management (Bloetzer et al., 1998: 139).

Three sub-sectors were defined, and for each sector a team was hired, composed of six to seven engineering firms, mostly specialized in civil works, hydraulics, and ecology. Despite the repeated statement that all measures should be integrated in a coherent landscape structure, designers from urban planning or landscape architecture disciplines were broadly absent from the selection. In 1997, the propositions were delivered and discussed within the canton's administrations. A private engineering firm, was hired to adapt and harmonize the various results, and in June 2000, the Road and Waterways Department could finally present the synthesis of five years of explorations, entitled 'Third Rhône Correction: safety for the future' (Hydronat, 2000), which laid out the basic principles for the future planning process and was officially approved by the Canton Council a few months later.

The conclusions of the preliminary studies in terms of flood risks and exposure were alarming: levees were unstable in many places; discharges higher than the one seen in 1993 should be expected; about 7000 hectares could be flooded by a 100-year flood. Even more worrying was the level of exposure reached at the turn of the century: the same 100-year flood could potentially cause six to ten billion francs of damage. Besides flood risks, the ecology of the river had been dramatically damaged by the channelling, and in that regard, the 2000 report took a new stance, stating that 'lasting solutions with limited maintenance' would be favoured and 'should as far as possible restore the river's natural dynamic'. Furthermore, the social and economic potential of the river in relation to the valley (farming, tourism, and hydro-electricity) had remained untapped and should be developed.

1992: the Sionne stream floods Valais' Capital city and paralyzes the railway traffic (Robert Hofer)

Breaking with past Promethean optimism, the report acknowledged the impossibility of completely preventing flood risk and the need to mitigate the effects of extreme discharges through controlled flooding: 'Residual risks ... must be addressed with the creation of a drainage corridor that avoids inhabited areas' (Hydronat, 2000). Retention was also a full aspect of the chosen strategy: 'In areas with low potential damages ... in order to manage extreme discharges between Brig and Martigny, and prevent aggravation downstream.' Despite an estimated budget of 800 million Swiss francs, it was then made clear that the Third Rhône Correction wouldn't provide total protection to the whole valley, but could limit the damage—a realistic view but a potentially unpopular approach.

Three types of possible measures were listed by the report: adaptation of existing flood defence; widening of the river space; and the creation of secondary discharge channels, distinct from the Rhône. The first type of measures was considered as incompatible with ecological restoration, but also presented as inevitable within urban areas, where the river could not be widened. The second and third types of measure allowed for a combination of more safety and reduced clearing of riverbanks, and was to be preferred for 'less constrained areas, particularly when extensive agriculture is thinkable'. Priority measures followed a parallel path, independently of the long-term and large-scale planning procedures.

The potential conflicts inherent in this differentiated approach were to be resolved within a global vision for the valley: 'Only a full integration of various domains from the start of the design process will allow us to achieve the goals and a sustainable development for the Canton' (id.). From the very beginning, the Third Rhône Correction's ambition was

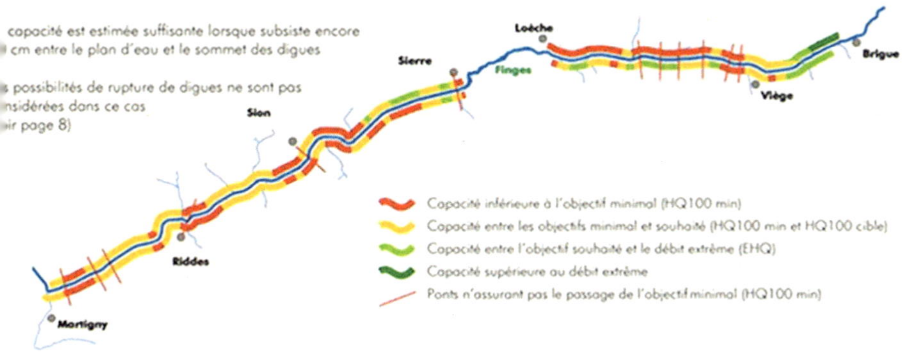

Preliminary hydraulic assessment: Only few segments appear to have a sufficient discharge capacity (in green) (Hydronat, 2000)

for a global and integrated transformation of the river, a territorial Gesamtkunstwerk that would incorporate all active components present in the valley, its past, and its future. A 'landscape concept' was to be developed 'simultaneously' and yet, strangely, not presented as an intrinsic part of the general plan but as a separate study to 'develop in parallel to the Third Correction'.

Dynamic river management and controlled flooding, although conforming with Federation policy, then represented for Valais a break with more than a century of damming and channelling measures, for historian and sociologist Gabriel Bender a 'true Copernican revolution: levees do not protect, they threaten, the solution promoted by common sense is the problem' (Bender, 2004). Local politicians remained more cautious about the application of the 'new philosophy'. In the report's foreword, State Councillor Jean-Jacques Ray-Bellet praised the heritage of the 'many generations that have done all they could to tame [the Rhône] and made the fantastic development of the valley possible' (Hydronat, 2000). He remained indecisive about the future flood protection that should 'adapt to current and future land use, valorize natural assets, and be integrated into economic and social life'. This general statement neither opposed nor supported the federal motto of ecological restoration and horizontal flood defence, and suggested that the project should adapt to the present situation, which was the direct result of the channelling of the Rhône and the full drainage of its floodplain.

The official name chosen for the programme, 'Third Rhône Correction', was also ambiguous. Federation and cantons seemed to agree on the enlargement of the river, but the name suggested a repetition of the two previous corrections that had deprived the Rhône of the space needed for seasonal flooding and ecological development. The drawings inserted in the 2000 report also illustrated the difficulty of the exercise. Showing concrete intervention proposals, the plans already seemed to address local conditions and reinforce the Rhône corridor rather than look for an overall landscape structure. More explicit than the text, the plans clearly indicated a 'constrained area', where the existing channel could not be modified but only adapted, generally corresponding to built-up areas, and an 'unconstrained area', where the channel could be widened or doubled with a secondary branch, corresponding to farmland, a classification that soon would appear controversial. Despite these conflicting realities, the *Grand Projet* took off, partly thanks to the Rhône itself....

RENEWED FLOODING AND NEW MOMENTUM

'The danger [of flooding] is not perceived in everyday life along the Rhône, as the memory of the last severe floods is lost to the inhabitants of Valais', lamented the report presented in June 2000. In October 2000, heavy rainfall poured over the south of Switzerland, and massive flooding occurred in Valais and Ticino Cantons. The Rhône sent a painful wake-up call. The river swelled again, and the discharge measured in Sion surpassed the 1993 record. The floods took the lives of 16 victims and caused 470 million Swiss francs in damage in Valais alone, 62 percent of which involved public infrastructure (OFEG, 2002: 16–17). The catastrophe confirmed the conclusion of the first studies and the need to engage in a new project that would this time address both the risk and the exposure of Valais' inhabitants and the infrastructure that had been built in the former floodplain, sometimes metres below the high-water level.

The scope of the project was widened, and an extended study was commissioned that included the Canton of Vaud, on the right bank from Lavey down to Lake Geneva; three large research programmes were also launched by the Valais Canton and various federal institutions, focusing on specific aspects of the mega-programme: flood response and emergency (programme MINERVE), the design of spillways (programme DIFUSE), the multi-functional potential of the correction, and possible development of hydro-power plants (SYNERGIE). In parallel, the European Interreg III programme funded a large research project on the Upper Rhône basin in Switzerland and France, comprising an extensive cartographic study of the Swiss Rhône Valley between 1800 and 2000 (Dupont et al., 2008, and Staüble, 2009).

In 2005, a preliminary version of the 'Plan sectoriel' was communicated as a recommendation to all concerned parties, to which environmental organizations reacted with disappointment. They endorsed the general concept but considered the proposal insufficient regarding the ecological ambitions set by the initial project and by federal laws. Four organizations, including the World Wide Fund for Nature WWF and Pro Natura, joined forces to elaborate a common proposal to ask for 'more space for the Rhône' and a global ecological restoration of the valley through a network of biological corridors. The counterproposals addressed the width of the enlarged segments as well as their amount. While the Canton proposed to locally double the width of the Rhône to achieve a basic ecological value, the organizations considered a factor of three to five as necessary to allow ramifications in the river course and 'the reestablishment of the biodiversity of desired species and their vital biotopes'. The document included zoning plans that illustrated their demands in terms of widening and land use adaptation, this on 24 sites along the Rhône (WWF et al., 2005: 5–6).

The final 'Plan sectoriel' published by the Canton Valais in 2006 integrated the federal guidelines and the new discourse on flood prevention more strongly in the project description. The document advocated

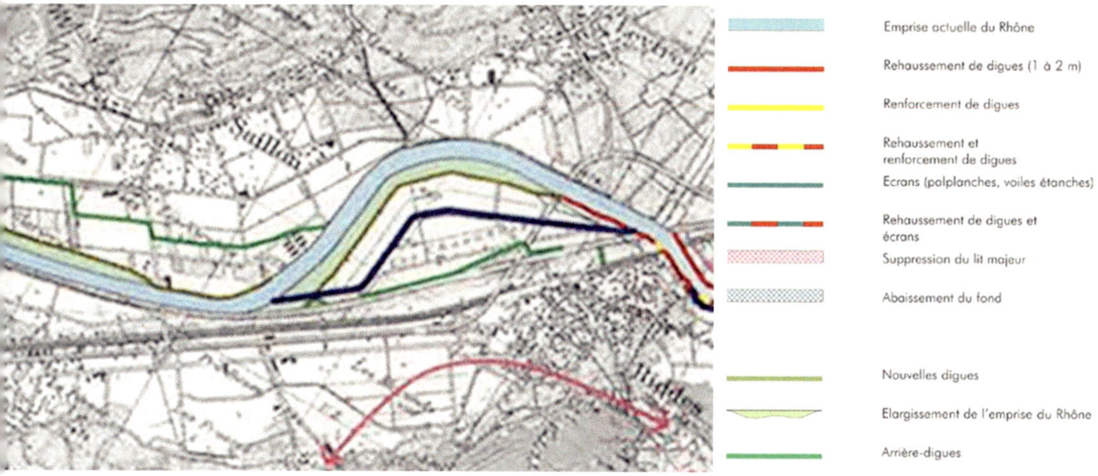

Preliminary proposal for the Third Rhône Correction, detail Riddes-Saillon area (Hydronat, 2000)

an approach through three key notions, illustrated on its cover: 'Territoire' was written on an aerial picture of Sion; 'Danger' showed areas of high and moderate flood risk in light and deep blue over the same picture; and 'Espace' showed a fluctuating corridor clearly related to the stream, but not limited to the existing channel, nor covering the full risk area: the newly define 'river space' or *Espace Rhône*.

The document clarified the definition of the Espace Rhône: 'This space was defined according to safety requirements; inside this safety space, environmental and socioeconomic functions can also find their place. It generally includes the current Rhône [channel] as well as a surface equivalent to the current width of the Rhône (on each side), but it respects important constraints existing in the valley and is therefore reduced to ten metres from the base of the levee, when close to built-up areas' (Projet Rhône, 2006: 5). The document stressed the fact that this river space (*espace*) differed from the space actually occupied by the stream (*emprise*): the Espace Rhône was a multi-functional space, defined but negotiable, depending on the priority given to the different contextual or project-related elements in place—a flexibility opposed by environmental organizations.

The law on hydraulic engineering passed by the Grand Conseil du Valais in 2007 again confirmed the integration of federal orientations into regional policy, although trying to counterbalance environmental requirements with an impressive list of interests to be taken into account in the maintenance and transformation of rivers. As stated in its first article, the law aimed at ensuring flood protection, fixing potential flood damages and 'maintaining, restoring, or developing streams into a state as natural as possible' (Grand Conseil du canton du Valais, Loi sur l'aménagement des cours d'eau, 15.03.2007, art.1). Its second article mentioned 'active and passive flood protection', to be achieved following no less than 14 principles enumerated in Article Five. These included: the 'preservation of [free] space to facilitate flood protection'; the 'conservation or restoration, as far as possible, of streams' natural character through renaturation measures';

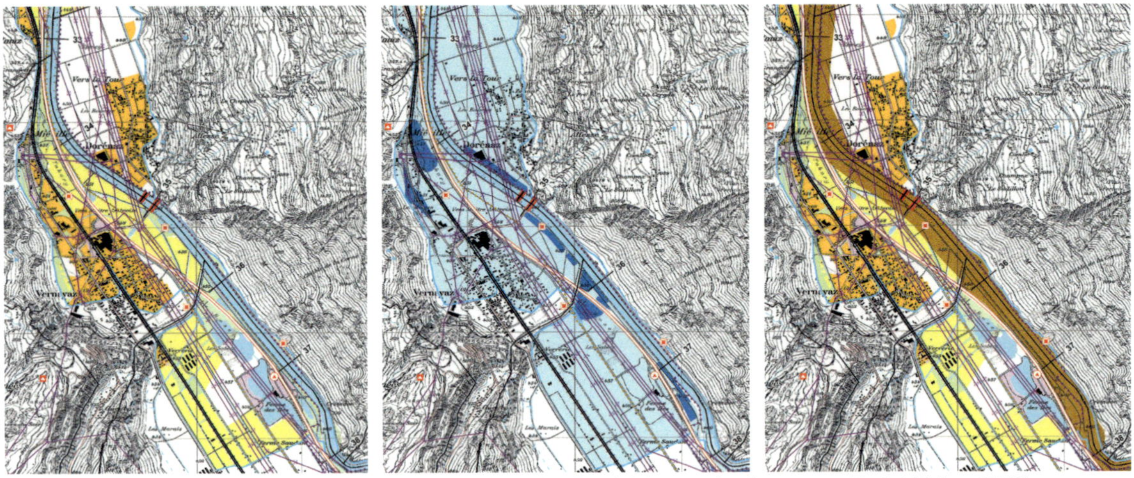

The three-steps approach of the *Plan sectoriel*: the territory, the flood risk area, the river space (Projet Rhône, 2006)

the application of measures allowing the installation of a diversified fauna and flora along the streams; the maintenance of riparian vegetation, and, if necessary, its replacement by a more adapted vegetation; and 'taking into consideration the requirements of spatial planning, of the protection of the environment, nature, landscape, as well as of fishery, agriculture and forestry ... the interests of navigation or the use of hydraulic power' (id., art. 5e, g, h, i, k, l). This new law indirectly framed the Third Rhône Correction, which was then already in the planning phase and from 2008 on could count on the Canton's funding for its implementation.

The mission described in this law was at the same time clear in its integrative ambition and complex in its implementation: the list of interests to be taken into account suggested a holistic and consensual approach, which in the dense context of the Rhône valley seemed, in retrospect, hard to reach, if even achievable at all. As the architects of the correction liked to stress, the valley floor was intensively exploited and inhabited; the state had no land buffers to compensate eventual losers and absorb the shocks of the re-allotments induced by the goals set for the correction. The discussion around the project logically quickly focused on the quantitative aspects: each party counted the hectares lost and won, arguing over the legitimacy of the opponent's territorial claims. Willing to 'make room for' ecology, safety, farming, recreation, construction ... the programme created high expectations and brought spatial competition to the foreground, at the expense of spatial coherence or possible synergies.

The *Rapport de Synthèse* published in 2008 specified the quantitative choices made during the design process: a widening factor of 1.9 would be necessary to ensure an 'ideal profile for the lasting securing of the valley' (Projet Rhône, 2008: 42). However, it was thought that the Rhône's river space could eventually be only 1.6 times larger with local enlargement of a factor two or three, this being considered as the minimum for 'lasting safety' (a term, however, partly tempered: the profile did not take into account the 'possible effects of climate change on high water discharges'). Eventually, twenty segments were planned to be enlarged, at more than 1.6 and up to 2.5 times the existing width, over an accumulated length of 25 kilometres (Veuve et al., 2008: 6).

The project had been discussed and detailed in local consultation trajectories. The Rhône was divided into five sectors led by steering committees—Commissions régionales de Pilotage or 'CoRePil', representing 53 municipalities. Each committee was composed of local representatives and organizations, in charge of envisioning the local application of global goals and measures predefined by law and by the Rhône authorities and of developing a *Concept de Développement de la Plaine*. While some of the CoRePils were already anticipated in 2003 and mobilized many actors in a genuine and broad reflection on territorial issues (notably in Sierre), the results were eventually narrowed down to a choice between limited options and evaluated according to quantitative aspects rather than dynamic visions (Largey, 2013). The authors of the 2008 landscape study would later praise the results of the local workshops, but lamented the physical impossibility to translate these ideas into projects (Veuve et al., 2008: 22).

The work done by the local CoRePil committees indeed resulted in 14 'plans of territorial and landscape insertion' ('Insertion territoriale et paysagère du Rhône dans la plaine') added to the general report. Contradicting their denomination, the various plans consisted of a general zoning scheme, combining factual information ('urban fringes', 'transition spaces'), specific functional orientations (possible location of recreational routes on the levee, re-allotment measures or hydropower plants), and areas vaguely classified as 'challenging zones'. In the end, the plans avoided any structural proposal, and appeared to be the result of quantitative negotiation rather than of a shared vision on the valley's future landscape, combining a sum of local compromises rather than securing the expected Concept de Développement de la Plaine.

Despite the consultation procedure that surrounded the definition of the *Plan d'aménagement*, many farmers still fiercely opposed the plan, contesting the very necessity of river widening. They joined forces in 2007 and establish the ADSA, or Association for the Preservation of Farmland. Two main principles guided their action: flood safety, they believed, could be better obtained without returning space to the river, but instead by lowering its bed; arable land was too precious to sacrifice to natural restoration, and should be secured and protected just as urban areas were. The first argument was developed in an alternative plan just like environmental organizations had done in 2005. This plan consisted of a minimal project supported by technical studies commissioned by the ADSA on the effects of dredging: a plan 'three times cheaper, three times faster and using three times less arable land', claimed the ADSA during its public campaign. They supported a prolongation of the two previous corrections, both focusing only on discharge capacity and the management of alluvial materials. The best alternative for the ADSA was thus to dredge the riverbed to lower the water levels, to keep the channel narrow to ease the evacuation of deposits, and to reinforce and locally heighten the levees where needed. The second argument reached further into local politics and psyche: keeping the Rhône in its channel and defending arable land was a case of 'respecting the work of our ancestors and the needs of future

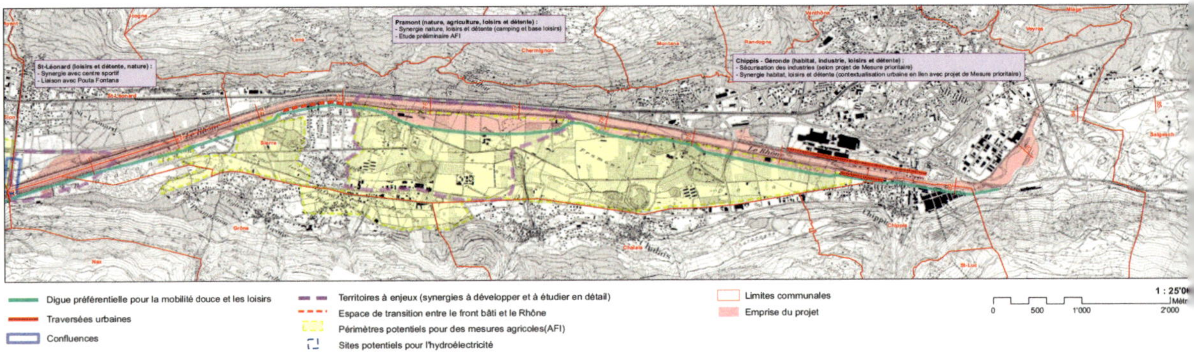

'Territorial and landscape integration of the Rhône into the valley', region of Sierre (Projet Rhône, 2008)

generations' (ADSA, 2008). Valais lacked arable land, a land that was 'indispensable to provide the country with food'. Only safety concerns could justify the sacrifice of this precious land, they argued, implicitly rejecting the environmental requisites contained in federal laws on river management and the expectation of environmental organizations in the Rhône valley.

The land argument, however, turned out to be a reversible one, as the debate organized by local TV channel Canal 9 showed in 2011. The delegate from the ADSA pointed out the contradiction between promoting a local and self-sufficient agriculture and widening the Rhône precisely at the cost of local farmland, and accused environmental organizations of illegitimately claiming the Valais territory for ecology alone. Thierry Largey, representing Pro Natura Valais denounced, in response, 'a great paradox': 'The greatest [land] consumer is the 'concrete', but they [developers] always manage to oppose nature to agriculture, and never get involved in the debate' (Canal 9, *Le débat*, 08.06.2011). The statement was supported by land use statistics that showed that 'every 18 months, the equivalent of the farmland needed for the Third Correction is being built on in Valais'. For Largey, farmers and environmentalists were not fundamentally competitors but faced a common enemy: urban growth. Largey repeatedly suggested uniting in a common fight against the building industry and inefficient planning, but lamented the fact that farmers instead preferred to 'fight for the leftovers', remaining prisoners of their own interests in rising land prices and for some selling their land to developers while fighting the Rhône Correction.

The proposed alliance never took place. In 2015, the conflict between farmer's organizations and the Rhône 3 office went further: the ADSA asked and obtained the holding of a public referendum on the project budget, hoping to paralyze its implementation. Supported by the conservative party UDC and by the Valais' Chamber of Agriculture, the initiative contested the allocated sum (considered excessive), as well as the widening of the river and its environmental restoration (although its scope had been reduced). The ADSA also confronted the building estate sector, who supported the flood defence plan to secure future land for housing. In that regard, the conflict went beyond the traditional opposition between farmers and ecologists, confronting the two pillars of Valais' economy, agriculture and construction industry. Jean-Pierre Jordan, from the Federal Office for the Environment, noted the limited scope for action: 'Either you meet the requirements for flood protection and the minimal ecological requirements, or you maintain and renovate the existing structures and take planning measures to take the remaining flood risks into account', implying less land available for future development and construction. (*Le Temps*, 6.05.2015). After a long dispute over figures and conflicting interests, the budget of the Third Rhône Correction was eventually approved by 57% of voters, providing a stable perspective for the megaproject.

LANDSCAPE, AT LAST

Contrary to what the first 2000 report recommended, no landscape structure was shown in the 2008 synthesis to support a coherent development of the Rhône correction within a wider context, local or regional. Landscape as such was introduced as a subsidiary aspect of the plan, relegated to the position of second-last 'specific theme', below hunting, fishing, and energy, and occupying two pages out of the 160 pages of the report (Projet Rhône, 2008: part 6.9). These two pages consisted of a dense summary of a specific study entitled 'Landscape Guidelines for the Rhône and the Valley', carried out between 2007 and 2008 in parallel with the local planning consultation process. Its main goal was to 'define landscaping principles applicable to the Rhône Valley in general and to the Third Rhône Correction in particular' (Veuve et al., 2008). It promoted two main ideas: first, the Rhône levees should become a 'life corridor', a continuous and car-free recreational route along the whole valley, giving access to its various natural and cultural assets; and, second, 'it is the river's natural dynamic that [would] create its own landscape' within the river space. The authors stated for the Rhône ambitions similar to those expressed earlier for the Isar in Bavaria and for the Isère in the French Alps.

The study further detailed guidelines for the transformation of the historical levees, taking into consideration their large-scale perception as well as their small-scale practicability, spatial quality, and maintenance. With a surprising openness, the authors partly turned the report into a pamphlet against the lack of long-term, transdisciplinary planning in a situation where the new spatial mitigation policy should provide an opportunity to create a coherent structure and prevent further degradation. They blamed the fragmented structure of the planning authorities and its consequences for the valley's landscape: the Rhône 3 Project might benefit from its own dedicated administration, but 'all the aspects relating the project to the rest of the valley and its future are diluted among the canton's various departments, regional organizations without political power and a profusion of public and private local interests'. The present weakness, they pointed out, 'is in striking contrast with the expectations of the Canton's various departments, which like to think that the Third Correction Project should deal with all the coordination issues present in the Valley'. As for the authors, the effort made so far to connect flood defence and regional planning would remain vain without structural evolution: 'The structure of territorial organization needs to change ... in order to compensate for the valley's growing fragmentation and lack of legibility through large landscape interventions (*gestes paysagers*).'

The authors seemed, however, disillusioned by the limited scope of their own assignment. They lamented the fact that 'for reasons of time and resources and because the problem is sensitive, the spatial translation of the goals and measures could not result in a concept for the development of the valley, leading to a vision for the valley's future' (id.: 22). 'Shall we continue with a piecemeal development fulfilling local and immediate needs?' asked the report; 'Which degree of inter-municipality cooperation can we accept? Everyone his own industrial zone?' With this rhetorical question, the authors openly criticized the lack of cooperation among Valais' municipalities, which were following competing strategies that accelerated urban sprawl and landscape fragmentation instead of coordinating development. 'This way of considering land as a consumer product, amorphous, indifferent to place or history, leads to a dead end' (id.: 23). The landscape report concluded with a vibrant call for more regional ambition and coordinated planning, underpinning the 'close interdependency between the Third Rhône Correction Project and the Valley'. However, if the Rhône project stimulated reflections outside the Espace Rhône, it was not its task to develop them, they concluded, an ambiguous statement calling for a large and inclusive analysis but a limited project in its practical applications.

Still, the report reaffirmed that 'the Third Rhône Correction provides an extraordinary opportunity to think about the organization of the valley', an opportunity that called for greater commitment and resources: 'It would be unfortunate, harmful even, to let the [river] project go further without securing equivalent means and quality of work for the valley' (id.: 32). This conclusion was echoed in a milder way by the author of the 2008 synthesis report, who admitted that 'the development concepts worked out by the CoRePil committees show the limits of an approached restricted to municipal borders, and highlight the necessity of pursuing the planning reflections above municipal level', yet without suggesting any further concrete proposal to remedy the lack of structural coherence (Projet Rhône, 2008: 79).

SION-SUR-RHÔNE: LANDSCAPE STUDIO AS ICEBREAKER

Urban and landscape design, kept apart from the studies engaged for the Third Correction, finally entered the planning process through the back door with an academic design studio held in Sion by the Federal Institute of Technology–ETH Zurich. In 2008, the state capital city of Sion still relied on the measures proposed by the Canton's Department of Roads and Waterways. The Sion planning department and chief architect were aware, however, of the river's potential for urban renewal, lamenting its inaccessibility and unattractiveness, but an initial attempt to organize a design competition was rejected by the City Council. At the time, the Council saw the Rhône Correction foremost as a flood safety issue, as a prerogative of the Canton authorities, and therefore not eligible for municipal funding. Considering the high level of flood risk in the low part of the city, the Council thought that the Correction would be implemented as it was sketched

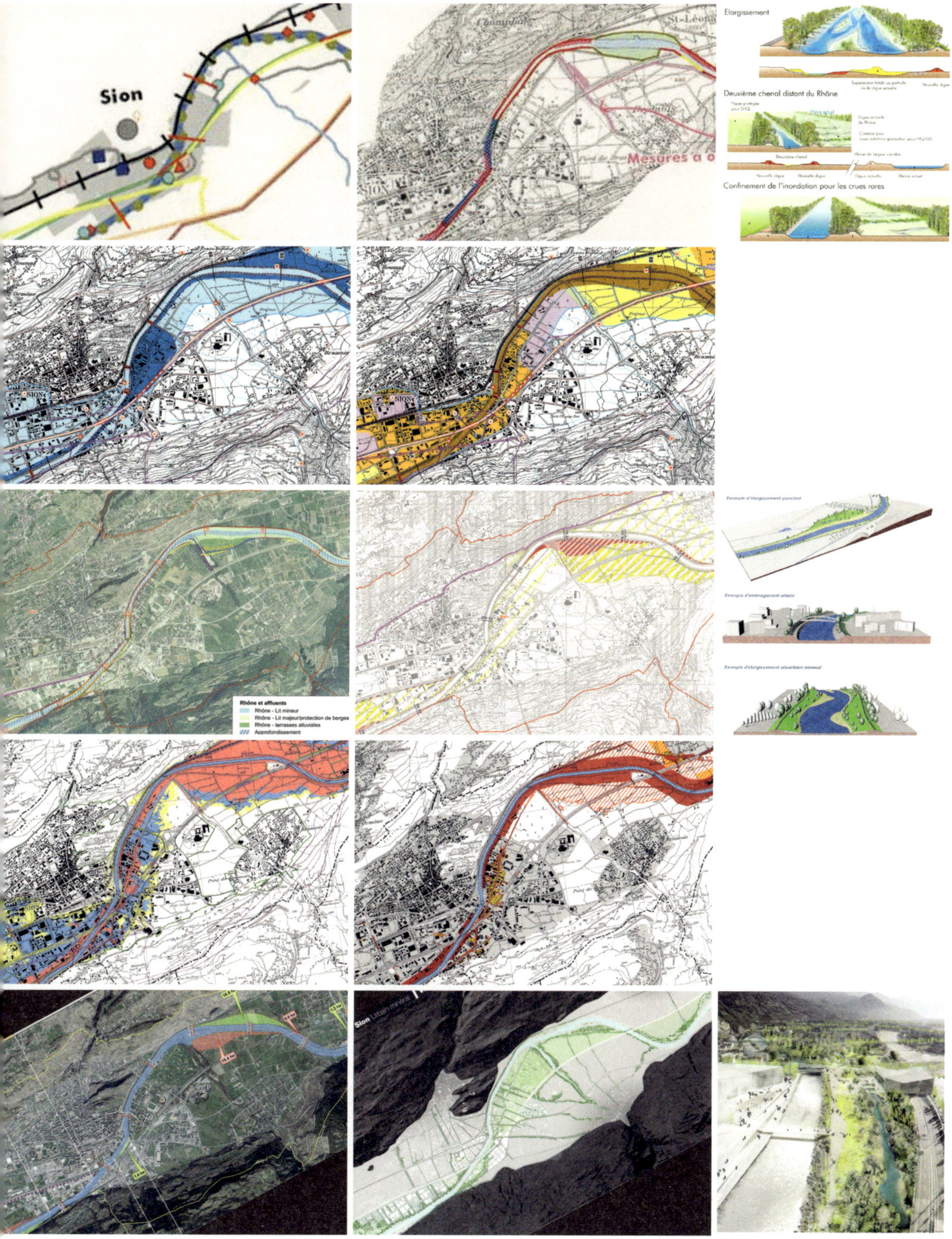

Main steps in the planning chronology of the Rhône Correction from 2000 to 2012, illustrated with the example of Sion and surroundings • 2000: identification main constraints and types of interventions • 2006: pre-defintion flood risk zones; Definition of river space 'Espace Rhône' • 2008: emergency measures; expected residual risks; preliminary design • 2011: flood-risk zones; high risk in built areas and building zones • 2012–2015: adapted layout; landscape design (design competition Paysagestion)

First Design workshop in Sion, discussion with the hydrologist

in 2000, by mechanically widening the river in 'unconstrained' (farming) areas, on the one hand, dredging the bed and reinforcing the levees in the 'constrained' (urban) area, on the other.

At the end of 2008, Professor Christophe Girot and his team from the chair of Landscape Architecture of ETH Zurich (then including the author) brought up the idea of the organization of a design studio along the upper Rhône. The Chair had been focusing for a long time on riverine sites and on flood mitigation, and the challenges posed by the Rhône Correction seemed to provide a rich and instructive base for a semester studio. Isabelle Duner and myself were responsible of setting up a programme for a design studio entitled 'Braided Streams', in reference to the multiple linear infrastructures that competed within the narrow valley floor. The studio brief challenged the students to 'elaborate a global vision for the Rhône Valley in Sion, integrating flood protection and urban redevelopment along the river' (Girot et al., 2010: 22).

Between February and June, the students developed six visions for the reconfiguration of the river, involving the whole water network of tributaries, drainage canals, and artificial lakes left by gravel pits, which they closely associated with the city's future development. Rather than channelling the Rhône further by raising the dikes, the projects suggested enlargement and diversions through the lower city's loose structure, creating opportunities for developing meaningful urban space and qualitative housing along the waterways, which had so far been neglected and hidden behind mediocre buildings.

This first design study perked the interest of the city, and was followed by a second studio entitled 'Captive River', as the enlargement of the river space within the city had become the key issue. Where the Canton's technician planned to reinforce the levees, the students suggested giving the river a secondary branch within the city, lowering the flood levels and creating a unique urban island out of industrial leftovers. This reminiscence of Geneva and Thun urban islands could form a new heart for the whole agglomeration of Sion and stress its riverine identity.

The resulting design study was presented to the city councillors in April 2010, and to the public with an outdoor exhibition shown during five months at various locations along the Rhône. 'Another vision, uncertain but carrying an indispensable dream element, is the one given by the students from the ETH Zürich, who imagined what the city of Sion-sur-Rhône could become tomorrow', wrote Jacques Melly, State Councillor and responsible for the Valais Department of Transport, Infrastructure, and Environment (*Rhône*.vs n.17, Sept. 2010). Simultaneously, a pamphlet entitled *Sion-sur-Rhône* was published. This apparently descriptive title was meant to seize the historical moment that could see a city, which always defined itself by its alpine landscape and elevated position, as discovering its riverine side, as the city born on the foothills had now extended down to and across the river. It offered a first urban and landscape vision to the

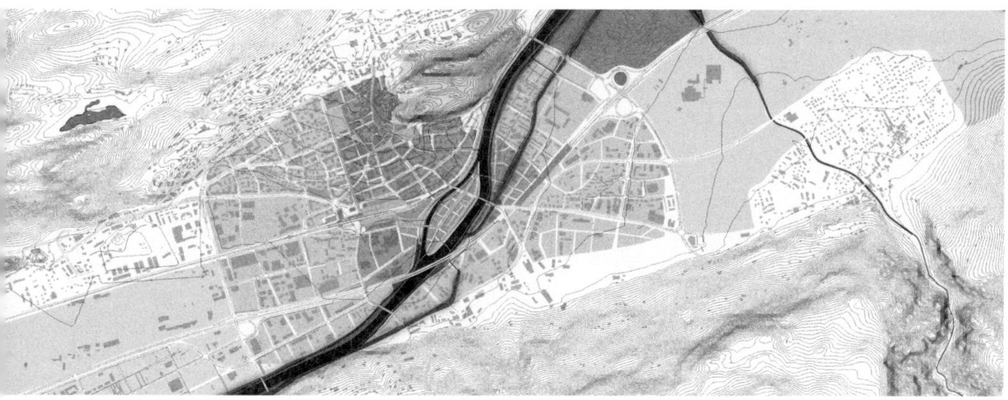

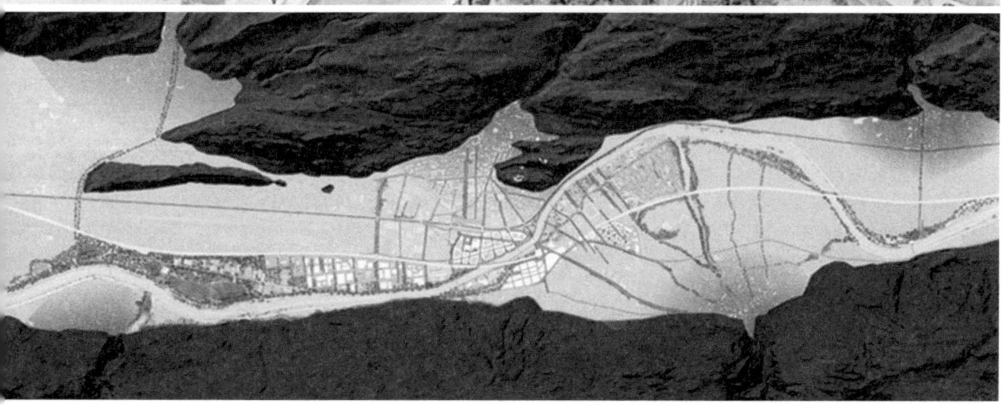

Sion-sur-Rhône design studio: The Rhône river as agent of urban redevelopment. Perspective onto the proposed second branch and new urban island (ETH Zürich, 2010) • Winning entry for the Sion-sur-Rhône design competition, bird eye view and general plan (Paysagestion, 2012)

people and their representatives, suggesting that the city could embrace rather than just endure the transformation of the river.

Taking advantage of this momentum in the planning process, city and canton decided to take the Rhône Correction to a higher level of ambition, and jointly organized a design competition for the reconfiguration of the Rhône River and surroundings within the borders of Sion. The brief took the title 'Sion-sur-Rhône', which has remained associated with the project to this day. A team led by landscape architecture office Paysagestion, co-author of the 2008 landscape report, won the competition in 2012. The chosen design did not literally take over the urban island envisioned by the Zürich studio but developed the idea to turn the 'problematic' river into an element of urban redevelopment. Paysagestion proposed a new park in the flood zone and accessible embankments that could be gradually flooded at high water. A substantial construction programme, capable of extending the city centre down to the river, was also proposed for the right bank. The suggestion to re-use former gravel pits and drainage canals to improve peripheral neighbourhoods was integrated into the new project as well.

For Sion and for the Rhône correction, this unsolicited intrusion of a teaching and academic institution into the planning process has proved so far to be a useful icebreaker. The city was able to facilitate the contribution of the ETH Zurich without necessarily endorsing its conclusions, using the school as an 'agent provocateur' to bring elected representatives to commit themselves to a more ambitious design than the original problem-solving plan.

ENGINEERING, ECOLOGY AND LANDSCAPE: A CONVERGENCE IN SIGHT?

On a broader scale, the master plan adopted in 2012 and the synthesis presented in 2016 didn't fundamentally change the philosophy of the project, still focused on technical objectives and land policy. The final changes were mainly aimed at easing the conflicts between ecology and agriculture: the area returned from arable land to river space was reduced with about 70 hectares, down to 310 hectares in total. New proposals were made for the levees, that allowed for farming activities on the slopes. Some nature and forest areas could be returned to the riverbed, providing that they would benefit to the river's ecology.

Together with the 2016 synthesis came a new landscape study, again entitled 'Landscape Guidelines' (Nomad, 2016), carried out by a local architectural firm. In between toolbox and pre-operational study, it established the basis for a systematic transformation of the river. Possible measures and corresponding authorities were proposed for three types of space: the space of the river, the 'top of the levee' and the space of the valley. The space of the river, delimited by the levees, would be shaped by the river dynamics, and the cantons would define the initial lay-out. The space of the 'top of the levee', literally the narrow and flat surface at the top, would also be designed and built by the Canton in collaboration with the municipalities, following as set of recommendations. For the space of the valley, municipalities and local stakeholders would be in charge of linking the project with the surrounding areas stretching from the river to the foothill, preferably through the adaptation of local roads and paths, engineering works, or the reshaping of river junctions, natural and agricultural areas. These guidelines were thought to facilitate the division of the whole programme into sections and lots, for a construction expected to last thirty years.

Although the document proposed a modular and pragmatic approach that divided the Rhône landscape among different zones and stakeholders, it still stressed the fact that 'the Third Rhône Correction [gave] the opportunity to redraw the landscape of the valley as a whole'. Significantly, the synthesis report acknowledged the fact that 'the development and optimization of the project' had revealed 'the necessity of a good integration with the neighbouring territories, and the exploitation of possible synergies with the municipalities along the river and with large infrastructural projects' (Projet Rhône, 2016). Sion and Sierre were given as two examples of the recommended planning process. Indeed, the most advanced sections—Sion, but also the Chablais region that put water at the heart of its regional planning (Chablais Agglo 3, 2016)—demonstrated the transformative potential of the Rhône Correction for urban, suburban as well as rural areas.

In 2018, the conflict between farmers, ecologists and developers seemed calmed if not solved, and more than half of the urgent measures were engaged or completed. The *Office Cantonal de la Construction du Rhône*, appointed by the Vaud and Valais cantons, finally commissioned a global design study for the Rhône public spaces. The programme management acknowledged the fact that a classical development section by section according to general guidelines wouldn't meet the expectations of local communities, and opened the door to a global landscape design.

Five teams were selected to make a design proposal based on the competition brief, which asked the competitors to 'draw a global project for the Rhône banks ... resulting in a global image for the whole length of the river, from Gletsch down to Bouveret, ensuring coherency among the various sections'. The design would be detailed for two test sites, Viège and the Chablais region (Office cantonal de la construction du Rhône, 2018).

In December 2019, a jury composed of 26 members, among which seven landscape designers, architects and planners, awarded the first price to the proposal made by the office of landscape architecture BASE, a project 'tailored for the region' according to Tony Arborino, director of the Third Correction. The winning team had succeeded in being creative within the numerous constraints, and the jury welcomed the fact that most of the project was 'built upon agreed elements', showed a 'good understanding of the landscape issues', addressed with 'pragmatic' and 'sparing' proposals.

Among the five entries, BASE' proposal stood out for its narrative and time-based approach, that sought to develop step by step new relationships between the river, its immediate environment (valley and villages) and its wider context (foothills, transverse valleys and mountains). For Bertrand Vignal, landscape architect and co-director of BASE, 'the issue isn't to compete for the design of riverbanks, but to bring people together

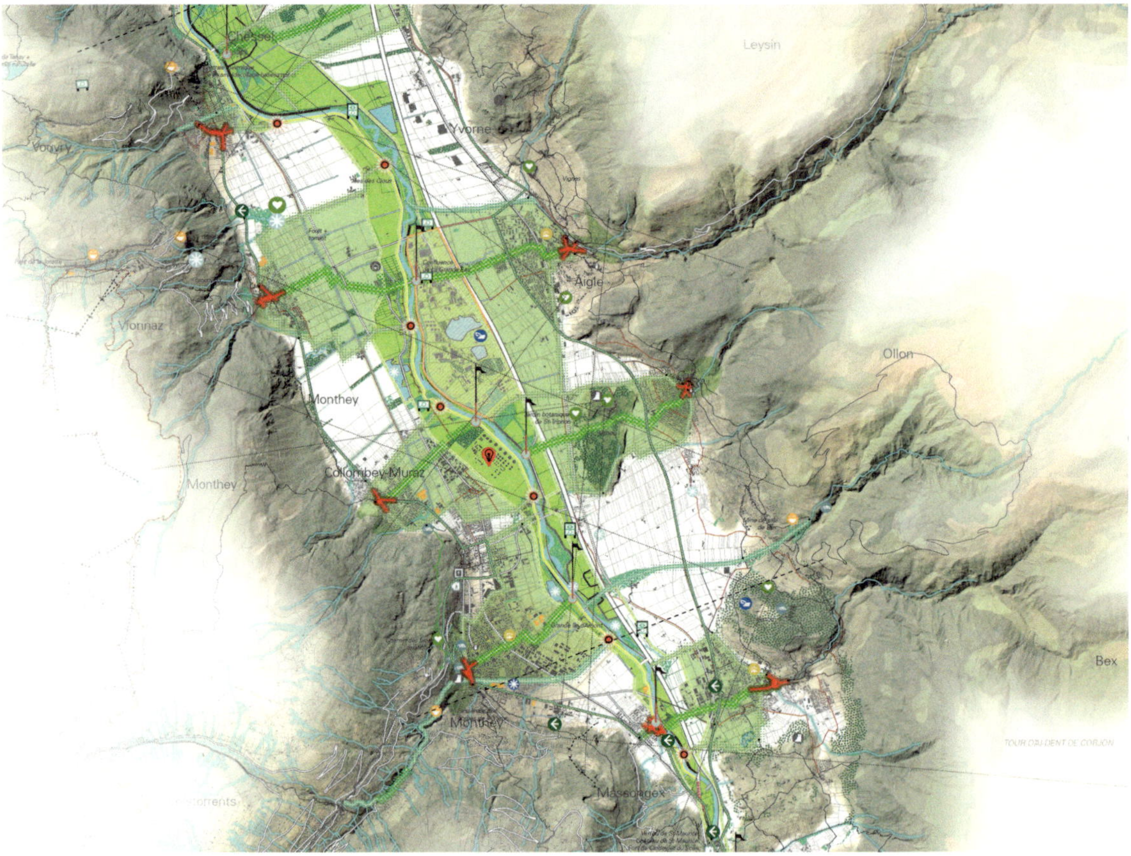

Design proposal for a better territorial integration of the Rhône river in the Chablais region (BASE, 2019)

around a common story'. He distances himself from the original linear division in parallel zones (riverbed, levees and valley floor) to promote transversal connections. These could be made through renovating, planting and reconnecting the side paths and tributary streams. On top of these spatial links, cultural links should be created as well. BASE suggests to built temporary pavilions in the midst of valley villages, meant to be moved and brought down to the riverside whenever the local correction works will be completed—which should take decades in some areas. The transport of these 'ambassador' constructions to the Rhône echoes local traditions, as several processions still take place in Valais, and will eventually create a string of 'hamlets', unique stopovers placed along the Rhône, drawing the attention towards the neighbouring communities.

These planting and building proposals are meant to announce the landscape that will be transformed over several decades. Bertrand Vignal underscores the 'long-term, transgenerational' nature of the project, which will first see the river stripped of its existing vegetation before it turns into a renewed, wilder landscape. 'It takes thirty years for a natural system to settle, this will impact a whole generation', Vignal stresses. 'In order to manage this time scale, we imagined the concept of "memory of

Designers' views for the banks of the Rhône river in Viège: 'beach' and belvedere (BASE, 2019)

Designers' view for the Rhône banks, a 'beach' for Viège (BASE, 2019)

the future". How to initiate a new narrative in a town that must wait another ten years to see the start of the construction works? By preparing the works and slowly developing the road and path system leading to the Rhône.'

Time is also a key aspect in what the designers call the 'acclimatization' of the Rhône valley, in other words its adaptation to global warming, particularly acute in the southern Alps. The project brings climate change again to the foreground, adding to sustainable flood management the mitigation of future hotter summers. At the competition stage, the proposal includes revegetation and the use of natural wind corridors which, together with the river's cool air, could refresh the valley in the future. For BASE, these options need to be further investigated and could become a new field of innovation for the region.

In the spring of 2020, at the time of writing, designers and commissioners are defining the contours of a long and complex mission. Constraints are numerous and sometimes contradictory, echoing lasting controversies and ambiguities. While the title 'Third Rhône Correction' has been criticized for being backward-looking and technocratic, the title 'RHONATUREPARC' chosen by BASE is now rejected for its supposed reference to a nature reserve or enclosed parks, which could upset local stakeholders, in particular farmers who tend to see themselves as victims of a project that, in their eyes, favours nature and recreation over production. The proposal to create connections from the Rhône river towards transverse valleys, appreciated by the competition jury, must first overcome serious obstacles: the administrative and jurisdictional boundaries that differentiate the Rhône riverbed from the levees and the surrounding land, in terms of land use regulation, responsibilities and maintenance. Visual and physical contact with water, celebrated in the competition's visuals, is another controversial issue, as regulations and laws on water and streams are numerous and sometimes ambiguous or contradictory.

For the Rhône valley as a whole, the winning team is mainly expected to supervise and to guide local developments with design guidelines, a masterplan and a tool box. However, the team will not provide a full and detailed design for the whole river. The Rhône correction has now reached a critical stage, as project managers and designers must agree on the scope and level of ambition for the river as a whole, and its future application in subsections. For the designers, the connection from the river to neighbouring towns and valleys is the most promising, in particular within the framework of municipal and inter-municipal development plans, some of which are now being revised and updated. They see the local scale as crucial for the future of the Rhône Correction.

Only the final result, expected in 2030, will tell whether the planning process will have succeeded in creating a coherent and sustainable river landscape, the large Gesamtkunstwerk that was envisioned at the turn of the century. Since 2000, the project has taken shape in hydrological, ecological, and economic terms, but the landscape value will be largely defined by local initiatives, good will, and 'happy accidents' that allow for concrete and mobilizing projects. Paradoxically, the success of this vast programme lays mainly in its ability to promote and articulate various local projects, which are as much subparts as driving forces behind the Grand Projet.

The Dutch programme 'Room for the River', the other European giant among river restoration projects, embraced this 'glocal' approach from the start under a title almost opposite to the Rhône Correction, but with strikingly similar means and ambitions.

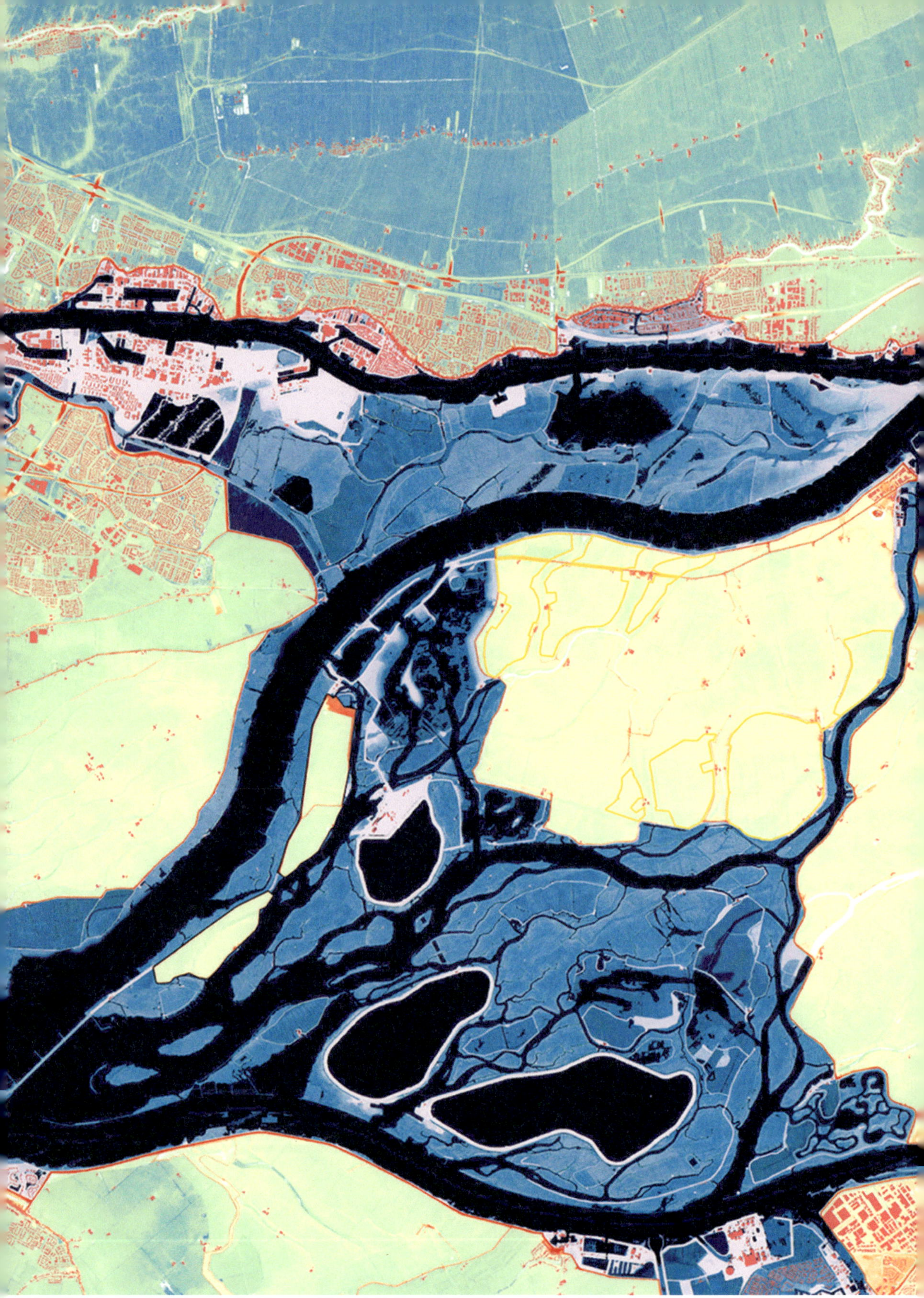

Biesbosch region, gate of the Rhine and Maas estuary

FROM RAISING DIKES TO MAKING ROOM FOR THE RIVER

After the sea storm that submerged part of the Dutch coast in 1953, hydrologically quiet decades followed and flood protection along the rivers remained in the background in the Netherlands. While the country was building the famous 'Delta Works' against the North Sea, the reinforcement of old river levees went on intermittently, mostly along the great rivers, Rijn (Rhine), Waal and Maas (Meuse). However, as in the rest of the Western European countries, concerns grew about the environmental and living quality of the river areas, challenged by intensifying agriculture, urban sprawl, new infrastructures, and the increasing degradation of valuable cultural landscapes.

NATURE ASSETS AND THE CULTURAL LANDSCAPE

In 1986, the today prestigious Eo Wijers competion for regional planning was organized on the theme 'Netherlands Land of Rivers' ('Nederland Rivierenland'). The Eo Wijers Foundation was then drawing attention to the future of the River Region, in the centre of the country, where the Meuse and Rhine rivers met and parted again before reaching the estuary. Out of 34 designs, the jury gave the first prize to the 'Stork Plan' ('Plan Ooievaar'). The Stork Plan proposed a reorganization of existing functions, acknowledging the rift between intensive agriculture permanently adapting to the market, and riverine nature that needed time to establish itself and develop. It also acknowledged a stronger differentiation between intensified agricultural areas behind the levees, adapted to the rapid pace of a changing economy, and the river's floodplain where slow processes should be granted space and time to recompose a rich, diverse, and dynamic environment—one not necessarily free from human intervention, in the eyes of the authors. Within this differentiated landscape, competing functions were isolated, while activities that could take place in the same areas and even reinforce each other, could be combined. To give the floodplain more legitimacy, the author proposed to create a new territorial status, the 'waardschap' going 'from dike to dike', a new entity that could turn the river space from 'administrative border and backside to an administrative backbone' (Eo Wijers Stichting, 1986: 16).

The Stork Plan attracted the attention of several non-governmental organizations, which saw 'a challenge and an opportunity to change their policy from a defensive to a more offensive and developmental strategy' (De Jonge, 2009: 187). The plan avoided the common opposition between production activity and nature protection—proposing to link, for example, the clay mining industry to the creation and management of wetlands, a proposal to which the industry reacted 'positive about cooperation as they were part of the win-win strategy'. Another original element was the interdisciplinary composition of the team, which included ecologist Frans Vera and river expert Dick De Bruin, along with four landscape architects then working together at the Forestry and Nature Agency Staatsbosbeheer. One of them was Dirk Sijmons, then head of the Landscape Department, who later with his team explored the construction of new, reinforced dikes in the Maas and Waal area from the point of view of a landscape architect.

At the time, environmental aspects were integrated into Dutch planning procedures, in particular with the MER assessment report ('Milieueffectrapportage'), mandatory for large projects and civil works, including flood defence. The concept of 'L.N.C. assets', the abbreviation used for Landscape-Natural and Culture-historical Assets, was then commonly used to summarize and evaluate environmental aspects, and had dominated the planning discourse since the 1960s (De Jonge, 2009). Established to provide assessors with objective criteria, the LNC system left little room for an in-depth, integrated analysis and project-oriented choices, making it for Sijmons an 'accounting system' rather than a concept useful to a designer. Engineers focused on the preservation of 'LNC-assets' rather than on global quality and transformation potentials: in the 1990s, 'it was always about the integration of LNC assets ... as a generic term: as long as it is well integrated' (Van der Grift, 2013). As a result, old, plain narrow levees were turned into new 'enormous levees swirled from one compromise to the next', a situation Sijmons decided to address with an alternative approach.

In 1988, Yttje Feddes and Frans Halenbeek, Sijmons' collaborators at Staatsbosbeheer, published a guide for the design and the renovation of levees, entitled 'a clear border' (Feddes and Halenbeek, 1988). 'Our advice was: you are forgetting your most important task, your cultural assignment, and that is building beautiful levees' (Sijmons, 2015). While 'the proposed profile was a nondescript plum pudding', the designers stated that a good levee should form a clear border between the protected and open areas and enhance the experience of 'floating over the landscape' that characterized the old narrow levees, both goals that could be achieved by giving the new dike a smooth trajectory, a clear, legible footprint, and a narrow top. More generally, the landscape architects rejected a pure protective attitude regarding supposedly permanent natural elements: 'Cultural history should take precedence over nature—nature is a dynamic system; you can "make' it" (idem). Yet a proactive design-oriented approach within the Forestry and Nature Agency was not welcomed by its directors, although the Dutch landscape was 'exhausted by a series of re-allotment programmes', and in need of a vision for its future (Sijmons, 2015).

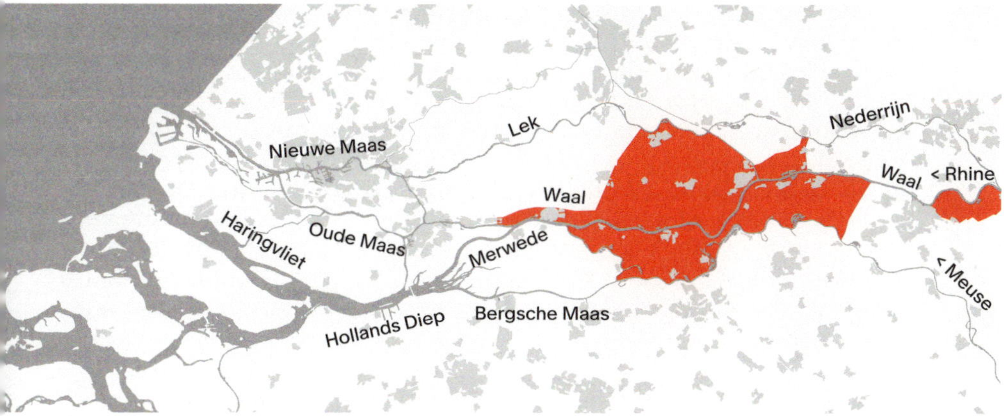

Areas evacuated during the 1995 high waters

However, with the start of the plan 'Living Rivers' launched in 1992 by the World Wildlife Fund and largely inspired by the Stork Plan, the idea was growing stronger that riverine nature did not consist of permanent and fixed habitats to be defended at all costs, but rather of dynamic processes that could also be created and enhanced by human intervention. Nature development was then already practiced by the Staatsbosbeheer Agency [State Forestry Bureau], next to traditional nature conservation, 'the only difference between the strategies [being] the length of time during which human interference is accepted in the system after the initial management measures' (De Hullu, 1995). In the world of water management, attention had slowly shifted from purely defensive strategies,—more water calls for higher levees—towards more dynamic visions, integrating alluvial nature as an important and malleable component of the river landscape.

1993 AND 1995: TWO WARNINGS ON THE RIVER FRONT

In 1993, forty years after the great sea storm, the country again endured a major flood event, this time from the Meuse River that flooded 6000 dwellings and about 7 percent of South Limburg province, generating damage estimated at 250 million guilders (Heezik, 2007). Although the inundations claimed no human victims, the disaster created the 'same shock-effect as in 1953' (Nienhuis, 2008: 548), laying bare the limited capacity of man-made channels, the crucial role of the river levees, and the exposure of the densely built river regions. The Dutch Government quickly created the Maas Flood Committee to investigate possible flood defence in the Meuse basin (the Maas Watersnood Commissie, known as Commissie Boertien II after its chairman). In its final report, 'The Meuse Back' (*De Maas terug*, 1994), the committee made a recommendation largely inspired by the Stork Plan and by the plan Living Rivers: combine material exploitation and nature development—already planned in the region—in order to restore the discharge capacity of the floodplain; raise the dikes only where necessary; prevent further construction in flood-prone areas (Heezik, 2007: 100).

Before the Meuse could be widened and secured, water rose again in the Rhine and Meuse Rivers in January 1995, after days of heavy rain in both catchment areas. Low parts of the Dutch Limburg region were again inundated. With the 1993 flood still in mind, 250,000 people were evacuated from the downstream River Region out of fear for a larger catastrophe. The situation eventually remained safe and people returned home again after less than a week, but the massive evacuation left the country in shock. After that second, traumatic event, levee strengthening was seen by the public as an absolute priority, far above 'LNC assets' and environmental concerns, held responsible for delays in maintenance works. Previous environmental protests against levee reinforcement were made null and void, as 'people who had previously opposed the reinforcement [were] given the cold shoulder and treated like wartime collaborators, so to speak' (Sijmons, 2015). The public fear and uproar led to a massive increase in public funding for flood defence, channelled through the Delta Plan Large Rivers and validated by Parliament in April 1995, only three months after the events.

The plan created the financial and procedural conditions for the accelerated implementation of maintenance and reinforcement works. This acceleration was made possible through 'the shortening of delays and neutralization of dozens of laws' (Min. van Verkeer en Waterstaat 1997b: 2). Another accelerating factor was the Flood Defence Act of January 15, 1996, which made it possible to proceed with legal validation procedures before the necessary environmental assessment reports (MER) were finalized. The Flood Defence Act also shortened expropriation delays from fourteen down to three months (idem). Besides being procedural, this mobilization was also financial: for the year 1996 alone, the State spent 225 million guilders on the Delta Plan Large Rivers (about 112 million euros), and in 1997, all emergency measures were implemented. Hundreds of kilometres of levees were restored and reinforced during large infrastructural works stretching from 1996 to 2000.

Unlike the earlier Delta Works, landscape and ecological concerns were integrated into the new campaign, but remained limited to general principles, such as 'the limitation, as far as possible, of the degradation of Landscape, Nature and Cultural (historical) values' (Min. van Verkeer en Waterstaat, 1996). However, the surge in financial resources allocated following the 1993 and 1995 high waters indirectly benefited landscape redevelopment: although there was haste and less participation, there was also more money, which made it easier for the proponents of stronger levees to respect the 'LNC assets'.

Parallel to a growing interest in an alternative, more 'natural approach to flood defence' (Heezik, 2007) that had been brewing since the 1980s, the shift from coercive measures towards spatial strategies reached the political foreground, nationally and internationally, partly activated by the failure of traditional containment measures. In February 1995, the Environment ministers of the five countries crossed by the Rhine and the Meuse rivers—Germany, France, Luxemburg, Belgium, and the Netherlands—met in Arles and signed a joint declaration. The declaration called for 'concrete measures', to 'be taken, in several fields, including spatial planning and land use (e.g. agriculture, forestry, nature management, urbanization and recreation) and water management (e.g. buffer zones, dikes and embankments, and flow management)'. Their recommendations specifically targeted two types of measures: adaptation of the rivers' morphology, and minimizing exposure in the floodplains. The five ministers also recommended 'taking spatial planning measures to permit greater water storage in the whole river basin, such as changes to land use, afforestation, returning riverside areas to nature, the development of overflow zones and impounding basins, and the de-canalization of streams'; they also recommended 'preventing further urbanization in vulnerable areas along the Meuse and Rhine (possibly including a ban on further construction). Such measures will not only reduce the risk of flooding but will also have a positive impact on natural features in the areas along the rivers'. The Environment ministers thus shifted the discussion from the fields of hydraulic infrastructures and technology to the field of spatial adaptation and regional planning. The declaration indirectly pointed the finger at states and regions regarding the consequences of their past *laisser faire* attitude concerning building in rivers' natural floodplains', concluding that 'In the long run, more emphasis must be placed on spatial planning'.

In reaction to this call, the ministers of Spatial Planning from the same countries met in Strasbourg in march 1995, and made a joint statement. The 'Declaration of Strasbourg' supported the proposals made in Arles, and decided to create a workgroup to engage in all measures needed in spatial planning, in particular the prevention of further constructions in risk areas. The ministers also insisted on the need to integrate flood defence policy with other fields, such as nature protection and water management.

In the following years, specific work groups for flood management and prevention were created within the existing Rhine and Meuse international commissions. Meanwhile, Dirk Sijmons' private office for landscape architecture HNS, together with DHV engineering firm, was commissioned to apply the principles previously developed along the Waal and the Meuse Rivers in 1995. Between Afferden and Dreumel, 50 kilometres of levees were transformed according to new design principles, which for Sijmons, marked 'one of the moments when technicians and designers learned to work together'; it also announced the return of spatial quality within the scope of flood management, a dimension that would gain importance in the Room for the River programme, as the emphasis shifted further from flood defence infrastructure to long-term spatial measures. Sijmons would soon be appointed State Landscape Architect, and would eventually join the management team of the Room for the River programme in 2006 as chairman of the Quality Team.

FROM POLICY TO PROPOSAL: OPENING THE SCOPE OF POSSIBLE MEASURES

Between 1995 and 2000, the Room for the River concept was steadily developed and engraved into governmental reports and policy documents. The 1996 governmental Room for the River Policy Guideline, jointly signed by the Minister for Transport and Shipping and the Minister for Housing, Spatial Planning and the Environment, drew new conclusions

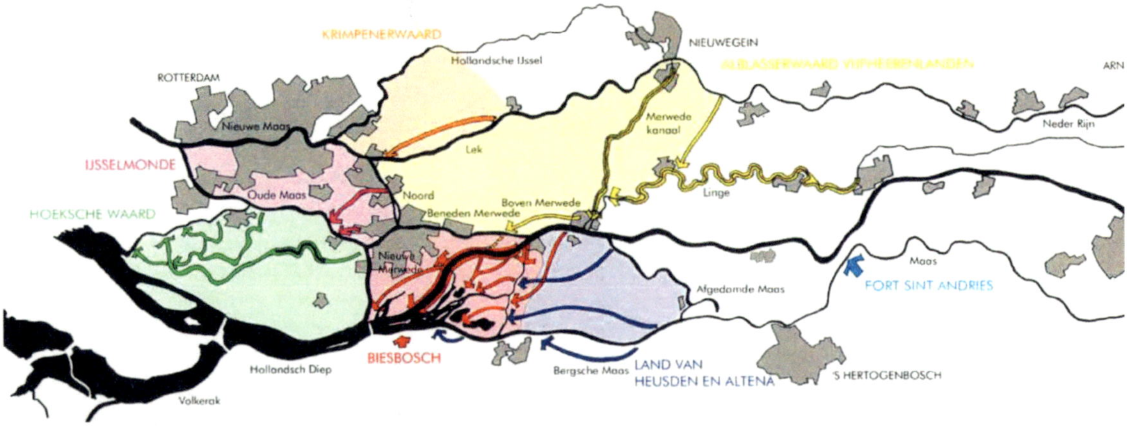

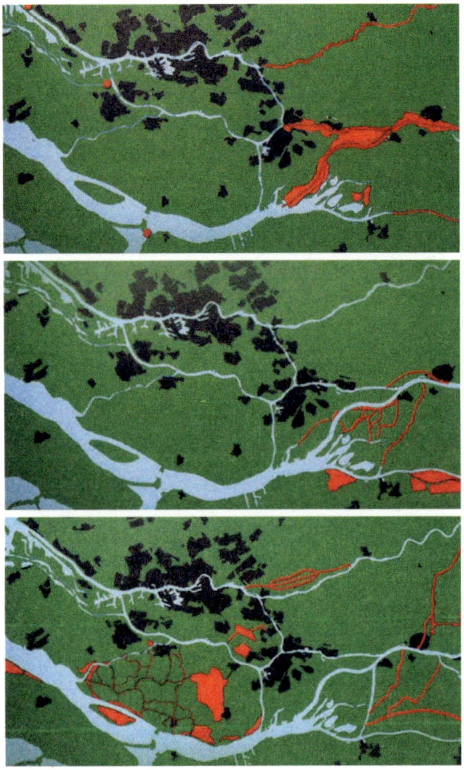

Possible spillways and bypasses investigated for the lower rivers (RIZA et al., 2000) • Three combinations based on 31 possible measures for the lower rivers (Min. van Verkeer en Waterstaat, 2000b)

from the 1993/1995 floods and expected sea level rise, a phenomenon already established in the 1980s, and it set the tone for the future decades: 'For the short term, the Delta Plan Large Rivers allows for the accelerated reinforcement of levees and the construction of riverbanks along the undiked Meuse. In the long term, sustainable protection will not come as a result of continuous heightening of the levees, but much more from measures intended to give more space to the river in order to accommodate higher river discharges' (Min. van Verkeer en Waterstaat et al., 1996: 9). The document was particularly critical of past diking and compensation policy that permitted narrowing or occupying the river's winter beds and floodplain. Despite compensating measures, 'we still must admit that the possibility of accommodating future higher river discharges was in fact diminished over time. With each activity [taking place in the floodplain] and its related compensation, a piece of spare space for the river was eliminated'. This observation led to practical applications: floodplains should be preserved in order to accommodate future water discharge levels and should only be used for activity directly related to the river; extra discharge capacity should be created through the extension and lowering of the floodplains, in combination with the creation of nature areas; and exposure should be limited in the areas along the Meuse that were outside the primary levees (idem).

From then on, governmental reports consequently presented river widening as preferable, if not inevitable. Furthermore, a close link was made between river management and spatial planning, breaking the tradition of centuries of land expansion at the cost of water bodies: 'The near-floods in the Rhine branches and the floods in the Meuse basin make it clear that rivers should get more space. To achieve this goal, more coherence is needed between water, spatial, and environmental policies' (Min. van Verkeer en Waterstaat, 1997: 3). The Governmental Statement of 2000 again confirmed the new strategy and launched the planning process that would lead to operational Room for the River proposals for the different river branches and regions (idem, 2000).

AMBITION AND PRAGMATISM: EXAMPLE OF THE WAAL AND BIESBOSCH

The Dutch Government addressed regional situations with a series of specific reports at the turn of the century. Following the 1995 Delta Plan Large Rivers, a broad preliminary study involving various disciplines had been engaged in 1997 by the Rijkswaterstaat agency vis-à-vis possible interventions along the lower rivers, including one specific study dedicated to the Biesbosch region. The general report, entitled 'Integral Foresight Lower Rivers' (Min. van Verkeer en Waterstaat, 2000b), was compiled by the Rijkswaterstaat's South Holland Direction, under the supervision of representatives from the Provinces of South Holland and North Brabant, from the Union of River Municipalities, and from the Union of Water Boards. It explored various ways to increase the discharge capacity on the lower branches of the Rhine and Meuse Delta (Lek, Waal-Merwede, Bergsche Maas-Amer, Old Meuse, and New Meuse) that all flowed through densely populated regions, including the scattered Rotterdam-Rijnmond agglomeration. Here again, the report stressed the necessity of a new approach: 'While it was a common idea in the past century to add a bit more on top of the dike to keep water out, future solutions will be sought after in terms of width' (Min. van Verkeer en Waterstaat, 2000b: Samenvatting). A cost-benefit analysis study commissioned that same year concluded that, in many cases, 'room for water' measures appeared indeed beneficial compared to heightening the levees or to no intervention at all (Stolwijk and Verrips: 2000).

In 2001, the official discharge reference for the Rhine was heightened as 'a result of the integration of the 1993 and 1995 high river discharges'. This significant step had immediate implications for the entire Dutch dike network. Dutch law indeed guarantees the safety of each hydrological region (or 'dike ring') up to a certain flood probability. This probability is set at 1/10,000 for the dike ring No. 14, which includes Amsterdam, the Hague and Rotterdam, but only 1/2000 for the dike ring No. 22, which encompasses the Island of Dordrecht, a less populated area located in the estuary. With the rise of this reference peak discharge, the government committed itself to take the necessary measures to maintain the same safety level. The reference level was thus significantly increased, but this time with the new Room for the River Policy Guideline: the new safety goal should, as far as possible, be achieved through spatial measures.

At the end of 2002, a synthesis for the whole river system was published by Rijkswaterstaat in association with ministries, provincial authorities, and water boards, entitled 'Adaptability of the Rivers Region' ('Spankrachtstudie Rivierengebied'). The report explored and combined various options available to reach the new discharge reference. In accordance with medium estimates for climate change in 2100 made by the IPCC (Intergovernmental Panel on Climate Change) and the KNMI (Dutch

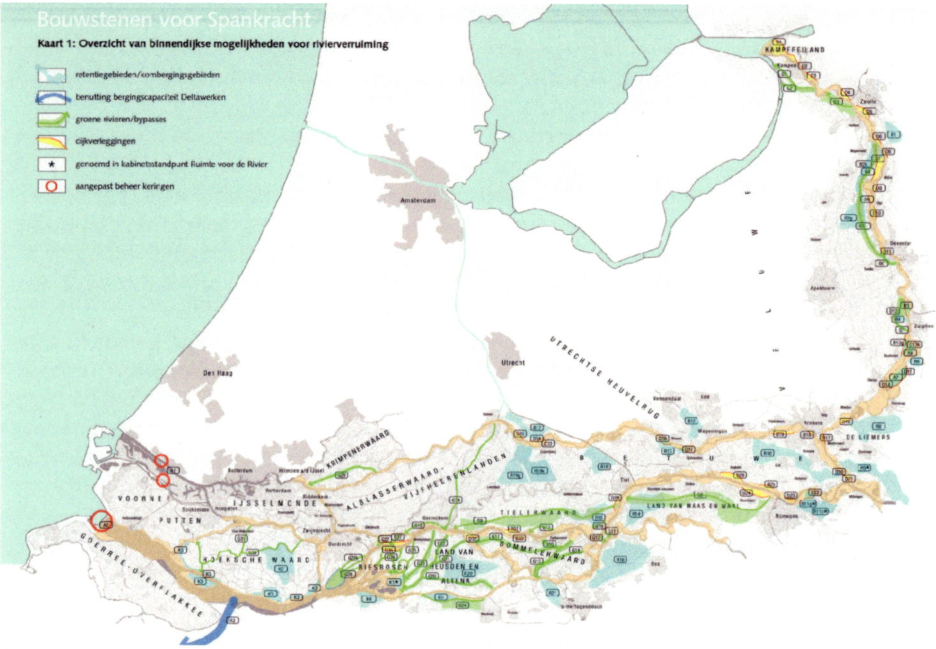

Overview of possible river enlargement measures within reclaimed areas. In green: spillways/bypasses, light blue: retention areas, dark blue: retention potential offered by the Delta Works, red: adapted use of existing barriers (RIZA, 2002)

Royal Institute of Meteorology), the report took as a base for the 21st century a 60 cm rise of sea level, and a further rise of reference peak discharge in the rivers, +25 per cent for the Meuse and +20 per cent for the Rhine (RIZA 2002: 17). On that basis, the report developed, combined, and measured various options available to achieve 'lasting safety and spatial quality'. From all possible options, 31 were sketched on the map and grouped into three thematic approaches as to how the river landscape could be adapted in a coherent way: 1) create retention and storage areas, 2) create new bypasses (or 'green rivers') in low areas, and 3) relocate the river levees to create a wide and continuous river zone.

The various sub-regions and rivers branches were further analyzed individually, and for each a global proposal was made. The study foresaw a growing role for the Waal/Merwede channel as long-term navigation and discharge route, where the 'intensification of river-related economy and the maintenance of navigation depth can be perfectly combined with the development of highly dynamic nature' (RIZA, 2002: 8). As a counterpoint, it set as a general goal 'the realization of a large-scale nature area from the banks of the Waal River through the Biesbosch, the Island of Dordrecht, the Hollands Diep, and Haringvliet down to the fore-delta'. The nature area envisioned 'can form a robust spatial buffer between the Brabant sandy lands and the Randstad' (the Dutch main agglomeration which hosts about eight million inhabitants). Anticipating a possible reopening of the Haringvliet dam, the report added: 'The long-term perspective is the realization of a fresh-salt water transition, and to give space to the tidal and river dynamics' (id.: 24). Sketching a both ambitious and pragmatic scenario for the southern delta, the report thus clearly hinted at a combination of intense river economy on the Rhine-Waal axis, and a greener and more dynamic future for the Biesbosch and Hollands Diep to the south.

Still a preliminary step on the way to a concrete programme, the 'Spankrachtstudie' or 'adaptability study' did not, however, deliver a plan. Instead, various possible measures and combinations were listed, evaluated, and shared through an ad hoc computer programme: the so-called 'blokkendoos' (literally box of toy building blocks, or 'tool kit'). The software allowed its user to choose and combine various measures, and check in real time the efficiency and the financial implications of the chosen combination. The tool kit automated and accelerated the preliminary studies, as 'a huge number of measures (and a corresponding number of measure sets) had to be evaluated. Nevertheless, the Spankrachtstudie came to conclusions within about a year, due to intensively using the tool kit' (Kors, 2004), and allowed for a relatively short planning process considering the scale of the programme.

Although the Water and Public Works Agency was then leading the planning process, the tool kit allowed the technicians and engineers to outline the goals and the scope of possible interventions, yet remain in a detached position when it came to global choices and spatial policy: 'It is very important that Rijkswaterstaat clarifies the safety task within the Key Decision, something that society cannot just make up, and says clearly: "If you do not act here, we will have to heighten the levee there", and we do not want that' (Van der Grift en Der Nederlanden, 2013).

The tool kit demonstrated for provinces and municipalities that various options were then still open and combinable, and all projects eventually integrated in the Room for the River programme would indeed be a combination of several types of measures. Furthermore, the tool kit also allowed Rijkswaterstaat to 'involve various parties in the exploration phase', although gradually, as 'there [were] so many different interests and priorities that one [had] to stay in control of in this phase of generating support' (Sijmons, 2015). The chosen strategy was indeed not fully 'bottom-up': the Room for the River team opened the discussion at a preliminary stage, but kept a firm hand on the sequencing of the planning process to avoid the plan being bogged down in endless discussions. As it would be stressed in retrospect, 'there is strong evidence that collaborative approaches can be only successful when they are complemented by more hierarchical, top-down arrangements' (Warner et al., 2013).

THE PROJECT AS TOOL OF INVOLVEMENT

Rather than elaborating a global programme to be later submitted to national and local representatives, the Room for the River agency tried to facilitate the planning process by involving local parties in the preliminary design, years before the programme was actually made official in 2006. Possible measures were exposed, discussed, and adapted during a large 'consultation tour' across the *waterschappen*, the powerful Dutch Water Boards, municipalities, and provinces of the two river regions (Upper and Lower Rivers).

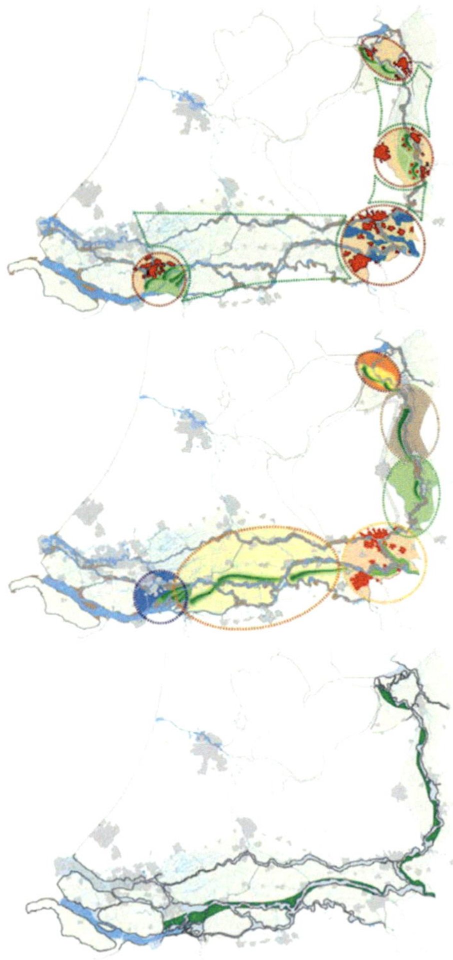

'Pearls on a string', 'Old and new river courses', 'Widened river ribbon': Three landscape strategies to increase the global discharge capacity of the main rivers (Projectorganisatie Ruimte voor de Rivier, 2006)

This preliminary round of consultation, mostly informal, generated local willingness and interest for the national programme. The combinatory approach also implicitly suggested possible advantages for the Dutch Treasury: 'The Water and Public Works Agency had limited financial means ... and maybe the measures could be combined with local developments and so turn out to be much cheaper, this at a time when real estate development was a profitable national "hobby"' (Sijmons, 2015).

Out of the 300 possible measures initially conceived, 30 projects eventually emerged, often combining different measures that could count on local support and allow for a good distribution of the measures over the River Region—as 'concentration of all measures along one river branch, for instance, was not an option' (idem). On the other hand, proposals that were not positively perceived by local authorities, such as a partial de-poldering of the Island of Dordrecht, were soon dropped in favour of less conflictual options—in that case a shift towards the reopening of the Noordwaard polder, a measure that proved to be extremely effective in reaching the hydraulic targets. Above all else, this informal consultation process allowed projects to emerge that were 'welcome in my backyard' and seemed 'ready to start as they were already granted the green light at the local level' (idem). Although more detailed and complex negotiations with inhabitants were still to come, this first round with local representatives and managers later proved to be 'one of the extraordinary success factors for the Room for the River programme' (idem).

The measures that seemed most promising, efficient and consensual were so filtered and assembled to create three main scenarios. The scenario 'Pearls on a string' looked for solutions through the implementation of large water landscapes close to or within urban areas. The scenario 'Old and new river courses' proposed the creation of by-passes for flood waters through low rural areas. The scenario 'Widened river ribbon' proposed to set the levees back all along the rivers and broaden the winter beds. All options appeared feasible, but the second and third options implied a major loss of farmland. The first option thus received the most attention and governmental preference, since it identified areas that provided 'opportunities to combine urban development with robust nature cores and recreation possibilities' (Projectorganisatie Ruimte voor de Rivier, 2007, 27).

Besides this first consultative approach, the Room for the River Agency imagined a second way to generate willingness among local parties: the so-called 'exchange rule' (*omwisselbesluit*). This rule gave local authorities the possibility to propose alternative measures, as long

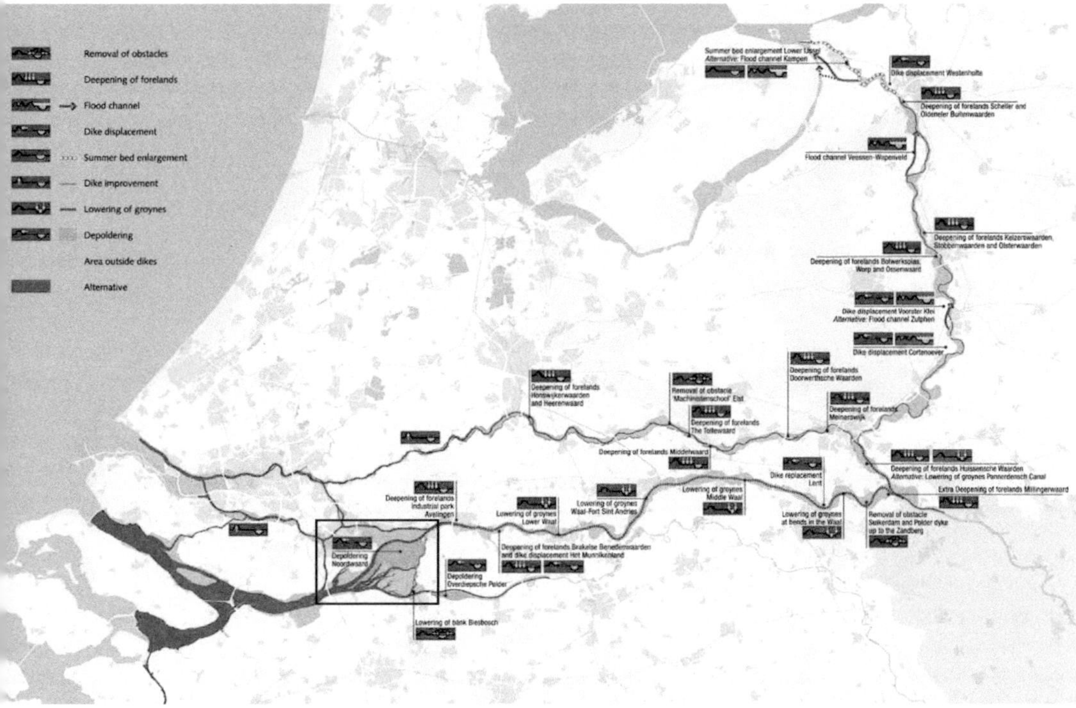

'Room for the River', location of the 39 planned interventions, Biesbosch area marked in red (Projectorganisatie Ruimte voor de Rivier, 2006)

as these measures could reach equivalent results and would not lead to higher costs for the government. Kampen's bypass and the 'Terpenplan' in the Overdiepse Polder are examples of locally initiated plans that have now been implemented. For Kampen, a large lowering of the river floodplain was first proposed, but the city envisioned a new by-pass between the River IJssel and the lakes south of Flevoland, which could be combined with future urban extension. The Overdiepse polder was initially planned to lose its agricultural function, in part or in full, in order to expand the Bergsche Maas floodplain, but farmers proved to be attached to their land and ready to exploit a polder that could be flooded at high water level, as long as the farms could be settled on safe grounds. A third spectacular 'exchange-project' would be the Waal river in Nijmegen, which saw the creation of a second river branch, a new island, and waterfronts, to the benefit of the city's urban ambitions. In all three cases, the Water and Public Works Agency studied the proposed alternatives and engaged in an open discussion with their promoters, sharing its knowledge rather than following traditional authoritarian strategies based on exclusive expertise.

Finally, the implementation of the projects, funded by the central government, was largely decentralized to local authorities, provinces, municipalities, water boards and private contractors, who could better balance local interests and opportunities in contact with the stakeholders directly concerned by the various projects, developed under the aegis of Rijkswaterstaat and the direction of the Room for the River programme.

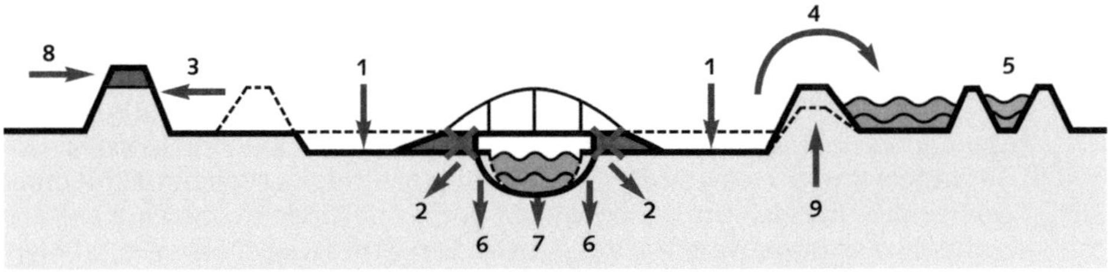

Nine types of possible measures (Projectorganisatie Ruimte voor de Rivier, 2006)

EMBEDDING SPATIAL QUALITY INTO FLOOD PREVENTION

With the Planning Key Decision Room for the River, first published in April 2005, adapted and validated in December 2006 and widely communicated, this change of philosophy within the Dutch Flood management was turned into a law that launched a colossal project meant to last until 2018 for an estimated budget of 2,1 billion euros for the state alone. Without explicitly criticizing past policies, the government clearly acknowledged the shortcomings of containment measures that had characterized the Dutch water management for centuries. In its introduction, the document depicts in clear terms the growing exposure of the river region: 'In past centuries, rivers were deprived of a lot of space; as a result, rivers were confined between levees that were heightened again and again over the past decades. In many places, the land behind the levees had sunk. Due to demographic development and economic growth, the assets to be protected have increased significantly. A flood would have huge consequences' (Projectorganisatie Ruimte voor de Rivier, vol.3: 7).

Considering the expected rise in reference discharge levels, the government chose to promote measures that prevented reference peak levels from rising further, meaning that the emphasis shifted from levee improvement to river widening, in addition to measures outside river channels that would also expand the actual floodplain. Furthermore, it was decided that 'levee improvement should be implemented only on segments where other measures are inappropriate or impossible to finance' (id.). Far from being presented as an ultimate solution, the programme was seen as a first step in an ongoing adaptation process: 'Because of climate change, the safety of the rivers region will need ongoing investment. This is anticipated in this Spatial Key Decision, through the reservation of a limited number of areas for the river-widening measures that will be necessary after 2015' (idem).

Discharge capacity was, however, not the only criterion: differing from the previous programmes implemented in a long history of flood defence, Room for the River set itself a second ambition: spatial quality. With this second goal, the River Region would 'be made more attractive and more liveable'. The notion of spatial quality had been introduced in the Dutch spatial policy during the 1980s, for example in the Fourth Memorandum on Spatial Planning (Dauvellier, 1991). Replacing the old 'Landscape, Nature and Cultural-historical assets' criteria, the new concept of 'spatial quality' was brought forward as a main requisite, thus committing the state to qualitative results but leaving room for local interpretations, variations and combination. As made explicit in the official programme validated in 2006, Room for the River should 'contribute to the improvement of the river region's spatial quality and simultaneously strengthen its economy, ecology, and landscape', it should pay attention to the 'preservation and the development of protected natural assets' (Projectorganisatie Ruimte voor de Rivier, 2006, vol.4: 9). Ideally, projects should seek to combine water-related measures with other functions, making it clear that the governmental decision was no carte blanche for unlimited wetland creation and floodplain restoration, but rather should associate the retention function with other profitable land use—whether new or pre-existing.

Rather than a systematic and uniform river widening, the Room for the River programme presented itself from the beginning as a 'tool box' consisting of various measures that could be taken and combined locally, and identified 39 locations where these measures were thought to be feasible and effective, adding a few locations where spatial reservations should be made for future adaptation measures.

Practically speaking, a first list of eligible measures was communicated, entitled 'six new ways to safety' and internally nicknamed the 'six-pack of spacemakers' (Sijmons, 2009). It included six ways to literally give rivers more space: removing obstacles from the riverbed, lowering floodplains or winter beds, lowering groins, creating high-water channels, displacing and relocating river dikes, temporarily storing water in existing lakes or on farmland. However, the original toolbox was eventually expanded with more traditional measures, as 'people found that it might be too dogmatic to put in only measures that were literally making space' (Sijmons, 2015). 'Strengthening or heightening dikes' was added, and it did play a role in the implementation. 'Water storage' and 'summer bed deepening' came along, and the toolbox expanded in terms of its initial definition.

As it appeared, the notion of 'room' in this context could be ambivalent. For the advocates of a radical change from containment to spatial flood management, the concept implied that more space and more land had to be made available for river fluctuation. Partisans of traditional flood defence technique preferred to interpret 'room' as a general quantitative concept including any measures that could increase the discharge capacity of river channels and floodplains. Dirk Sijmons retrospectively perceives these additions as an attempt on the part of the promoters of containment strategies to reintroduce exactly the sort of measures that were meant to be avoided by the programme, a reaction he considers 'Pavlovian' within the context of Dutch water management (idem).

A key tool in the safeguarding of the Room for the River approach was the creation of the 'Quality-Team', or Q-Team, initially composed of Dirk Sijmons (chairman), Frans Klijn, Maurits de Hoog, Dick de Bruin, and Sjef Jansen. Respectively landscape architect, geographer, urban planner, river expert, and ecologist, they together composed a multidisciplinary team that over the course of several years would supervise the development and implementation of the various projects. The Q-team however faced the difficult task of evaluating the project's 'spatial quality', a relative concept. In the 1980s Former Governmental Advisor for Architecture

Tjeerd Dijkstra had introduced Vitruvius' principles for good architectural design, *utilitas, firmitas,* and *venustas*, as an evaluation tool (Sijmons, 2008). The Q-team translated this triple criterion within the context of the Room for the River programme: projects should meet and combine hydraulic effectiveness, ecological robustness, and cultural meaning and aesthetics, with special attention given to the relations among the criteria. Hydraulic effectiveness was obviously already assessed at the time the projects were to be detailed and built, but the Q-Team chose to consider 'the relationship between hydraulic effectiveness and form, as well as all other functionalities of the site' that most impacted the lives of inhabitants (Klijn et al., 2013).

Being unable to judge the result of projects involving long-term processes, the Q-Team decided to also 'assess the planning and design process as such, so that we could give useful recommendations in early stages and thus influence the designs when still possible' (idem): more than a fixed preliminary framework or a subsequent judgment, the aim was to provide active and continuous guidance to the designers and to the authorities in charge of the projects.

FROM FIXED TO FLEXIBLE: REMODELLING THE RHINE-MEUSE DELTA

The Dutch Rhine-Meuse Delta is famous for its tumultuous history of reclamation and dramatic floods. The Room for the River programme should therefore be seen in the context of an intrinsically malleable geography, shaped by strong natural processes and rapid human interventions. This is particularly visible in the Biesbosch region and its succession of dramatic turnarounds up to the recent de-poldering that turned it into a key project within the Room for the River programme.

In Roman times, the Southern Delta formed a vast peat bog, covered by a metres-thick peat layer, and isolated from the North Sea by a narrow dune ridge (Leenders, 1999). Without stable and permanent drainage, the lower delta region was barely habitable or exploitable, yet the first traces of drainage there date from 75 and 125 CE, with the oldest known levees and tidal gates built close to the Meuse, near Rotterdam, to drain the land at low tide and defend it against rising waters at high tide.

This first reclamation period stretches across the Middle Ages, a long period when hydraulic infrastructure developed and spread around the North Sea, and land exploitation slowly gained in scale and intensity. Behind the dikes, which increasingly isolated land from water, peat was extracted as fuel and for the sea salt it contained. This intensifying exploitation, mining, and drainage of the soil led to land subsidence, which in its turn made it easier for the sea to cover the land again, causing 'one of the best known examples of massive ecological damage at the countryside induced by man in the early modern times' (id.: 58). Human exploitation caused the breach of the dune barrier (id.: 40) and the loss of large land areas. The wide opening of today's Haringvliet brought seawater closer to the river area and to the future Biesbosch, and the Southwest Delta progressively took on the contours that we still know today.

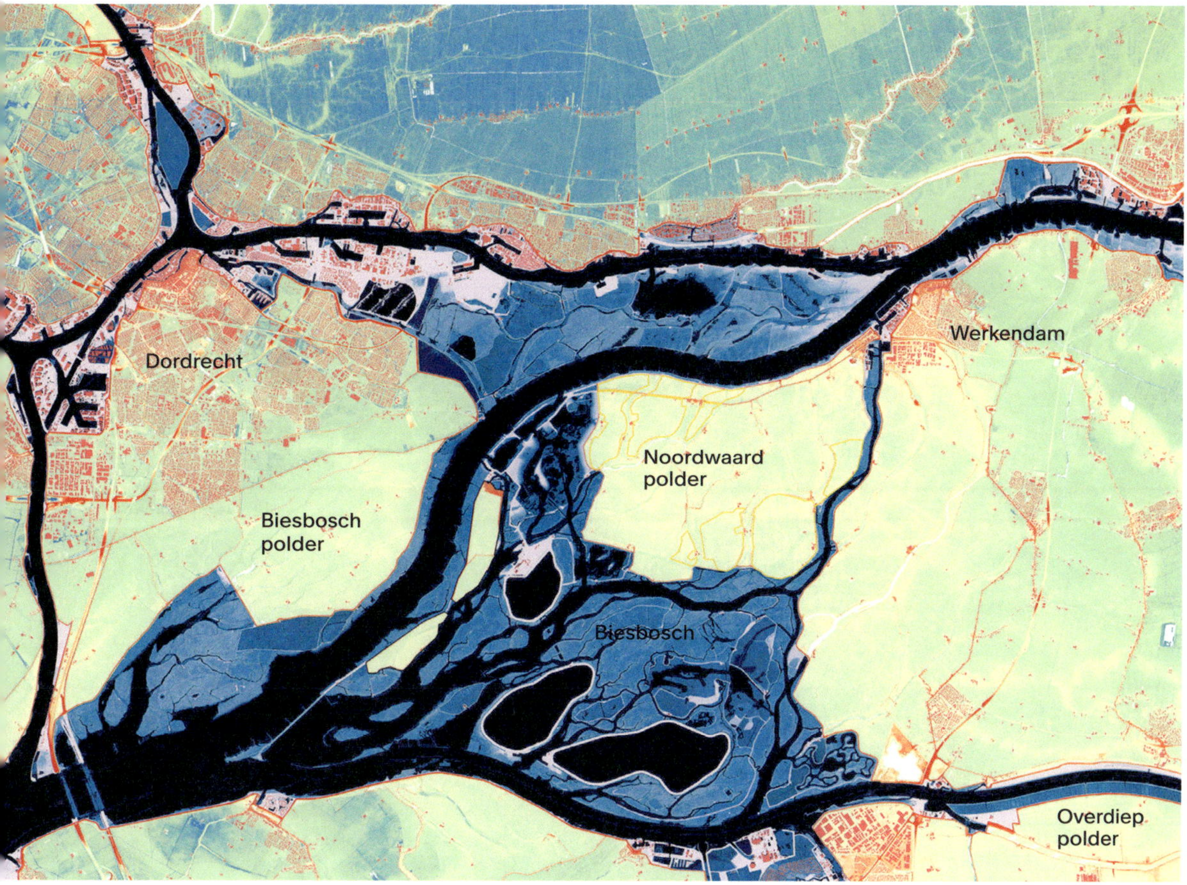

The Biesbosch region in the Dutch Southern Delta, with the recently reopened Noordwaard and Overdiepse polders

The period between 800 and 1200 saw the sea retreat again. The climate was relatively mild and bog reclamation expanded again on a broader scale, later through planned operations. In the centuries that followed, perpendicular flood-mitigation dikes were replaced by longitudinal defensive levees, progressively joined to form ever-larger enclosed areas. With the damming of the Meuse River in 1283, the Grote Waard or Zuid-Hollandse Waard polder was enclosed and became the greatest achievement of the Middle Ages, stretching over 400 km² from Heusden to Dordrecht (id.: 77). However, this permanent enclosure by dike-rings considerably reduced the space available for rivers to fluctuate and for water levels to rise within the narrowed channels. Combined with land subsidence, rising waters made gravity-fed drainage obsolete and weakened the levees, making the new, reclaimed land even more dependent on damming and drainage infrastructure and on continuous maintenance, thus increasing risk and exposure. A victim of the weakening of natural coastal defences, of the side-effects of its own reclamation, and of maintenance failure, the Grote Waard was inundated during the famous Saint Elisabeth Flood in 1421, which drowned numerous villages and most of

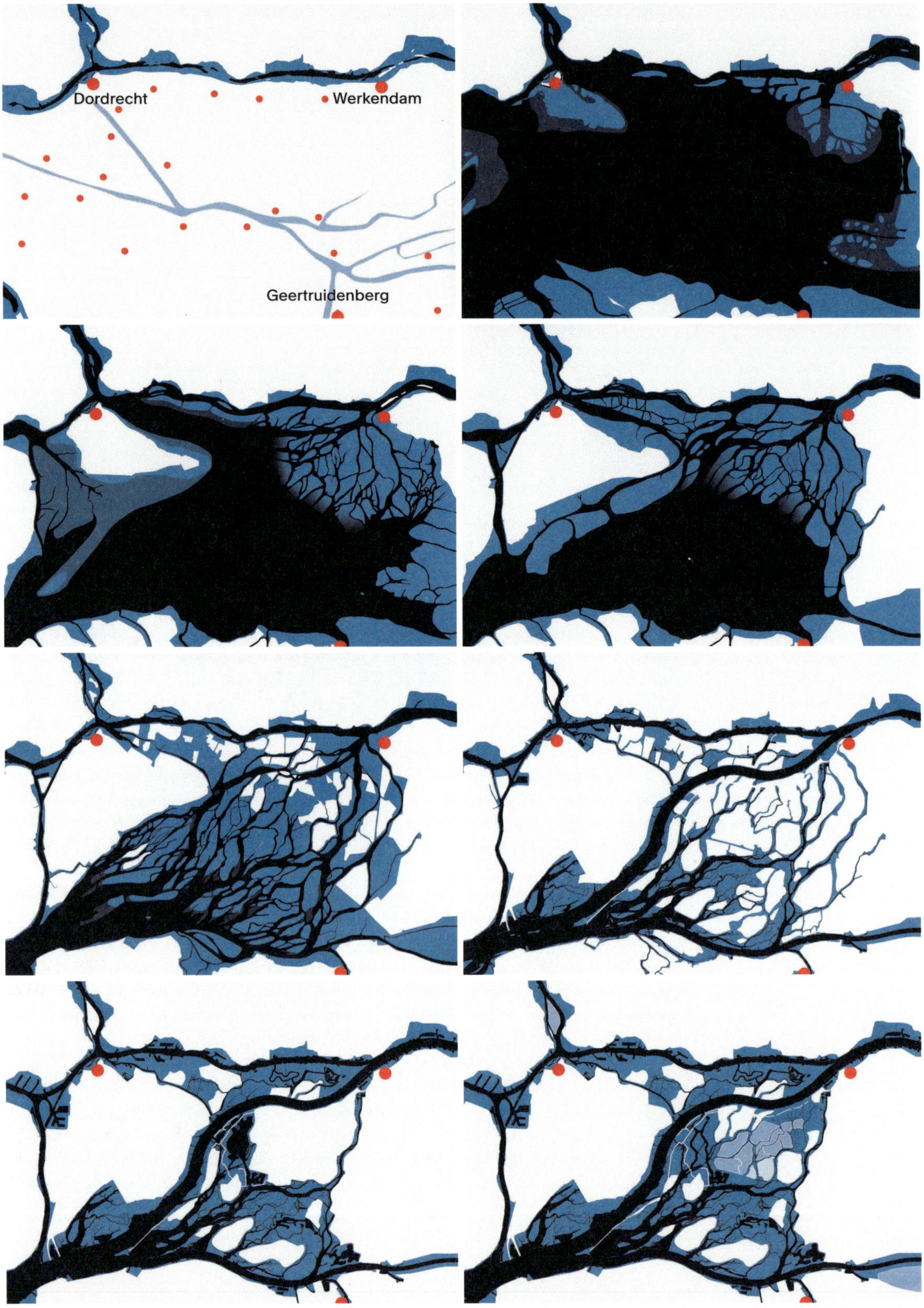

Flood and reclamation history of the Biesbosch area, in the early 15th, mid-16th, mid-17th, mid-18th, mid-19th, mid-20th centuries, and in 2010 and 2015, before and after the completion of the 'Room for the River' programme

the polder between Dordrecht, Werkendam, and Geertruidenberg. The three cities, once part of the same territory, were suddenly isolated by a vast and shallow tidal lake, with Dordrecht remaining isolated on a tiny island.

Despite uncoordinated efforts to rescue the lost land, the western part of the Grote Waard was largely abandoned to the waters after a series of disastrous storms, less than 150 years after its completion. This dramatic episode became a national trauma for the man-made lowlands, a constant reminder of Holland's vulnerability and dependency on permanent and coordinated water management; with the emergence of environmental history, it will later also be seen as a 'classic example of the devastating effects of human greed' (id.: 76).

The Biesbosch, from Land to Water to Land

In the centuries following the flood, the lost region appeared on maps under the name of 'Drowned South-Holland Polder' (Van Wijk, 2012: 15, 67). The new lake had extended the maritime influence further land inwards, up to the city of 's Hertogenbosch (id.: 125), and had radically upset the local geography—in land use but also in navigability, economy, and politics. The name Bergsche Veld appeared to designate the area that slowly emerged (*veld* meaning field in Dutch). As the lake started silting on its periphery, the lost land became gradually covered with wetlands and woods; the name Biesbosch, known before the flood as a small area east of Dordrecht (id.: 12) appeared as well to designate what slowly became a vast wetland—Biesbosch meaning literally 'bulrush wood', bulrush having the double advantage of being dried and sold for the handicraft and building industry and to accelerate natural sedimentation. Rivers and streams, tidal movements, seasonal floods, vegetation growth, silting, and erosion finished erasing the medieval agrarian landscape, and replaced it with a new, fluctuating archipelago of islands and mudflats, out of which old church towers still emerged here and there (Chrysotomus Neapolitanus, 1514, quoted in id.: 12).

From the mid-fifteenth century, it took about 400 years of sedimentation to raise the land again above water level, as shown by historical maps. The first areas that were diked and reclaimed again were the flats west and south of Dordrecht, today lying about one meter below sea level. Taking advantage of natural silting, side creeks were dammed and small pieces of land were again enclosed by small levees and aggregated. The reclamation works accelerated from the seventeenth century on—although the fear remained that narrowing the river could again cause a catastrophe (id.). In the eighteenth century, the eastern part of the Biesbosch, south of Werkendam, was recovered as well and incorporated into the Land of Heusden and Altena.

Rapid sedimentation, often 'helped' by adequate plantations, had allowed new land to arise and be exploited everywhere, but also caused headaches for the municipalities around the Biesbosch. The narrow pass downstream from Gorinchem, where the Meuse and Waal waters join the Merwede, formed a dangerous bottleneck. Further downstream, the many branches let the river spread and slow down; waters were so shallow there that navigation between the Western harbours of Rotterdam and Dordrecht, and the Rhine and Meuse watersheds became difficult. During harsh winters, the shallow waters occasionally blocked the ice floating on the river, creating ice dams that hindered the flow, leading to dike breaches and inundations upstream.

Facing this recurring problem, the States of Holland created a commission to investigate ways to remediate these multiple problems. The resulting plan of 1730, drawn up by Nicolas S. Cruquius, a surveyor and member of the commission, envisioned the first engineering intervention and flood mitigation on the Merwede, and advised the construction of a bypass for the Merwede to expand when high waters threatened to inundate the adjacent polders. Two options were drawn up by Cruquius (white dash on the map): a northern bypass located parallel to the river and temporarily isolating the village of Hardinxveld; a southern bypass creating a large short-cut for the Merwede to flow into the Biesbosch through the Old Land of Altena and occasionally turning Werkendam into an island (white arrows). For this second option, the Commission proposed the construction of a three-kilometre-long submersible dam on the southern Bank, between Werkendam and Woudrichem, a solution designed to let water overspill only when necessary. To implement these spillways, the river levees had to be lowered where indicated on the map, and new levees had to be erected further back inland.

The use of spillways and bypasses, or *overlaat* in Dutch, was not radically new at the time, as it was known that rivers could spontaneously use former riverbeds during high floods. Some of these temporary riverbeds or 'green rivers' were partly left open for extreme flooding. However, spillways were not yet developed as a coordinated strategy for flood defence as they would be after Cruquius (Kappers, 2000). Old river branches and lower forelands regularly flooded there at high waters, a natural process sometimes instrumentalized as part of a flood mitigation or military defence strategy. The Cruquius plan was welcomed by the western harbour cities, from Rotterdam to Gorinchem, all eager to secure their access to the hinterland. However, it raised worries and opposition from the dike masters responsible for the large polders boarding the Merwede, who feared that the planned dam would increase flood risks upstream (Van Wijk, 2012: 137). Due to persisting disagreement, the project was not built.

After the 1757 flood in the lower river region, caused again by the formation of ice dams, state engineer Jacob Pierlinck published a book in which he examined the circumstances of the catastrophe and advocated the use of 'green rivers'. Pierlinck wrote: 'In order to redirect pressuring waters towards the Biesbosch, there are ways to secure the region this memoire is about' (Pierlinck, 1757).

Several spillways were later built, often using old river branches. Along the Meuse and Waal Rivers came the Baartwijkse Overlaat in 1767 and the Beerse Overlaat in 1828 (Vandersmissen, 1998: 61–62). The Lobith-Tolkamer Overlaat was successfully used as recently as 1995 to mitigate the effects of exceptional discharge in the Rhine River. The 'Cruquius Plan' has therefore remained to this day an inspiring attempt to address a major hydraulic challenge through 'soft power' and controlled flooding, as a necessary complement to dikes and dams. Furthermore, its visionary mix of discharge repartition and disaster mitigation through submersible levees

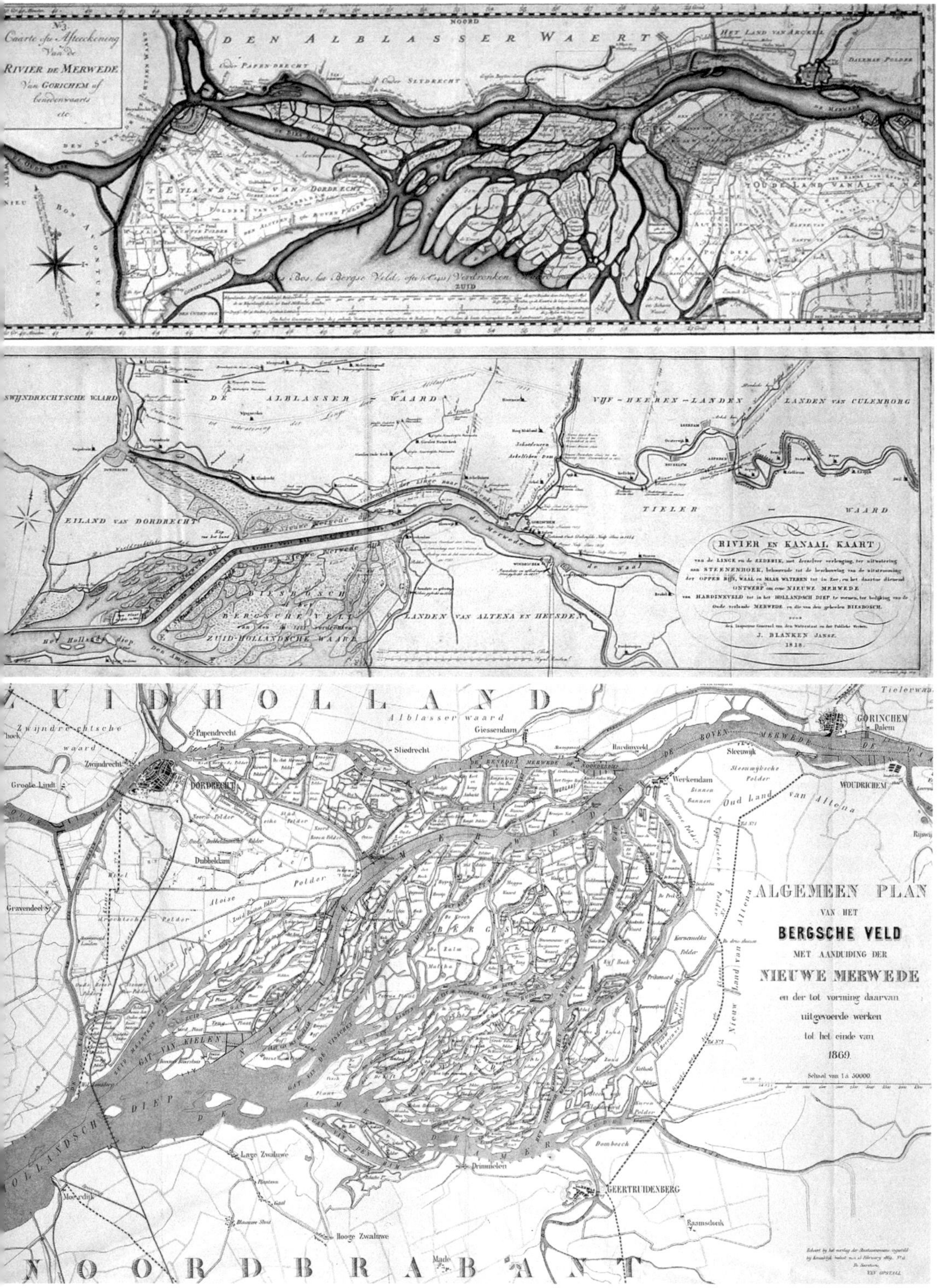

Three designs for the Merwede River: Cruquius 1730 plan for new by-passes • Jan Blanken 1818 radical canalization plan (Regional Archive Dordrecht) • 1869 implementation plan of the Commission Van Opstall (University of Groningen)

shows a striking resemblance with the current transformation of the area by the Room for the River programme implemented almost 400 years later.

In the mid-nineteenth century, the whole area lost in the fifteenth century had turned into a complex and fluctuating archipelago of flats and islands, separated by numerous river branches and tidal creeks. Most islands were not yet enclosed by levees and were still regularly submerged by tidal and river floods. Although not very suitable for field agriculture, the Biesbosch at the time hosted an intensive activity: reed was used for construction; bulrush was harvested for handicrafts but also planted to accelerate silting and land creation; willow coppice was exploited along the streams; and fishing took place everywhere. This period might be considered as the ecological climax for the Biesbosch in terms of diversity of habitats and species, just as it is today generally considered for the countryside of the river region, then still partly open to seasonal floods and not yet 'shaved' from its extensive network of hawthorn hedges by massive re-allotment programmes.

The plans made in the nineteenth and early twentieth century, however, changed this situation, with the implementation of two major 'river improvement' projects: the Nieuwe Merwede and the Bergsche Maas. In Holland, as in the rest of Western Europe, the population was growing and with it the need for arable land, trade, and navigation, putting more pressure on the rivers and their floodplains. As the floodplains were increasingly reduced upstream, lower rivers silted up, filling their own bed and rising over the surrounding lands. As the Biesbosch rose again thanks to river and tidal sedimentation, its shallow wetlands carried two major disadvantages lamented already since the seventeenth century (Van Wijk, 2012: 136): poor navigability due to changing, narrow, and shallow streams; and the abovementioned formation of ice dams, blocking the river's course and increasing flood risks in the area. In order to ameliorate both the navigability and the discharge capacity of the lower rivers, two new river channels were created along the Waal (the southern branch of the Dutch Rhine) and the Meuse Rivers before they joined the waters into the Biesbosch, radically changing the landscape of the area.

On the north side of the Biesbosch, various options were investigated to create a navigable channel through the shallow swamps, and facilitate the discharge of the Waal River into the estuary. A radical plan drawn up in 1818 by the Governmental Water Agency Waterstaat and presented by General Inspector Jan Blanken, envisioned a cornered canal dug through the Biesbosch that would not only fully channel the Waal's southern branch but also redirect most of its waters into the Hollands Diep inlet and split the Biesbosch into two areas ready to be drained and exploited.

The plan that was eventually implemented half a century later fortunately showed a bit more concern for the pre-existing landscape: the Nieuwe Merwede, dug between 1861 and 1874, basically connected and enlarged existing streams into a regular curved channel with a standard width of 325 metres and a length of 18.2 kilometres. The chosen track might indeed be more adapted to the initial morphology of the Biesbosch, but still mercilessly sliced the region into three isolated patches: the Sliedrechtse Biesbosch (North-east), the Dordtse Biesbosch (South-west) and Brabantse Biesbosch (South-east).

Following the completion of the New Merwede, in 1926 a new polder was dammed on its right bank, along the Island of Dordrecht. The 'Biesbosch Polder' took off another 1100 hectares from the Merwede floodplain, allowing Dordrecht to expand its surface within the primary dike by 40 percent in a few years (Vis, 1936: 11). Besides the reclamation of arable land, fighting the post-WW1 economic crisis was another priority. All over the Netherlands, large urban civil works were undertaken during the inter-bellum period to put unemployed workers to work, in exchange for a modest salary or allowance, resulting in new large city parks (the Amsterdamse Bos in Amsterdam and the Kralingse Bos in Rotterdam, among others). For the Biesbosch polder project, 1316 unemployed Dordtenaars worked at damming, clearing, and draining the new land. The polder was opened up for farming by 1927, and the reclamation was celebrated as a final victory: 'A work like the reclamation of the Biesbosch Polder will honour the city of Dordrecht until the end of time' (id.: 111). In the meantime, on the left bank of the Merwede, many inlets were obstructed by the construction of a winter dike (bandijk) alongside the new channel, hindering the connections to the Biesbosch, and closing many stream branches off at their upstream end. A full reconquest of the Grote Waard seemed nearer.

A similar scenario developed south of the Biesbosch. A crucial factor for the overload of the Merwede River was long identified: the confluence of the Meuse and Rhine Rivers into a bottleneck in Gorinchem. Meuse and Rhine waters used to flow in separate beds to the estuary, until the Meuse was dammed from Hedikhuizen in the thirteenth century, directing its waters to the north. From the old Meuse there remained a few branches and tidal channels of small importance, with most being dammed during the reclamation of the Grote Waard, although some were later reopened to tidal influence by the Saint Elizabeth Flood. The main track remaining from the former river was later named the Oud Maasje (the 'Little Old Meuse'), a silted tidal channel used to drain and fertilize the surrounding hay meadows—and after 1766 occasionally used as a spillway to offload the 'real' Meuse through the Baardwijkse Overlaat (Bink et al., 2011: 54–56). The Overdiepse polder was increasingly diked and gained upon the river banks of the Oud Maasje in the early years of the eighteenth century, although it was still flooded in the winter to benefit from the so-called 'fat water' (*vet water*) used to fertilize the 'flow meadow' (*vloeiweide*) (see chapter 'Flood as Land-maker').

This situation changed again radically in the nineteenth century. Due to persisting floods along the Rhine, Meuse, and in their overloaded common branches, the Waal and Merwede, plans were made to redirect the Meuse towards its former route—the Oud Maasje. The first serious attempt was made in 1823, with Lieutenant Generaal Krayenhoff's test design ('Proeve van een ontwerp') for a new river channel from Heerewaarden towards the 'Bergsche Veld', another name for the Biesbosch (Van de Vorstenbosch, 2011). Several plans followed, and the creation of the Bergsche Maas was eventually decided by the Dutch Government in 1883. The 24-kilometre-long channel was dug and bordered by new levees between 1888 and 1904 under the aegis of Cornelius Lely, civil engineer and Minister of Waterstaat. Although the new riverbed mostly followed the old bed of the pre-medieval Meuse, it curved to the north in order to protect the Overdiepse polder—then protected by a strong levee, 150 years later reopened to flooding.

The Biesbosch Drained—and Flooded Again

The twentieth century saw the Biesbosch come close to disappearing. In 1927, the area was still under the strong influence of tides and river dynamics, but its water network was diminished, amputated from many branches by damming works and compartmented by rigid borders. With the construction of the Biesbosch polder in Dordrecht, and of the Nieuwe Merwede and its winter dike, a northern limit had been set for the wetland archipelago. To the south, the new Bergsche Maas channel already marked a possible edge, although not yet materialized by a continuous levee. In between was left a reduced Biesbosch, where many tidal creeks and silted channels were dammed, and where small polders were joined to form larger arable areas.

However, contradictory events and concerns were at play. A research department for sea inlets, lower rivers and coast (Studiedienst van de Zeearmen, Benedenrivieren en Kusten) was founded in 1929 within the Rijkswaterstaat Governmental Water Agency, initially to study and ameliorate the navigation network. It soon also investigated the flood safety of the lower rivers and coastal regions. Initially asked to study the effects of the young Biesbosch Polder on flood risks, the department drew an alarming conclusion for the safety of the whole region, as the levees appeared to be too low to resist a major flood (Sanderse et al., 2004). As a result, the city of Dordrecht became reluctant to see more land being reclaimed in its surroundings, but its reserve did not discourage the supporters of further drainage in the Brabantse Biesbosch. Plans were made to finalize its piecemeal reclamation with one great gesture embracing the whole area from Merwede to the Bergsche Maas. In 1940, one large polder was drawn that would achieve the final reconquest—and the end of the 'Bulrush Wood' as such. This Grand Projet was, however, soon frustrated by the Second World War, which turned the Biesbosch into a battlefield between the German occupiers and Dutch resistance fighters, and was further delayed by the reconstruction period that drew all energy towards the cities and the Ijsselmeer polders.

The 1953 storm surge and giant flood, although long feared by the Rijkswaterstaat Research Department, paradoxically hit a poorly prepared country. The poor state of the levees and the need to better protect the land against combined sea and river floods had been identified before the war, but no action was taken due to the economic crisis, the occupation, and the reconstruction that followed. The loss in human lives, cattle, and land abruptly redirected the nation's attention towards the southern Delta, leading to the construction of the world-famous Delta Works, a series of Herculean sluices built to protect the Netherlands from more sea surges and storms. The combination of dams and storm-surge barriers drastically lowered the chance of sea floods for the coastal provinces, by reducing about 700 kilometres of coastline. The closing of the Delta and the reclamation of the IJsselmeer which reached its peak in the 1970s

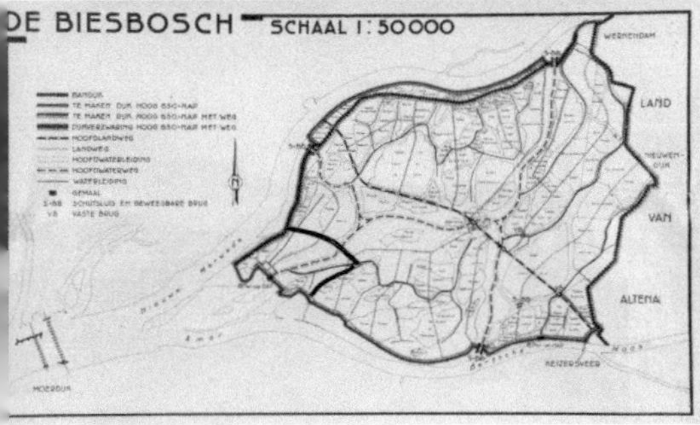

Plan for the reclamation of the Brabantse Biesbosch, ca. 1940 (Biesbosch Museum)

marked the culmination of the 'Art of a Nation: Dredge, Drain, Reclaim' (Van Veen, 1948).

As a downside, the former sea inlet formed by Biesbosch, Hollands Diep, and Haringvliet became the downstream chemical depot for the Rhine and Meuse rivers, having received more than 150 million m^3 of polluted mud (Nienhuis, 2008: 288), a process that persists today, although with reduced pollution levels and at a slower rate. The drainage of the Biesbosch was on hold, but the Haringvliet tidal gates built in 1970 brought another blow to the area: the Biesbosch then largely lost its connection to the sea and its tidal influence, which only subsisted through the distant Rotterdam harbour channels. Environmental concerns would only become decisive six years later, with the construction of the last Delta Work, the Eastern Scheldt storm-surge barrier. Under pressure from environmental groups, this last work included an opened part to maintain tidal exchanges between sea and estuary. Unfortunately for the Biesbosch, these concerns came too late to save the Haringvliet/Hollands Diep inlet from enclosure, and the threat of sea salinization carried more weight than the repeated calls to let sea water in again, although the ecological consequences brought about through the 'hard' approach soon became visible. Only in 2019 were the sluices partly opened again in order to restore the ecology of the estuary.

As shown by the 1995 flood, the Merwede bottleneck addressed, among other issues, by the 1730 Cruquius plan, remained problematic. The area was not exempt from dramatic events: in January 1995, 9000 inhabitants of Gorinchem had to flee the threatening flood risks, and a thousand were temporarily hosted in an evacuation centre for several days (de Gast, 2004). After this near-catastrophe, options were again sought in the surroundings to lower water levels along the sensitive urban shores of Gorinchem, Dordrecht and upstream. Considering the larger scale, it also appeared preferable to concentrate the measures on the Meuse and Waal branches rather than on the Lek River. The latter had the disadvantage of streaming through densely populated areas, offering less space for widening and a narrow outlet channel through Rotterdam. By contrast, the Meuse and Waal branches offered a long-term opportunity for spatial measures. They could also divert future additional discharge towards the wider Delta inlets, including the Volkerak, where extra capacity could be made for emergency retention whenever sea dams had to remain closed. At the crossroad of the Meuse, Waal, and sea inlets axis was the Biesbosch region. With its significant nature areas, its low building density, and its long history of frequent conversions from land to water and water to land, the Biesbosch was naturally seen as a place of opportunities for Room for the River measures.

The 'tool kit' software made it possible to model the hydraulic effect of various possible measures such as the diversion of the Merwede waters to the south through the existing Steurgat stream, to the north through Dordrecht's Polder, and to the southwest by lowering the Merwede levees. The first one appeared to be the most effective to reach hydraulic and spatial quality goals, followed by the second and third measures, classified as average (RIZA, 2002: 125). In the following design session, the creation of a large flow corridor through the Noordwaard polder appeared to have a strong short-term and long-term effect on the peak level in the Merwede River; it also appeared to be the favourite option of the session participants—although declared controversial by the steering group (Veerschelling, 2003: 12). But despite the possible controversy, it was again the vision of a wide-open Biesbosch that aroused interest and was elaborated further towards an operational plan.

As a part of the elaboration of the 'Integral Foresight of the Lower River', landscape architects Bosch & Slabbers were commissioned to study possible landscape adaptations in the Biesbosch Region, in collaboration with the Planning Department of the National Water Agency Rijkswaterstaat, which resulted in the report 'River widening and landscape development in the Biesbosch' (RIZA and Bosch & Slabbers, 1999b). The team explored the history of the region and analyzed the different compartments fragmented by modern channels and land use specialization. Not only had the Biesbosch been divided by the digging of the Nieuwe Merwede and Bergsche Maas channels, it had also lost its coherence through changes and various concentrations in land use. The Sliedrechtse Biesbosch consisted of small spaces unified by a dense, organically grown water network, just as the south of the island and the west of the Brabant side. South of the Nieuwe Merwede, the landscape was characterized by numerous small polders surrounded by informal, ancient streams bordered by linear vegetation. The east and south of the region had lost most of their former wetland character and now formed continuous lands with a few, wide waterways. Finally, a singular, vast and strongly engineered landscape characterized the Dordrechtse Biesbosch Polder.

While stressing the various qualities of the Biesbosch landscape, the authors lamented the fact that 'in the current situation, it is impossible to imagine that the whole area stems from the same initial situation, that it once was a united system. Spatial coherencies are lost.... The Biesbosch was for centuries a mythical landscape, a labyrinth of changing mudflats and channels, the domain of poachers and pirates, a haunted domain where the threat of water still existed, water that hosted sturgeons almost three metres long'. That myth had been brought back to 'a limited nature area surrounded by crop polders' (id.: 67). The study advocated the reopening of historical streams that had been successively dammed since the nineteenth century. It also suggested reshaping the farming polders and returning them to a scale corresponding to an earlier stage. Consequently, the authors supported the reopening of the Noordwaard polder to old creeks and saw it as an opportunity to emphasize the historical archipelago structure of the Biesbosch and use it as a framework to shape a more dynamic water landscape. They stressed the possibility offered by the option to 'reinforce lost coherence' and the 'creation of new spatial quality' (RIZA and Bosch & Slabbers, 2000: 53).

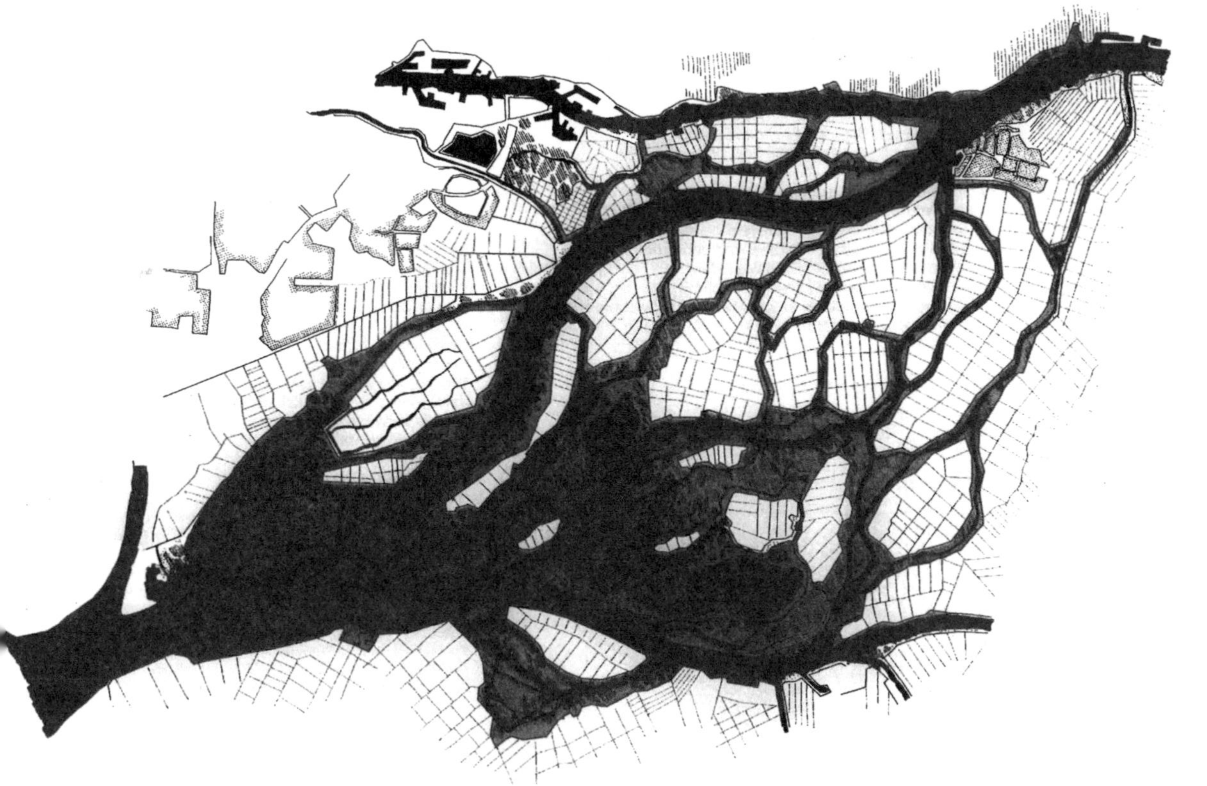

Study for the 'de-poldering' of the Biesbosch (from RIZA, Bosch & Slabbers, 1999)

NOORDWAARD GRADUAL LANDSCAPE

Before the landscape study could lead to concrete results, there had already been a plan to turn the western part of the Noordwaard area into wetlands and expand the National Park De Biesbosch by another six hundred hectares. This local plan was made in 1992, as a contribution to the 'EHS' project—a new nature corridor through the whole country as part of the governmental programme Nadere Uitwerking Rivierengebied (NURG), meant to develop 'new nature' along the large rivers.

Led and designed by the Rural Areas Department and the Governmental Water Agency, this initial project progressed rapidly, and is now known as the Kleine Noordwaard (Small Noordwaard). The process was accelerated by a ten percent 'sale bonus' offered to farmers and inhabitants, along with a possible land exchange with neighbouring parcels owned by the Department of Rural Areas (Van Herwijnen et al., 2003: 33). Arable land was removed, topography was lowered and unformalized, new channels were dug to let waters flow in, and the area was completely transformed into a large wetland punctuated by a few heightened islands and later a land art piece at the northeast end. However, this first 'renaturation' project, today celebrated as a 'bird mecca', paradoxically accelerated

First 'Nature Development Noordwaard', plan area in 1950 and 1988 (Kadaster) • Project approved in 2002 (Rijkswatersaat/Grontmij)

the disappearance of the Biesbosch cultural landscape. The 'Small Noordwaard', born out of ancient streams and mudflats, and characterized by small and rounded patches of heightened land and a fine, reticulated network of tidal streams, had already been altered by the 1950s reallotment. Its historical landscape was further erased in the 1990s and replaced by a vast wetland fed by a formless water network, unrelated to either past or present surrounding landscape, despite the warnings of governmental departments for archeology, heritage and water.

Conceived before the 1993–95 near-floods but only validated in 1998, a third ambition was added in the meantime: the reduction of river flood risks. The project was consequently adapted to offload the Merwede channel during peak discharge and contribute to the safety of the neighbouring cities. Eventually, the clay gained from the project also came in useful for the construction and reinforcement of new levees within the Deltaplan. The final design presented by the engineering firm Grontmij was validated again in 2002, and completed in 2008.

The 2000 'Integral foresight large rivers', however, had envisioned a wider plan, whose priority was to allow the Merwede to flow right through the Noordwaard polder during high discharge, from Werkendam to the estuary. The regional development plan already integrated the potential transformation of the area in 2002, although the Room for the River programme was not yet officially validated and the polder remained officially dedicated to stock farming. The 2006 Key Decision Room for the River confirmed the long-term plan and defined the ambitions: a lowering of the river level by at least 30 centimetres in Gorinchem at peak discharge of 16,000m^3/s at the Dutch-German border, along with increased spatial quality. The proposed de-poldering of the Noordwaard area was presented as 'the first step in the direction of a sustainable long-term solution for this area', combined with a 'positive impulse given to its spatial quality' and the materialization of 'an area around the Biesbosch shaped in a more natural and more dynamic way' (Projectorganisatie Ruimte voor de Rivier, 2006, vol.3, Nota: 50).

During the elaboration of the Room for the River programme, the Dutch population was made aware of flood risks through a nation-wide campaign, 'The Netherlands live with water' (Nederland leeft met water, 2002–2006), but local reality has been less harmonious, and the inhabitants were less enthusiastic about another de-poldering operation. The first Noordwaard Nature development project had not yet been completed, when a new de-poldering project was planned right next to it. Furthermore, many felt that rural areas were again paying the

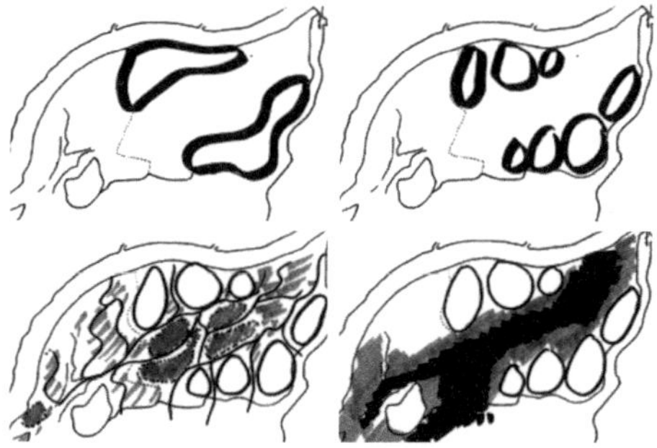

Work drawing overlapping actual levees and 1905 topographic map (Detail, ca. 2005, Robbert de Koning) • Study drawings for the Noordwaard Polder: Central area open as emergency spillway, two option for the periphery: large or small compartments (from Bureau Noordwaard, 2006)

price for the safety and recreational needs of urban regions. During the workshops organized by the 'Lower Rivers Agency' (Bureau Benedenrivieren), many locals opposed the new plan and blamed public authorities for the incoherent policy and for the continuing uncertainty inflicted on the inhabitants and businesses. They joined forces within the local action group Alternatief Platform Behoud Noordwaard, which proposed a set of alternative measures that they considered sufficient to reach the goals set by the government without actually reopening the polder to (high) waters (id., vol. 3, Nota: 118). This alternative was, however, rejected by the government and later by the Council of State (Raad van State) for the reason that 'the proposed alternative seems to reach the goals in the short term, but de-poldering remains necessary in the long term' (id.).

Robbert de Koning, a landscape architect from Arnhem, was commissioned to shape the project, in collaboration with the engineers and advisors of the Governmental Water Agency, working under the aegis of the newly opened 'Bureau Noordwaard'. A preliminary plan was soon presented (Projectbureau Noordwaard, 2006), followed by an adapted version refined during a regional consultation round (Projectbureau Noordwaard, 2007). All existing constructions were surveyed in terms of their resistance to future floods. Constructions that could hinder the flow, as well as those positioned too low to resist future water levels, were bought out by the state at *binnendijkse* value—meaning at the price one would pay for the same construction were it protected by a primary levee. This allowed farmers and inhabitants to move out of the area or plan a new construction within the Noordwaard, adapted to its projected situation. Out of the 25 exploitations, only 11 could eventually remain in the plan in a traditional form, since the central flow corridor could not accommodate regular crop fields that would be damaged at every flood, but only allowed for extensive livestock breeding. Since there were more candidates than places, a long and difficult negotiation process followed to redistribute land and building plots—a negotiation made more complex after the government granted all inhabitants the possibility to go on living in the area, committing the designers to providing new homes for those who wished to stay. New terpen (see chapter 'Early Adaptive Landscapes') were eventually built along the highest dikes to build replacement houses and barns at a safe level.

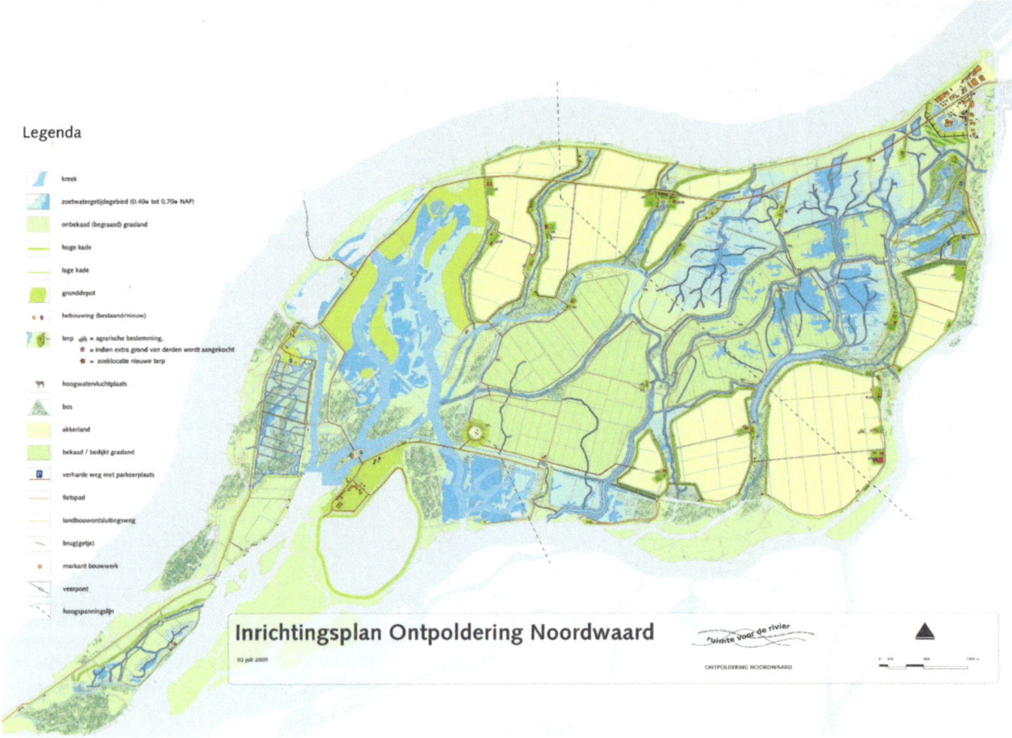

Development plan 'De-poldering Noordwaard' (2009, Room for the River, Robbert de Koning)

Following the direction suggested by Bosch & Slabbers in their 1998 and 2005 studies, De Koning took the 1905 topographic map as a reference and under-layer in order to shape the new 'flood-ready' landscape. For the landscape architect, the Biesbosch was, before that point in time, 'very wild, a barely cleared landscape', as 'the river ruled ... and the water was high'. The configuration in 1905 was 'more in balance. More embankments had been built; more land was cultivated. In terms of structure, the situation then better matched the current situation' (De Koning, 2013). Around this time, the Biesbosch had indeed largely been reclaimed, but the fine and meandering network of streams was still structural, forming, together with small polders, a complex but coherent cultural landscape.

Overlapping a CAD drawing of actual levees and the 1905 map, the landscape architect thus looked for remaining lines and structures that could help to recompose a Noordwaard open to water. Many traces were still visible in terrain maps: where streams used to flow the ground lay a few decimetres lower, since the organic materials present in the water had compacted. Former embankments could be traced from the soil map. The old traces, however, were not literally rebuilt, for technical and aesthetic reasons. The newly built dikes needed to be slightly higher and wider than the historical ones; instead, the landscape architects 'mainly sought to create beautiful lines and views' within a 'more differentiated space'. The differentiation between high-dike polders bordered by trees

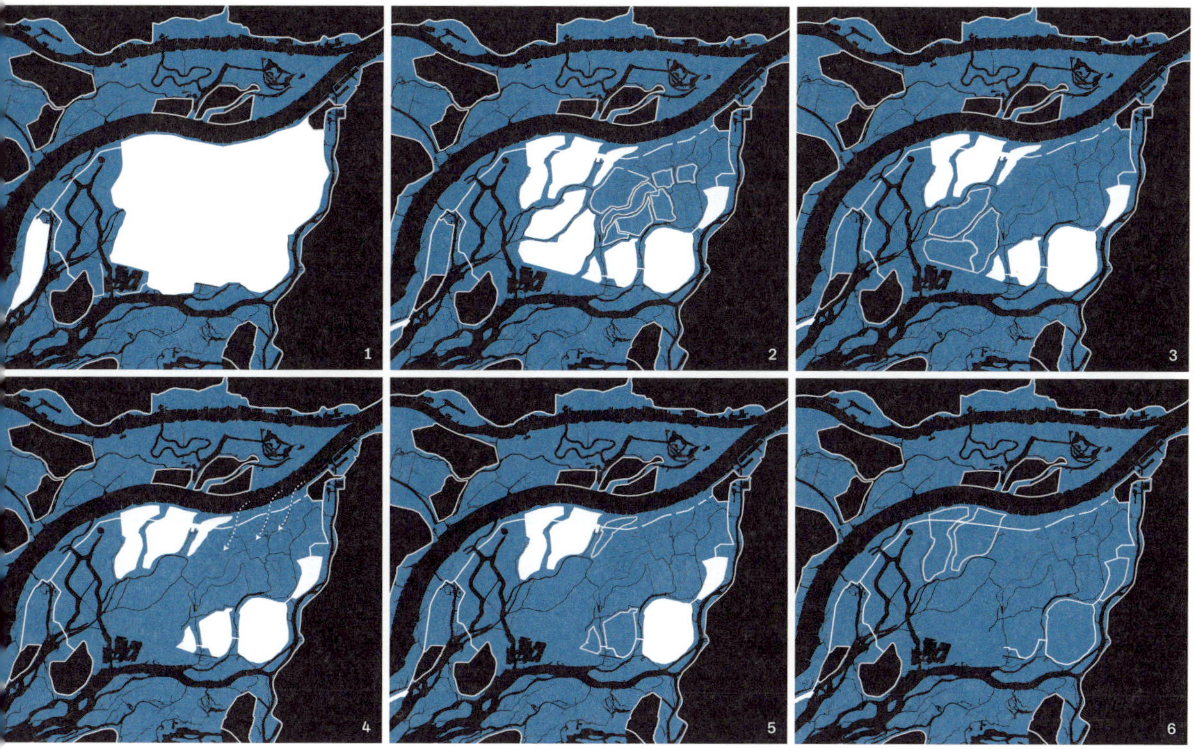

Noordwaard / Biesbosch: diked/open areas in 2010 (fig. 1), submersion phases after project completion (fig. 2–6)

at the periphery and low-dike polders with low vegetation in the central flow corridor created strong variations in openness.

Hydraulic studies showed that a wide, diagonal corridor was the shortest and most efficient way to let large amount of waters overflow from the Merwede towards the estuary. Furthermore, a southwest/northeast corridor better respected the orientation of existing creeks and avoided inhabited hamlets; a large corridor of about 1500 metres width, where vegetation was kept low, precluded the costly construction of a narrow channel, which would include important spatial disruption and heavy digging and damming works. Finally, the chosen option allowed the existing polder structure to be maintained even inside the emergency flood corridor (Projectbureau Noordwaard, 2007: 13).

Opponents regretted that this option also enlarged the surface that needed to be regularly flooded, and thus even further reduced the land available for arable farming. Legal appeals were rejected, however, on the grounds of the 'spatial quality' criterion: a large flood area integrating the historical creek structure was considered by the Council of State as more qualitative than a new canal dug through the Noordwaard (Beroep en oordeel Raad van State, 26.10.2011).

In terms of flood mitigation, the project allowed for various flood stages, from daily tidal flooding to extreme events, following a gradual transformation rather than a binary one. In the pre-existing situation, Noordwaard formed an enclosed polder with a 1/2000 years flood probability (fig. 1, p. 205). After completion of the project, five flood stages could occur. About 25 to 100 times a year, average high waters would come in from the estuary and gradually cover the areas surrounded by low dikes (fig. 2–3). Once a year, on average, the Merwede level would reach two metres above NAP (Dutch reference level), and the river would flow in through the inlets created in the winter dike (fig. 4). Small high-dike polders would face a one percent chance of inundation per year at peak discharge (fig. 5). Eventually, the whole Biesbosch could be flooded at extreme discharge, with a flood probability of 0.1 percent per year (fig. 6).

For the high-dike polders alongside the flood corridor, the designers chose seven small polders instead of two large ones in order to create more differentiation between the inhabited and cultivated areas on the one hand—making them more intimate and wooded—and the vast open horizons of the central portion on the other. Farmers opposed the idea, preferring larger and continuous farmland over small compartments, but the inhabitants eventually chose the proposed option, attracted by the perspective of having more streams and creeks around the area and along the houses (De Koning: 2013). Despite the differentiation among open, low-dike and high-dike areas, the whole Noordwaard officially changed status, from an 'inside-the-dike' area to an 'outside-the-dike' area

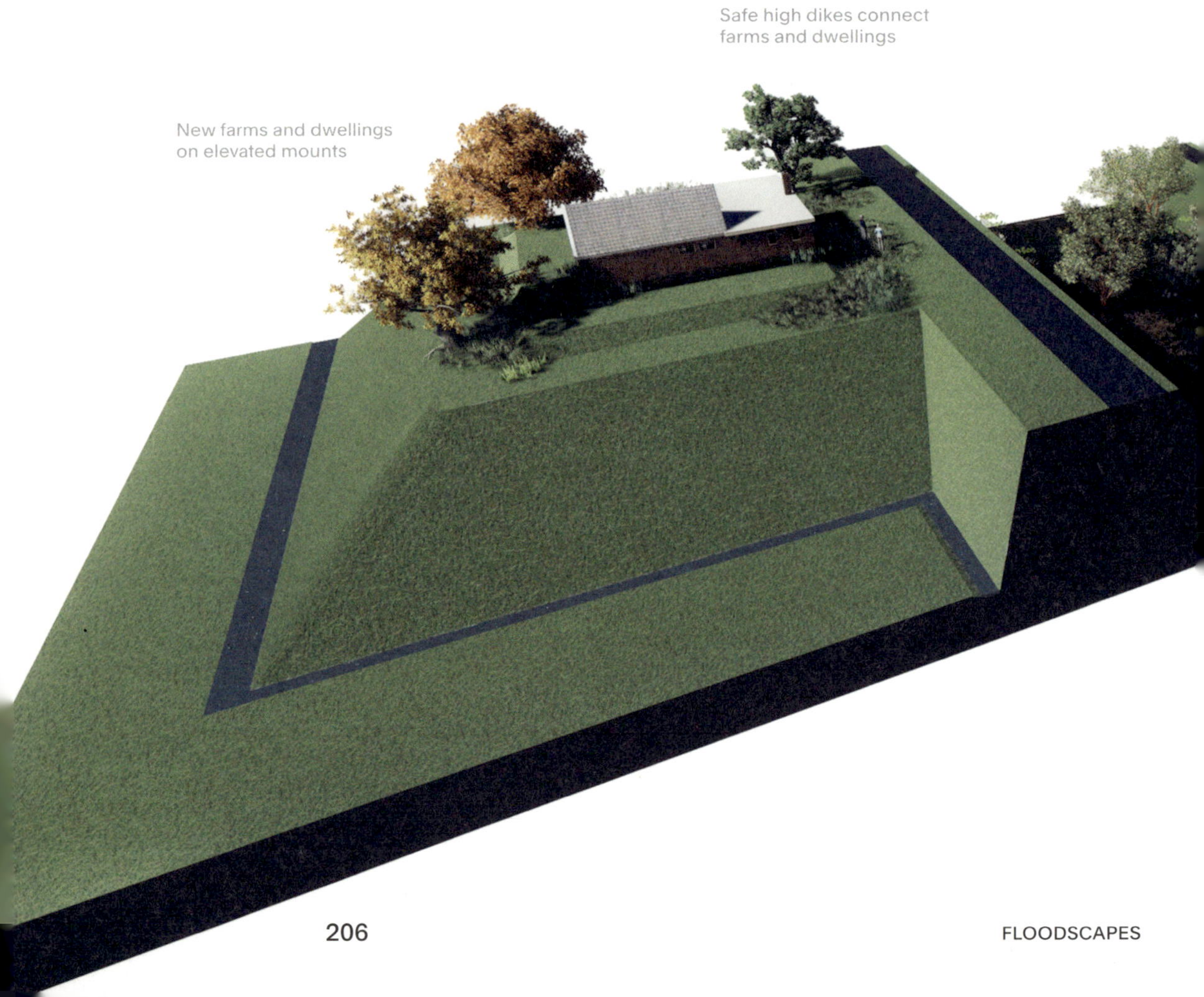

(two distinct statuses in Dutch law). Depending on their altitude, some of the pre-existing constructions were considered safe, while others were declared uninhabitable. In the latter case, inhabitants and farmers were expropriated and compensated. Some chose to remain in the area and move to one of the five new terpen that were built along the new dikes.

The project was also expected to change life in the area, with the development of new nature areas and recreational facilities. With its new network of streams and wetlands, the Biesbosch would soon be expected to attract not only more wild species but also more human visitors. Inhabitants who had enjoyed the Biesbosch for its nocturnal darkness and its silence, feared this permanent nuisance (*overlast*)—a word that appears no less than 27 times in the preliminary design booklet. To prevent future conflicts between ecological interests, residents, and visitors, recreational facilities were confined to two locations, with the rest of the area being restricted for car traffic. Eventually a 'high-water Noordwaard road map' was developed during the construction years, and later distributed by the municipality to help inhabitants anticipate flood situations (*Draaiboek Hoogwater Noordwaard*, 2015). Among other aspects, the guide detailed the risks of living 'outside-the-dike-area', the evacuation procedure in case of flooding. and the restoration procedures.

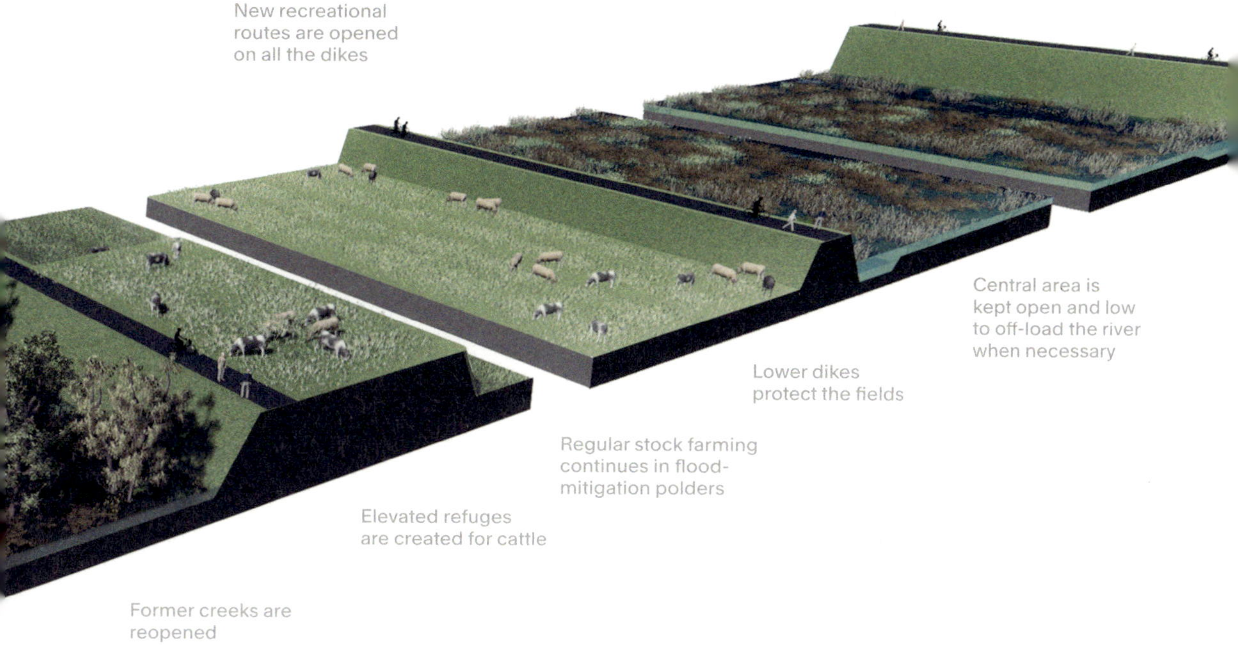

Schematic section of the flood gradient: permanent farmland within high dikes, new houses built on elevated platforms connected by high dikes, historical creeks reopened, low farmland diked but floodable by high water, nature area permanently opened to water fluctuation (Rossano/Orfanopoulos)

On August 27, 2010, the project was endorsed by the government. The construction works were outsourced to a contractor, under the supervision of the Bureau Noordwaard, and the landscape architecture office West 8 was commissioned to design specific civil works. The project progressed rapidly, with occasional verification on the part of Robbert de Koning, and today attracts world-wide attention. In October 2015, the project was completed in the allotted time planned by the Room for the River programme, and inaugurated by the Minister of Infrastructures and Environment.

Retrospectively, the architect did however conceal a few hindering factors in the design process. Spatial quality appeared to be a fragile concept and had to be fiercely defended, sometimes against farmers, wanting to secure large and functional farmland, and occasionally against engineers who had difficulty putting aesthetic choices over and above technical or more economical alternatives. A similar source of concern was the outsourcing of detailed design to a construction consortium which in turn sub-contracted an architect for the design of specific civil works. The subordinate position of the architect appeared to weaken the aesthetic ambitions that were initially claimed by the contractor, since the architect was in no position to argue for high (and sometimes more costly) design quality vis-à-vis his own employer.

A second frustration concerned future maintenance. In De Koning's opinion, this aspect was not discussed in timely fashion in the case of Noordwaard—a problem he considered recurrent in landscape architecture projects. The Q-team 'consistently required to consider the phase after implementation' (Klijn et al., 2013), since the hydraulic effectiveness of the project would later be defined by its vegetation; the maintenance regime thus had to be 'decided together with the responsible land manager already during the design stage' (idem). Future vegetation appeared particularly sensitive in the central area, where no obstacles should impede flood waters, meaning that no tree could be planted and that rigorous pruning of spontaneous vegetation was needed every year at the end of the summer to prepare the area for possible floods in the fall and winter. Maintenance work would therefore be crucial to the long-term effectiveness of the plan but could not be agreed upon during up-front studies: legal tender procedures for public commissions made it impossible to choose and discuss these issues with future maintenance contractors, since the selection could only be made on the basis of a finalized plan and a predefined assignment. According to De Koning, contracting the National Forestry Agency Staatsbosbeheer, by far the biggest maintenance company in the country, would for instance have allowed a useful dialogue to take place between designers and maintenance managers, a dialogue the law did not permit.

Despite remaining doubts as to the future development and effectiveness of the Noordwaard landscape in the coming decades and centuries, a successful design and construction process is evident so far. Through a precise differentiation in dike elevation and profile, the new topography provides a gradual answer to flood levels that range from daily tides to a near-catastrophic situation—a range of more than three metres in terms of water levels. Within a spatial framework restored to its past unique character, a strong landscape diversity in vegetation and cultures was realized, varying from traditional crop farming in the polders protected by high dikes and adapted agriculture for low-dike polders to semi-natural grazing fields with minimal maintenance and open wetlands. In february 2020, the Noordwaard proved its value: due to the high water levels in rivers, the area was successfully flooded for the first time, with no damage to its infrastructure (see picture on page 218).

CONTROLLED DISASTER IN THE OVERDIEPSE POLDER

As a counterpoint to Noordwaard's diversified landscape and gradual flood response, the nearby Overdiepse Polder project brought a straightforward and binary answer to the targets set by the Room for the River programme, illustrating its flexible approach when it came to local translation of the overall goals. The Overdiepse Polder also distinguished itself by the active participation of local farmers in the planning process. In 2000, the Overdiepse Polder was identified as well as a search area ('zoekgebied') for the Room for the River programme, then still in an initial phase. The above-mentioned report, 'Integral Foresight Lower Rivers', investigated various ways to increase the capacity of the Bergsche Maas channel opened in 1904, mainly by looking at those areas flooded during the great 1953 flooding (Min. van Verkeer en Waterstaat, 2000b, De Jong et al., 2000: 49). The southern bank of the Bergsche Maas provided various ways to reopen former flood areas subtracted from the floodplain, with the Overdiepse Polder being one of the biggest. In the various landscape studies, reopening the Overdiepse Polder was evaluated as 'very promising' (RIZA and Bosch & Slabbers, 2000: 91). Reopening the southern banks of the Bergsche Maas provided the opportunity to 'reinforce lost coherence' and 'create new spatial quality' by restoring some of the former flood meadows that had allowed the Meuse branches to fluctuate before they were channelled. Although still officially dedicated to stock farming, the polder was also soon identified as a potential area for water retention by Brabant's 2002 regional development plan.

Contrary to the Noordwaard area, the pre-existing situation of the Overdiepse Polder appeared rather simple, both spatially and in economic terms. The polder was delimited to the south by the old levee of the Oude Maas and to the north by the levee of the 1905 Bergsche Maas channel. It was located in a farming region with few recreational uses, exploited by 17 farmers, and counted 18 homes, all related to farming activities. The parcel structure along the riverbanks still reflected the premodern long and narrow units, while the 550-hectare inner land showed larger plots inherited from various re-allotments programmes, and for the most part owned by their users. The land elevation was close to the average level of the river.

With the Room for the River programme, the Overdiepse Polder was given the relatively simple but no less conflicting task of creating Room for the River from profitable farmland, while national and provincial policies explicitly favoured combinatory solutions upon the 'renaturation' of former floodplains (Van Rooy and Slootweg, 2003: 44).

When the polder was designated as study area for river-widening measures, the local community started, already in 2000, organizing itself and began to explore possible options that would secure their own activities, even before official planning procedures started. Supported by local representatives, in 2001 the initiative led to a special planning procedure: a so-called *Spiegelproject* (mirror project), led by semi-public organization Habiforum and by independent consultants mainly financed by the governmental agency 'Room for the Lower Rivers' (Ruimte voor de Benedenrivieren). This parallel planning process was meant to initiate innovative and yet feasible regional development plans, and to bridge the gap between public authorities and local stakeholders through an open and multidisciplinary design process. Spiegelprojecten were used in various regions of the Netherlands and generally focused on combinatory land use. The project's steering committee included national, provincial, and municipal authorities, the water boards concerned, as well as the interest groups formed by the inhabitants of the polder to represent them in the discussions. Five meetings of the steering team and seven meetings open to inhabitants allowed the participants to discuss and evaluate 19 possible interventions (id.: 11–63). Here again, the combinatory exploration of various hydraulic measures was made possible by the Water and Public Works Agency Rijkswaterstaat, which remained the main actor for hydraulic issues, while local farmers took the initiative in the planning process, in collaboration with the Provincial authorities.

The design studies were conducted between July 2002 and July 2003, and resulted in a comprehensive report on the social, economic, and geographical situation of the polder, exposing the conclusions of the meetings and workshops. In its introduction, the report stressed that 'dike construction and reinforcement has reached its limits', and that 'spatial measures are needed to face rising discharges'. Considering the density of the country and its history of land reclamation, it advocated multiple land use solutions and the use of water safety measures as opportunities for new developments, mentioning the Meerstad-Groningen project, then in its planning phase, as an example. The investigation confirmed that farmers feared the consequences of lasting uncertainty more than territorial change, and would therefore rather take the initiative than try to hinder the planning process. It also became clear that farmers' situations and perspectives for the future could be quite varied, and sometimes they had contradictory interests, depending on their personal, economic, and geographical situation within the polder (id.: 24–26).

Three main alternatives eventually came out of the discussions, each of them leading to several options. An initial alternative suggested the transformation of the entire polder into a nature area, consisting of wetlands opened to high waters. The second 'middle option' or 'split polder' envisioned a division of the polder into two parts by a new primary levee, with the northern part being open to 25-year floods and above, the southern part being permanently protected by a new primary levee. A third alternative went further: the 'terpenplan', named after the elevated farms built in ancient Frisia, proposed opening the whole polder to high waters by lowering the northern edge (or by creating an inlet and outlet in the northern levee), and to upgrade the southern levee and relocate

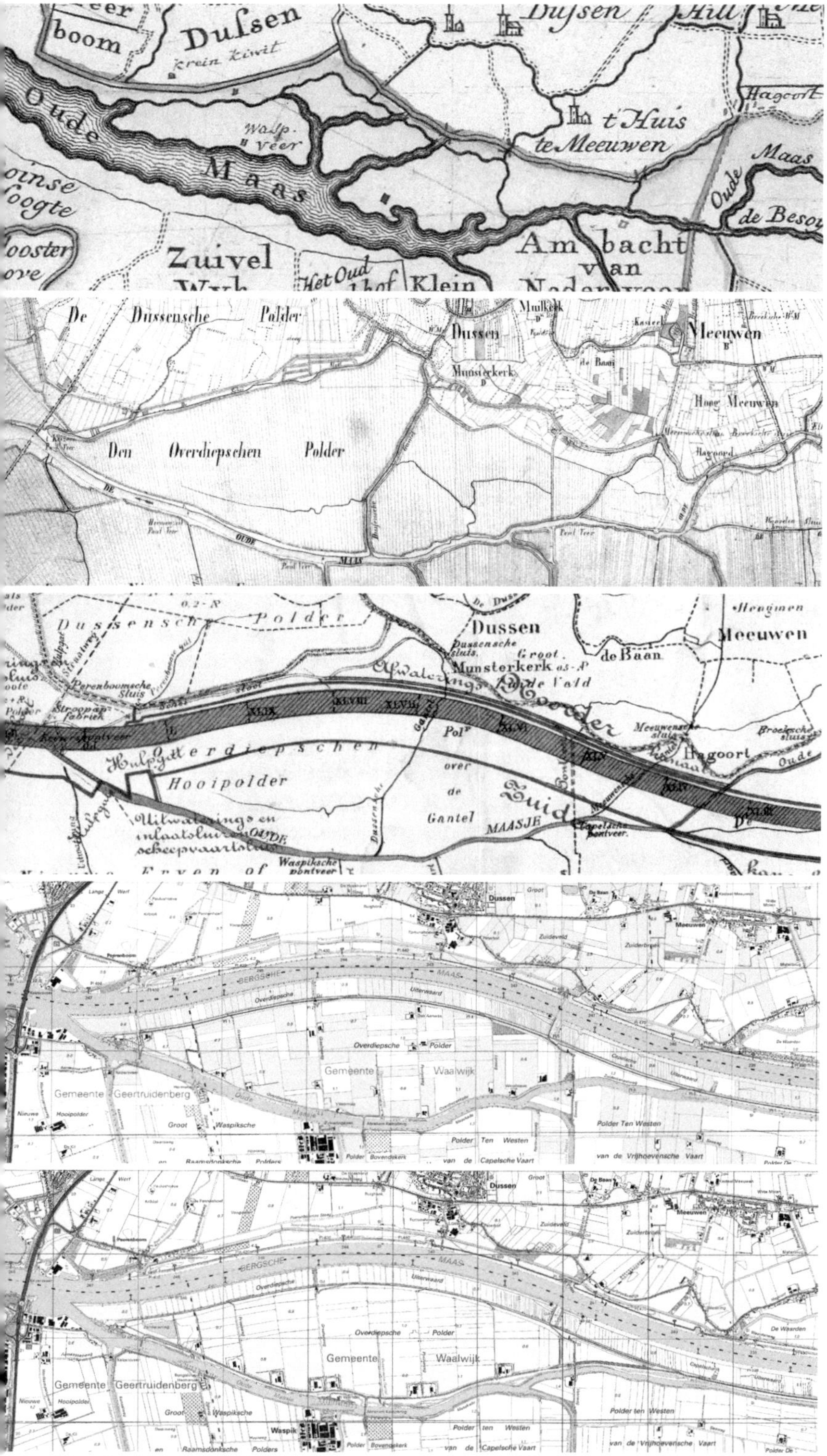

Transformation of the Meuse River upstream from the Biesbosch: The Oude Maas in 1789 before embankments were build (Nationaal Archief, The Hague) • After the reclamation of the Overdiepse Polder (Nationaal Archief, The Hague) • Governmental project for the Bergsche Maas in ca. 1900 (Brabants Historisch Informatiecentrum) • Situation in 2010 and in 2016, after the removal of the primary levee to the south and construction of elevated farms (Kadaster)

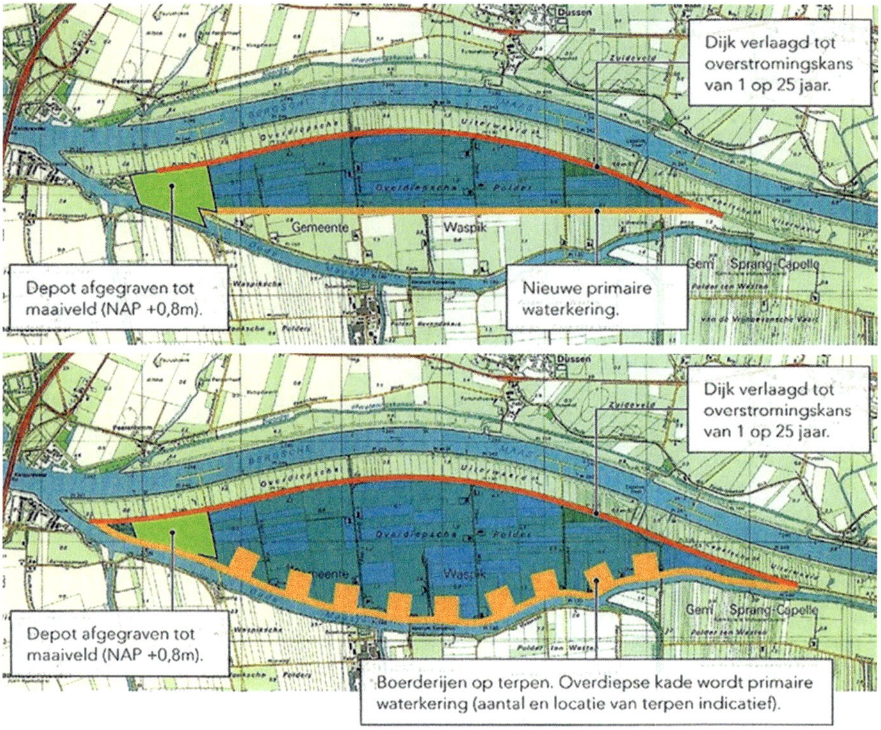

Overdiepse polder: 'middenvariant' and 'terpenvariant' (Bink et al., 2011: 157)

the farms onto a string of raised platforms leaning against it. Given the efficiency requirements of intensive farming, it appeared that this latter alternative would create enough space for about ten farms, including habitation and exploitation (id.: 11), implying the departure of seven exploitations and seven families.

The 'nature-option' was soon set aside as the most expensive, the most destructive in terms of local economy, and as contradicting the national and regional policy. In the 'middle option', most of the pre-existing buildings could be spared, but the central division was seen as very disruptive by the local community (Provincie Noord-Brabant, 2008: 27): smaller parcels with various statuses regarding flood defence would have to be distributed among the farmers, implying long negotiations and painful arbitrage. In the third 'terp-option', the full polder could be flooded. In both cases, all existing constructions needed to be removed, whether located in the polder or along the levees. The third option generated the most interest but also mixed feelings, partly because it appeared difficult for the inhabitants to imagine the terpen without visualization (Van Rooy and Slootweg, 2003: 26). Despite these doubts, this last option was eventually favoured for its strong effect on the river-water level, which could be lowered 30 centimetres at peak discharge, versus twenty for the 'middle-option'. The hydraulic inquiries confirmed that the Overdiepse Polder could play a significant role in regional safety as a retention area but even more

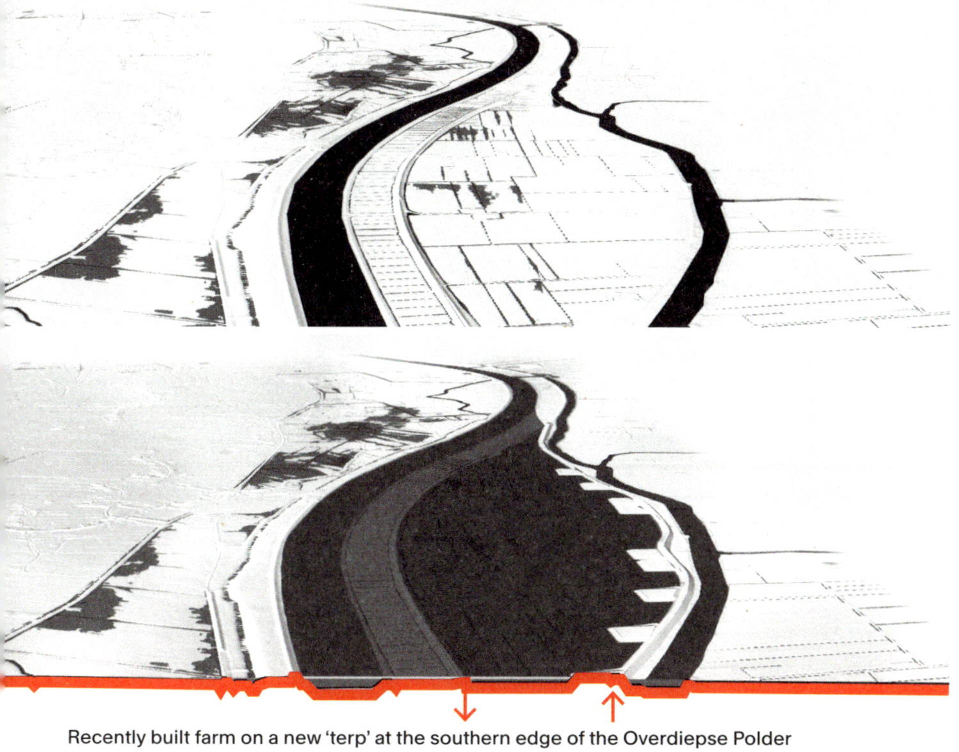

Recently built farm on a new 'terp' at the southern edge of the Overdiepse Polder

so as an extension of the river channel at extreme discharge. As it appeared, this second role, however, implied that the polder remained free from obstacles: In order to remain efficient as a temporary flow channel, its vegetation should remain low, and stock farming was thus preferred to other land use (Verschelling, 2003). The 'terpen' option also appeared to be three times cheaper than turning the polder into a nature area, with equivalent effects on water level (Van Rooy and Slootweg, 2003: 52).

For local farmers, the Terpenplan brought a perspective of lasting activity in a renewed landscape, and an opportunity to shape new, efficient, and secure farms on the future mounts. Paradoxically, the transformation of the Overdiepse Polder into an emergency flowing area brought hope of long-term and stable activity for the remaining exploitation—with the authorities being committed to financing the achievement of the project and to a once per 25-year flood threshold—for the discharge expected in the 2050–2100 period on the basis of the IPCC moderate scenario (id.: 43). Following the preliminary studies, the terpenplan was further developed at a high pace, anticipating the Planning Key Decision Room for the River that was officially validated by the Dutch government in 2006. At that time, the Overdiepse Polder project was already defined and the planning procedure largely engaged, making the project one of the 'frontrunners' of the Room for the River programme, completed and inaugurated in september 2015.

The vast agrarian landscape of the Overdiepse Polder seems today almost unchanged. The two existing massive levees have been transformed, one lowered to submersible levels, the other heightened to the level of primary flood defence and bordered by the new terpen and their modern farms. Although the local population took the initiative and shared the responsibility of the transformation, and managed to secure farming activities in almost the whole project area, the social impact of the plan appeared utmost sensitive. With the planning of the nine new farms, eventually seven, the project implied a radical re-allotment in order to secure a profitable economy, and forced more than half of the inhabitants to leave their homes and land. Since most families wanted to remain in the polder, this negotiated departure has been a hard time for a small, strong, and forward-thinking community that dared take the lead and accepted its responsibilities in the face of regional challenges.

The revival of the 'terpen' in the Dutch landscape was seen as an important event, as was the construction of a series of large new farms, not seen since the completion of the Flevoland polders in the 1970s. The Room for the River organization and its Q-team were therefore keen on reaching a high level of landscape and architectural quality and on putting spatial ambition into each project already at the preliminary stage, rather than exercizing control afterwards: it also used its influence to involve

Farmland is submerged 1/25 yr average to lower the water level in the Bergsche Maas river

Habitation buildings are located alongside the new main road/levee

designers in the various projects and at various stages—even when these projects were mainly directed towards local activity, as was the case for the Overdiepse Polder, where a competition was eventually organized for the design of the nine new farms.

Most of the functional aspects were strictly predefined by the future users, as the jury stressed after the first round. However, the wining architects, Onix, managed to create diversity within a coherent ensemble through a set of simple rules for plantation and volumes, combined with precise guidelines for construction details. A level of detail that would be perceived as intrusive by some of the future users, but eventually succeeded in giving the farms an elegant and familiar aspect without contradicting the large-scale, unsophisticated, and productive character of a polder devoted to intensive farming. The size of the 'terpen' was also a matter of concern for Rijkswaterstaat spatial quality experts. Intimate in the Noordwaard area, the 'terpen' in the Overdiepse polder were dimensioned for large industrial farm buildings, with a surface of about 1.5 hectares. Eventually the ensemble

Entwined river and infrastructures in the Rhône valley (Rossano, 2013)

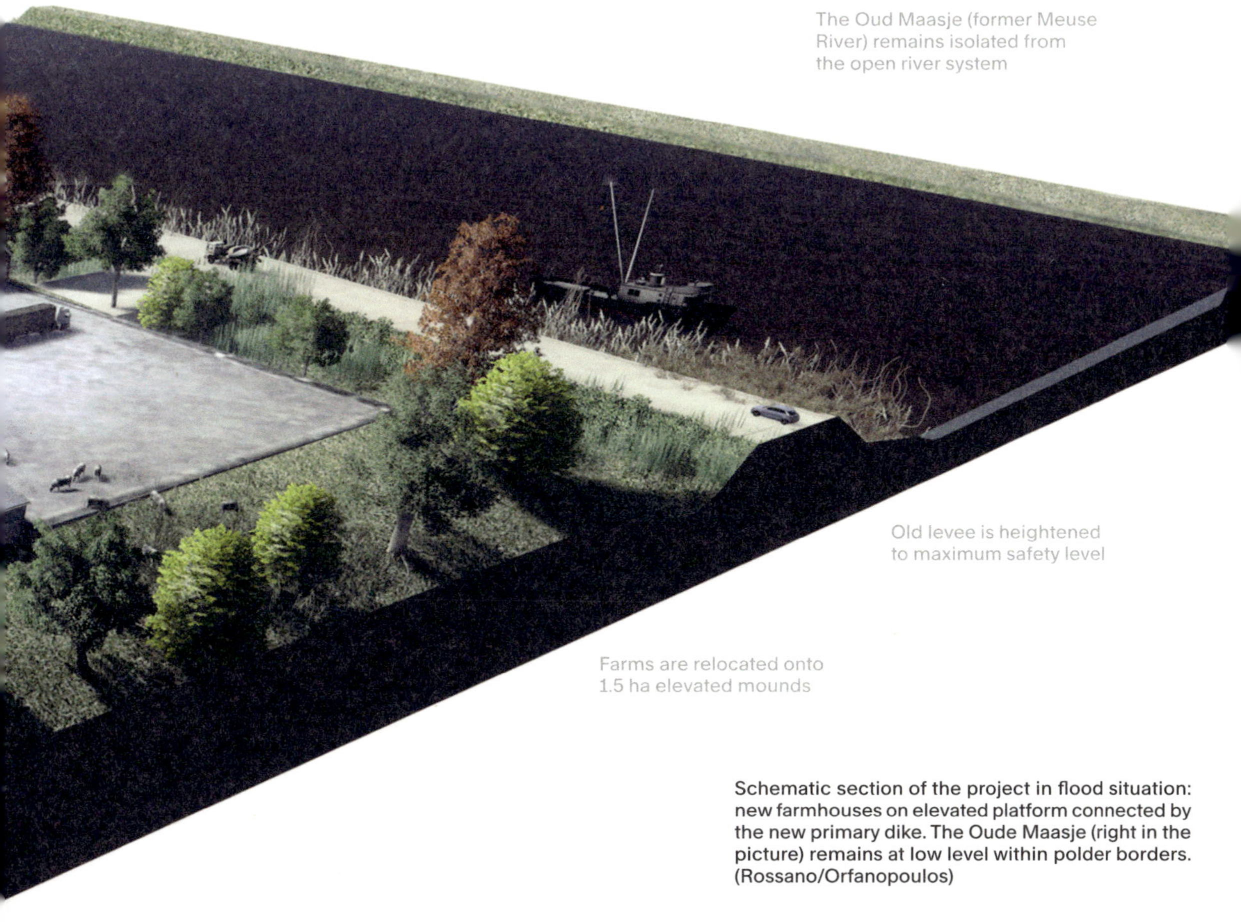

Schematic section of the project in flood situation: new farmhouses on elevated platform connected by the new primary dike. The Oude Maasje (right in the picture) remains at low level within polder borders. (Rossano/Orfanopoulos)

was considered successful also in terms of spatial quality, since the string of farms had its own logic and 'fitted within the scale of the polder (Van der Grift and Der Nederlanden, 2013).

Paradoxically, the most striking side of the project is its ordinary aspect: the polder looks just like many modern Dutch polders, as nothing indicates that it might be covered by up to nine million m^3 of river water for several days or weeks. Five years and 111 million euros later, the Overdiepse Polder has proven that 'floodscapes' should not necessarily be associated with extensive wetlands, heck cattle, and nature-like aesthetics, but could also take on or keep the aspect of a highly productive landscape.

4

DESIGNING ELASTIC LANDSCAPES FOR AN UNCERTAIN FUTURE

February 2020: High waters flow through the Noordwaard region (Rob Engelaar/ANP)

> 'Uncertainty is ineradicable, and there is every reason to see it not as a limitation but to approach it positively as a challenging fact that is inherent to social action.'
> (D. Sijmons, Hajer, Maarten, et al.: 37)

The projects described in the previous chapters illustrate a global change in flood management in the whole of Western Europe. They also show many similarities in goals and development processes. Past civil works were often built by authoritarian regimes, in under-used areas, aimed at safety and economic growth. Today's projects reach many objectives, sometimes contradictory, in developed regions. They involve numerous stakeholders in long negotiations.

This multiplicity of goals and stakeholders allows for new territorial deals, conditioned by two key dimensions: the space within which flood risks are managed and redistributed, and the temporality of the responses to various flood levels, which implies a great uncertainty. The drafting of these agreements generally follows a few essential steps: defining and communicating the overall framework; clarifying the risk levels and their possible redistribution within the area or region; creating favourable conditions for negotiation, deconstructing binary oppositions and clearing unfounded suspicion; valuing flood zones as useful and multifunctional spaces.

This complexity also calls for new design tools that go beyond pure engineering and involve other disciplines. On top of hydrological aspects, issues of functionality, land use, and aesthetics are addressed through various visual tools of communication, and discussed with more qualitative concepts, such as authenticity, beauty, or sustainability. Flood management uses diversified expertise, and involves more frequently designers of the living environment (landscape architects, urban planners and architects). However, these designers may encounter difficulties: their intervention can be either perceived as a source of conflict (and limited to a final shaping of a technical decision), or be integrated from the start as a true contribution to the whole project.

LANDSCAPE NEW DEALS

FRAMING A NEW FLOOD MANAGEMENT POLICY

Flood defence might seem a univocal concept, but its implementation in time and place makes each choice and each project highly specific. Experience shows that these choices must rely on a clear framing and a coherent communication—large projects can otherwise stumble over misunderstandings and frustrations and become bogged down in lasting conflicts.

The name chosen for the projects, often from the start, already reveals the angle from which authorities approach flood management. In France, the term 'amont' (upstream) added to Isère (the river) clearly set the priority on the protection of the city of Grenoble located downstream, although its name wasn't mentioned. In the Netherlands, the name 'Room for the river' advertised the new approach chosen by the government already in 1996. However, the name didn't specify the way room should be made, nor the locations of the projects, which were still being discussed. The sub-projects within the programme did clearly name the area meant to be transformed and reopened to flooding, undoing the excesses of the past ('Ontpoldering Noordwaard', which can be translated into 'depoldering' or 'undamming' Noordwaard, or 'Ruimte voor the Waal' in Nijmegen, meaning 'room for the river Waal', which was spectacularly widened and reshaped). Conversely, the Third Rhône Correction, envisioned in 1999, suggested the continuation of the previous damming, channelling and draining. As the project leader would later recall, the name was inspired by a 'sort of filial duty, respect for the heritage and the work made by our predecessors' (Arborino, 2019). However, the chosen name was, as we have seen above, both inaccurate and confusing.

Logos and diagrams, often used to communicate the projects at an early stage, are crucial as well, as we can see by comparing the Swiss and Dutch communication. The Third Rhône Correction used a naive hand-made drawing, showing a diverse landscape and suggesting that all types of land-use would be taken into account: river (a meandering line), transport infrastructure, farming, built environment, and nature (a bird, a tree). This logo came along with a before/after diagram, representing the enlargement of the riverbed at the cost of farmlands. Name, logo and diagram actually told different stories: the first suggested a reinforced

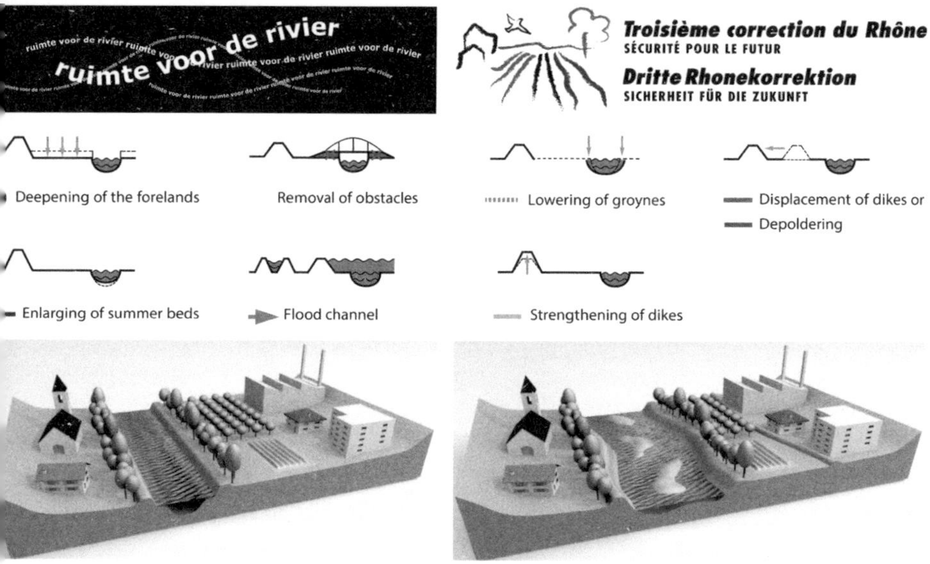

'Room for the River' and 'Third Rhône Correction': logos of the two programmes • Range of possible measures for the Dutch rivers • Schematic diagrams for the Swiss Rhône

channelling; the second a consensual one-fits-all solution; the third a priority given to nature and urban safety. In its local developments, the Rhône Correction appeared more complex and diverse than was communicated (and maybe originally thought), and the project suffered from of the original ambiguity: was it a river 'correction' as suggested by the name, or the 'liberation of a stream' as the diagram suggested? The debates were long defined by oppositions: safety for all versus differentiated risk management; food production and self-sufficiency versus environmental restoration; environmental restoration versus cultural landscape. Twenty years after the first studies, the 2015 referendum showed that the same dilemmas, expectations and frustrations still dominated the discussion, hindering the development of the programme.

At the other end of the river network, the Dutch authorities started to promote integrated measures over dike reinforcement from the 1990s on. The logo chosen to advertise the 'Room for the river' programme suggested a free meandering stream, and came with a set of schematic sections showing various ways to improve the capacity of the riverbed. This palette left a wide margin for interpretation and future development, and slightly varied in the course of the programme, at some stage incorporating more traditional approaches (heightening levees and deepening the riverbed). These variations reflected internal debates among the advocates of century-old hydraulic technique and the supporters of a strictly spatial and territorial approach. This debate however didn't reach the wider public: they received a stable and, in appearance, totally coherent message, which facilitated the acceptation of potentially controversial measures. The palette of possible measures also provided more freedom to local projects, a crucial advantage over the apparently standardizing approach chosen for the Rhône.

Protest sign in the Biesbosch Polder: 'Keep the polder, no nature nonsense!' (Rossano, 2012)

RISK TRANSFERS

Most landscapes recently transformed for flood mitigation relate directly to urban exposure and constraints, in particular to the limited resistance of existing urban embankments, levees, bridges, and railways that offer little flexibility to accommodate natural fluctuations. They imply an outsourcing of flood risk from city to landscape, a transfer legitimized by different land-use, the one use being considered more valuable and more vulnerable than the other, or, as the euphemism used for the Third Rhône Correction has it, carrying more 'constraints' (Hydronat, 2000), often resulting from a lack of foresight and uncontrolled growth. The Isère valley is now transformed in order to spread future floods before they wipe out the old bridges and embankments in Grenoble. The 'de-poldering' of the Noordwaard area was shaped in order to keep the level of the Merwede River below a predefined maximum in the city of Gorinchem: The river had to be lowered by 30 centimetres at peak discharge in order to avoid massive damage in the built-up areas. In Switzerland, the Third Rhône correction gave priority early on to the securing of cities and industrial areas even before the project was fully defined (Arborino and Perret, 2008: 10), focusing on the urban areas of Visp and later Sion. The density of inhabitants and workplaces was in all cases an argument put forward to legitimize a risk transfer from city to countryside, and to justify the level of investment. Communication around Room for the River concerned four million people living in the Dutch river area, while the Third Rhône Correction reported 100,000 inhabitants at risk.

Infrastructure too represented a crucial motive for flood mitigation, especially so in linear landscapes concentrating all means of evacuation within one single geomorphological corridor. Transport infrastructure networks involve larger population and broader interests, and help to mobilize national authorities and funding, if not international. For example, road and rail links were a decisive argument for the Swiss federal council to engage in the Rhône Correction already in the nineteenth century.

Among all types of land-use, intensive farming was the biggest beneficiary from past river engineering works, and the loser in the current redistribution of flood risks, often linked to river restoration. This reality should be acknowledged and anticipated to avoid feelings of betrayal or neglect being expressed, sometimes violently, by farmers: They make 'room for the river', mostly to ensure flood safety for urbanites, and may legitimately expect a form of compensation. This perspective should however be qualified by the fact that other types of land-use might lose space to river engineering: constructible zones have been cancelled when prone to high flood risks, and natural woodlands have been cleared to ease flood mitigation. Furthermore, areas transformed into temporary flood fields can still be fit for farming activities: flood fields aren't necessarily marshlands surrounded by steep levees, they can provide suitable land for farming or pasture and multifunctional infrastructure.

Combined uses in flood spaces: new lake for recreation and water storage in Meerstad, cultivated flood meadows in the Bergsche Maas' winter bed, along the Overdiepse Polder (photos Rossano) • Winter bed of the Isar River in Munich, transformed and opened for urban leisure (photo Kuenzel)

COMBINED AMBITIONS FOR MULTIFUNCTIONAL LANDSCAPES: FLOODSCAPES AREN'T LOST SPACES?

The main argument used against the widening of rivers and horizontal flood mitigation strategies is partly true and partly contradicted by old and new examples: floodscapes are indeed generally lost for cultures that require costly equipment and built infrastructures, and lost for regular urban extensions, which cannot withstand temporary flooding. Floodplain cultivation and flood-resistant urban spaces have however existed for a long time, and recent research and projects show that there are solutions for combining space for flood mitigation with profitable land exploitation, provided this combinatory approach is integrated at an early stage. As shown earlier for past flood-resistant civilizations, flood-prone areas have long been exploited, not only in-between floods but also through the exploitation of the flood itself for fertilization or land-making. For the farmers of the Overdiepse Polder, cultivating flood prone spaces wasn't new either: from a 730 hectare-large area, 180 hectares were located in the floodplain of the Bergsche Maas River, below high-water level and nevertheless cultivated since the Middle Ages. When the Bergsche Maas channel was dug in 1904, a winter bed was kept outside of the levee to allow for extra capacity at peak discharge. This winter bed remained exploited after the construction of the channel—the trees planted in the floodplain conveniently marking its edges for ships to remain on course even at high waters (De Bont, 2013).

In 2003, more than half of the farms active in the Overdiepse Polder were still renting land in this floodplain, using it mainly as meadow and, for a small part, for raising corn (Van Rooy and Slootweg, 2003). The majority of the farmers were thus used to exploiting flood-prone land, long before the polder itself was identified as potential space for flood mitigation, which was not the case along the Swiss Rhône and the French Isère. This familiarity with combinatory use may explain the fact that the farmers of the Overdiepse Polder actively participated in the planning process, even initiating an accelerated procedure to promote a solution that allowed them to sustain their activity in the future flood mitigation space. For the farmers of the Overdiepse Polder, just as for the

	BEFORE PROJECT IMPLEMENTATION	AFTER PROJECT IMPLEMENTATION
Rhône 3	Fruit culture Agroforestry Sand and gravel mining	Nature development Urban recreation (Sion-sur-Rhône) Recreational cycling / pedestrian route Farming on new shallow levees
Isar Plan	Unexploited meadow	Urban recreation Swimming / beach Recreational cycling / pedestrian route Nature redevelopment
Isère Amont	Crop farming Isolated riparian woodland Recreational ponds	Crop farming Riparian woodland reopened to river flood Recreational cycling / pedestrian route
Meerstad Groningen	Arable land Intensive pasture Rowing basin	Water recreation Sailing and rowing sports Nature development (wetlands) 'Sail-in' villas on islands
Noordwaard	Arable land Intensive pasture Farm buildings and dwellings	Arable land Intensive pasture Farm buildings and dwellings on terpen
Overdiepse Polder	Arable land Intensive pasture Farm buildings and dwellings	Arable land Intensive pasture Farm buildings and dwellings on terpen

Land use before and after transformation into flood expansion areas

farmers of the Isère Valley, the land is today no less productive than before, despite its change of status.

The case of the neighbouring Noordwaard Polder is different in that regard, since the area will function as a by-pass for the Merwede and should be flooded at a higher frequency and higher speed, making intensive farming more uncertain. As a consequence, farmland in the 2050-hectare flow area decreased from 80 percent to about 50 percent after completion. The Noordwaard Polder is, however, located in the heart of the Biesbosch and surrounded by several isolated nature reserves: its transformation was thus partly a step towards the restoration of a greater and more natural Biesbosch that made sense on a regional and national scale—although nature restoration was not communicated as a primary goal but rather as a side effect welcomed between two conurbations. This choice was also legitimized by the position of the Biesbosch nature area, in a densely urbanized region and in a country were nature occupied only three percent of the national territory in 2000 (CBS, 2004).

The re-balancing of nature and farming was more difficult to defend for the architects of the Third Rhône Correction, where river widening was explicitly related to the restoration of riparian nature, excluding productive land-use in a canton where more than 50 percent of the land was classified as unproductive. Although the ecological state of the Swiss Rhône was as poor as that of the Dutch Rhine, the global balance between cultivated and natural space was more often used *against* the renaturation

'Linking water and space': examples of combinatory flood defense from the Dutch governmental Water Plan (Min. van Infrastructuur en Milieu, 2015)

of the Rhône, opponents arguing that the local economy had little space available in the valley, and that the Alps left enough space for nature to thrive.

Paradoxically, floodplains preserved within the urban environment eventually appear to be profitable, since they provide large recreation space at relatively low construction and maintenance cost, which can partly be covered by the budgets allocated for flood protection. As the popular success of the Isar Plan shows today, there is a strong demand for large, informal, and open landscapes in cities, which can provide space for unforeseen and unprogrammed activities, a demand the city could meet thanks to the space available along the river, without buying new land. Furthermore, an increase in retail activities in the vicinity of the Isar floodplain showed that the transformation of the central section benefited the local economy, in particular ice cream parlours and cafés located near the bridges, and increased the neighbourhood's attractiveness for inhabitants and investors.

Still officially seen as flood-defence infrastructure, the Isar floodplain isn't subject to the limitations that apply to regular urban parks, meaning more freedom for visitors, but also more latitude for its maintenance: 'It's not a park, it's a floodplain.... If it was a park, we would constantly have to ask: "Are we allowed to take a tree out? Are we allowed to cut the grass?" It would be much more difficult for us to take care of this area' (Schaufuß, 2013).

Thanks to this hybrid status, the managers can rapidly take the necessary measures to ensure that the floodplain can fulfil its hydraulic function without facing long procedures, in particular for the removal of riparian vegetation that could hinder the flow. Due to the nature of flood mitigation spaces, the design of the Isar floodplain has sought to curb the cost and frequency of maintenance work, and ensure its resistance to flooding, based on both discharge statistics and erosion simulation.

The (re)creation of floodplain and mitigation space can also provide opportunities for urban and real estate development, as in Groningen with Meerstad and in Sion with the Sion-sur-Rhône project. The first case concerns a massive investment for a large and rapid programme of construction, while the second case is built around various projects along the river Rhône that improve flood safety and the attractiveness of the surrounding neighbourhoods. In both cases, water management plays a central role in urban development. The Meerstad project showed, however, that the close association between expected real estate revenues (10,000 new dwellings planned) and investment in landscape infrastructure (a 600-hectare lake good for three million m^3 water retention) meant both could fluctuate according to the market. Eventually, initial ambitions had to be scaled down as a result of the 2008 crisis, for the construction of the lake and the houses. Nevertheless, the combinatory value of the project remains in the long term, with housing construction taking place at a slower pace than initially planned, and the creation of the recreation and retention lake being phased accordingly.

Sustainable mobility can also benefit from the restoration of floodplains and the adaptation of riverbanks. Riverbanks offer a possibility to implement continuous paths, an opportunity seized in all investigated projects. The creation of slow traffic routes, when designed in an integrated approach, allows for a more efficient and healthier mobility infrastructure, and makes the city economically more attractive. It also allows a larger public to enjoy green riversides, appreciated during heat periods. Finally, public paths along the water bring a larger public in contact with the river's fluctuation, thereby generating more awareness for flood risks.

BUILDING UPON THE LEVEE: NOT SUCH A GOOD IDEA

The levee itself is increasingly subject to various combinatory proposals, associating functions as diverse as recreation, farming, parking, and housing. 'Minimizing mono-functional levees' and 'maximizing levees with multiple functions' were advocated for the Rhône Correction by the SYNERGIE research programme (Pellaud et al., 2005). The design studios held on the Rhône by the Institute of Landscape Architecture at the ETH Zürich also illustrated the potential of levees as support for urban, natural, and rural developments (Rossano, 2012). In its new 'Water Plan 2016–2021', the Dutch Government too is promoting the combination of flood defence with urban development in the form of 'integral area development with large-scale uses' (Min. van Infrastructuur en Milieu, 2015: 16).

The combinatory use of levees remains, however, controversial for aesthetic reasons, but also because of safety issues. Unlike mitigation spaces, levees are built to resist specific thresholds in water levels, thresholds that are and will be regularly updated in the future—most probably towards higher levels. The construction of fixed urban amenities on the levees makes future adaptation difficult if not impossible, depriving vulnerable cities from future adaptive capacity.

Historically, the development of water cities followed the construction of their flood defence structures (as shown by the suffixes '-dam' and '-dijk' found in the names of numerous Dutch cities and streets), and many of these structures were indeed used for traffic and building. However, urban extensions generally implied the construction of a new and higher levee in front of the old one, while old rivers were dammed and regulated, and often turned into artificial canals. This is the case for the rivers IJ and Amstel that cross Amsterdam, as well as the river Rotte that gave start to the construction of the city of Rotterdam, the name meaning literally 'the dam of the Rotte [river]'. As a consequence of every extension, riverbeds and floodplain were reduced a bit more each time; this process is however today compromised in most urbanized deltas and valleys, where river channels have reached their minimum width, while variations in discharge are expected to increase in the near future. It is thus important to consider river banks and levees as evolving structures, the future safety of entire cities and regions depending on their flexibility. The difficulties faced by the island-city of Dordrecht is in that sense characteristic: the northern section of its primary flood defence is formed by the ancient Voorstraat, a street and levee clamped together between buildings of great historical value; as a result, this crucial protection cannot be raised to a safe level without damaging the historic city centre, while building a new levee along the river would narrow the flow, raise the water levels and worsen flood risks. This double fixity leaves municipal authorities with an unsolvable dilemma.

Furthermore, the profile and the components of levees are today still the subject of research and improvement, and access to the levees appears to be a key issue in the implementation of reinforcement measures, an access that was in many places hindered by construction and improper vegetation. When needed, removing constructions will bring high costs for future maintenance, just as the construction of cut-off walls built in the Isar levees avoid cutting existing trees that were planted at the wrong place—right onto the levee.

Finally, aesthetic and societal objections can be raised against constructions that tend to hide or trouble the general aspect of main levees, since they form the essential cultural landscapes in floodplains and riverine regions. Their concealment could also worsen public awareness, if the only permanent reminder of flood risks were to disappear under various constructions and urban facilities.

Contemporary projects for spatial flood mitigation show that floodscapes are not lost for productive land-use, but should on the contrary also be seen as an opportunity to optimize and recompose a territory. Multi-use projects should therefore be fostered by combinatory approaches, but also maintain the possibility to change and evolve, a key aspect to future resilience.

FUSE LANDSCAPES

In the debates about the costs and the productivity of flood mitigation, time is the main issue, and more particularly time in land use and planning processes, and our capacity to think and plan our living territory according to different times and speeds. The development of temporary flood fields is even more revealing, as they are designed to be used according to various frequencies, at different flood levels, and for different horizons in terms of present and future climate.

These new landscapes play similar roles of hydraulic fuses, and are preferred to other measures for their large mitigation capacity and the possibility they provided to sustain productive land-use in the concerned areas. Less intrusive than the widening of the riverbed, they imply relatively limited and relatively cheap interventions (spillways, submersible embankments, low levees); more controllable, they can be set into motion only when necessary. Last but not least, they can store enormous volumes of water, and are therefore adapted to exceptional floods.

In the two Alpine cases, Isère Amont and the Third Rhône Correction, flood fields and flood corridors are thought to answer residual risks in the long term, including possible changes in climate and hydrological conditions. For the Swiss Rhône, the transformation of the river bed is meant to accommodate regular floods, but the creation of a 'dry'

	TOTAL MITIGATION AREA (ha)	MITIGATION / DIVERSION AREAS	CONCERNED STREAM LENGHT (km)	BUILDING COSTS (millions €)	IMPLEMEN-TATION	POPULATION PROTECTED
Rhône 3 (CH) Corridor of residual risk management	Unknown	15	80 / 160	Unspecified	…–2030 +	100,000
Isère Amont (FR) Controlled flood fields	3387.4	16	50	135 (estimate 2016)	2006–2016	100,000–300,000
Room for the River (NL) Total programme	4400	10	450	2300	2006–2015	4,000,000
Overdiepse Polder (NL) Retention polder	550	1	14	111	2008–2015	
Noordwaard (NL) Retention polders + Diversion corridor	2050	1 (7 sub-areas)	15 (70 km banks)	295	2009–2015	

Examples of mitigation/flood deviation areas (NB: project costs are global and preliminary estimations for ongoing projects)

	ACTIVATION THRESHOLD / EXPECTED SUBMERSION PERIODICITY	TIMESCALE TAKEN INTO CONSIDERATION
Rhône 3 (Corridor of residual risks)	≥ 100 years	Unlimited
Isère Amont (Flood fields)	1× 30 years	Unlimited
Overdiepse Polder (Emergency retention polder)	1× 25 years	2050 (levee) 2100 (inhabited mounts)
Noordwaard (Semi-open wetlands and emergency retention polders)	1× per day (tidal wetlands) 25% × per year (low dikes) 2/3 × per year (low dikes, direct upstream inlet) 1% per year (high dike polders) 0.1% per year (all polders)	2015–2050

Activation threshold and timescale of four systems of controlled flooding

corridor, consisting of a series of bypasses between riverbed and valley floor, will allow the stream to expand over vast areas of farmland in case of extreme flooding. 'Typically, it is about avoiding dam breaches with discharging systems, anticipating that "the worse can happen", and not "fighting to find out whether the system is perfectly dimensioned for the hundred-year flood"' (Arborino, 2011: 20). Olivier Manin, project leader of the Isère Amont project, also stressed the fact that the controlled flood fields created along the River Isère have almost no upper limit—provided that they are well maintained. As no specific long-term projections were available to him to integrate possible climate evolution, it was thus the system itself that was seen as more sustainable, since it wasn't based on a fixed upper threshold but rather functioned as a safety valve.

The timescale taken into consideration to calibrate the Room for the River programme was more precise and more limited: the interventions made along the Rhine and Meuse branches were dimensioned to resist a discharge corresponding to a 1250-year flood under the climate expected in 2050. In the alpine cases, flood diversion areas were seen as an ultimate tool to face long-term uncertainty, while in the Dutch case they were created to reach a precise and limited goal set for a limited time scale. The adaptations made today in the Dutch river area aren't seen as a final stage but rather as a step in a process of bringing flexibility back into the river system.

The activation of these 'safety valves' placed on European rivers also greatly differs in expected frequency, from daily tides to 200-years flood (see table), a modulation made visible in the various water level thresholds used to transform the Noordwaard area. An aspect that generally remains underexposed is the terms of the risk redistribution agreed upon between the collectivity on the one side, and users and landowners on the other. In the article 'Who likes to live in the "calamiteitenpolder"', the Dutch newspaper *NRC* echoed the debate following the proposal of the Luteyn governmental commission to designate several emergency retention polders, where waters from the Meuse and Rhine rivers could be diverted to in case of threatening high waters (*NRC*, 02.12.2002). House owners complained that the value of their properties had decreased, even though the proposal was still at an early stage. In the French Isère Valley, opposite protests were heard when regional authorities announced the creation of 16 'champs d'inondation contrôlée' closed for construction: local representatives saw land prices soar around one of the designated flood zones, threatening municipal housing policies. In both cases, local economies were influenced by the mere eventuality of a status change that would turn (potential) building land into designated flood areas.

While public authorities stress the collective advantages of controlled flooding compared to higher levees or river widening, stakeholders fear the individual losses implied by this strategy. The main disadvantages expressed in the various contexts involve: short-term uncertainty generated by complex planning processes; transfer of flood risk from urban areas onto farmlands; long-term uncertainty for future harvests, and the induced loss of reliability in the eyes of purchasers of agricultural products; restrictions on future types of production; finally, loss of speculative

NEGATIVE ASPECTS	POSITIVE ASPECTS
Short term	
Uncertain future for the exploitations Uncertain value of farm and land, restraining Investment, transmission or sale	Opportunity for optimizing land-use Opportunity for changes in land ownership within the area No heightening / enlargement of levees onto farmland
Long term	
Additional risk to existing constructions Additional risks of bad / lost harvest Untrustworthiness vis-à-vis buyers Restrictions on productions sensitive to flooding (erosion or submersion) Restrictions on productions involving infrastructures or constructions (greenhouse, irrigation) No possible profit on future change in land-use zoning Uncertainty on future flood frequency	Opportunity for optimizing parcels division Potential revenues from recreational functions when project is linked to nature development Secure agricultural zoning and protection from land speculation Investments hedged by compensation agreements Financial compensation for all flood events, including at discharges above reference levels Public guarantees for long-term maintenance

Advantages and disadvantages of controlled flooding for farmers and/or landowners

value for land which, in a normal situation, could possibly become building ground in the future and see its price soar, to the benefit of the owner.

The projects where flood areas are now agreed and defined (Noordwaard, Overdiepse Polder, Meerstad, Isère Amont) show, as a counter balance, that several positive effects can be expected from controlled risk transfer. The planning and design process, although generally experienced as stressful, opens up opportunities to re-organize and upscale existing exploitation. This was decisive in the Room for the River projects, through local land-consolidation schemes, as well as for Meerstad, where those who wished to give up their farms could do so, eventually for higher prices than when the land would have been designated for building.

For flood-prone land, the status of 'official mitigation area' can also provide supplementary coverage: since the land is purposely flooded for the common good, its users are guaranteed to receive compensation for their losses in all circumstances, which is not the case for regular exploitations located in a floodplain and exposed to high flood risks. The farmers of the Hunze-en-Aa's Water Board jurisdiction first rejected the proposal to lend their land to emergency flood mitigation, but eventually, during the negotiations, realized that they were better off exploiting an 'emergency polder' with a guaranteed compensation, than not having this guarantee and still exploiting flood-prone land (Van Hall, 2013).

A change in status from regular floodplain to controlled flood area can therefore have positive effects, and cannot only be understood in terms of risk catchers and beneficiaries but also in terms of traded and mutual benefit.

SPACE FOR FLOOD, SPACE FOR LANDSCAPE?

A concrete and visible effect of floodscapes is the establishment of lasting clearings within metropolitan areas, securing vast urban open spaces that can be crucial for the future quality of growing agglomerations. In the European Union, artificial surfaces have increased by 711 km^2 per year

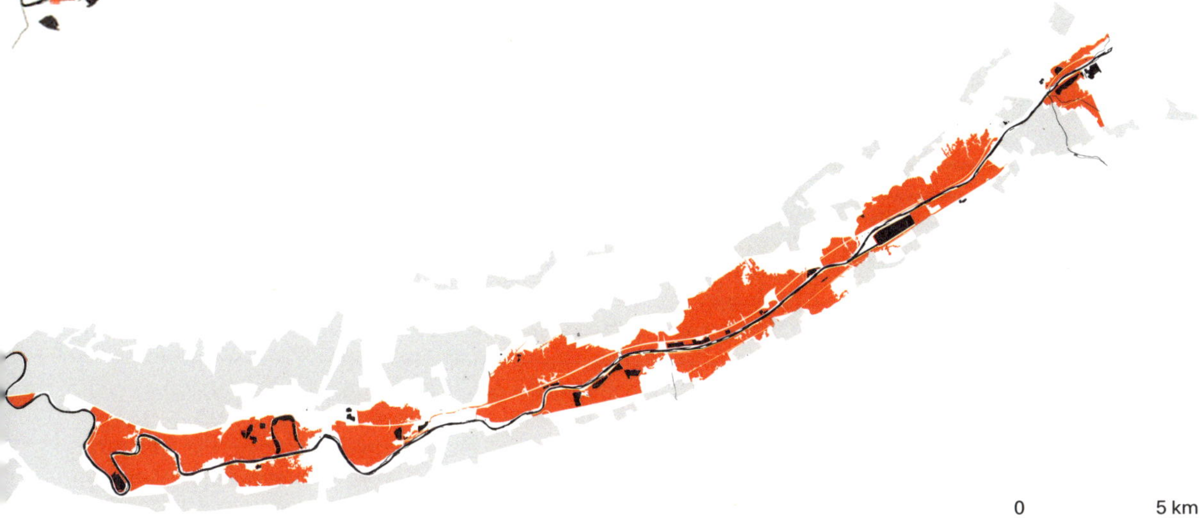

'Flood evacuation corridor' planned along the Upper Rhône between Brig and Martigny, urban areas in gray, flood mitigation areas in red • 'Controlled Flood Fields' planned along the Isère between Grenoble and Pontcharra

between 2012 and 2018 (EEA, 2019), about seven times the area of Paris. The Union seeks to curb this trend by 2050, while large metropolises seek to 'green up' to keep their inhabitants. In this context, preserving large and open floodscapes seems to be a sustainable and an efficient way of securing the necessary and meaningful voids within urban areas.

The Meerstad project was in that regard the most explicit, since a hundred percent of its mitigation capacity was to be implemented in the form of a large lake at the edge of the city, in a mixed suburban landscape. In the Southwest of the Netherlands, the Room for the River programme motivated the 'de-poldering' of the Noordwaard area not only with hydraulic goals in mind but also using the argument that 'developments in the Biesbosch reinforce its function of green buffer between the urban networks of "Randstad Holland" and "Brabantstad"', two major Dutch agglomerations (Projectorganisatie Ruimte voor de Rivier, 2006, vol.4, Nota: 27). More generally, the Dutch Room for the River programme voluntarily concentrated the projects near or in urban areas, in order to reach multiple goals and serve the quality of the city that had to be protected. This is particularly visible in the project implemented in Nijmegen, which nearly doubled the width of the river along the very centre of the city, creating a green island in the heart of the agglomeration.

The effects of controlled flood areas could in the future be even more significant in the narrow space of Alpine valleys. With the Isère Amont project, an open area of 3400 hectares is protected against urban sprawl, and secures a vast and steady green corridor of about one kilometre wide in the centre of the valley. This forms a precious reserve for an agglomeration counting about 550,000 inhabitants, concentrated in a valley plagued by frequent air pollution and long heat waves. The Upper Rhône Valley too could be positively affected by the future 'Flood evacuation corridor', which should protect large parts of the cultivated floodplain against uncontrolled urban sprawl. The positive impact of these projects on the quality of the living environment was however underplayed by local authorities, possibly to avoid controversies and the above-mentioned legitimacy conflicts. Securing open space for flood protection is generally accepted as legitimate, yet setting a fixed limit to urban sprawl remains controversial and frustrates many stakeholders. The subject was therefore for all intents and purposes kept out of the Isère and Rhône projects in spite of their obvious implications for future land use in regional planning, most probably to avoid having to engage in a broader and risky debate.

Judging by the growing activities developing in the newly opened floodplains in Munich or Nijmegen, flood areas are no longer seen only in terms of technical necessity and risk management, but are actively exploited and enjoyed. The definition of areas for controlled flooding could thus rightly be celebrated as a multi-purpose instrument of flood safety, sustainable urban planning, and landscape development. Flood fields, spillways, by-passes and sleeper dikes should therefore not be hidden and forgotten, but rather become a matter of landscape design and be affirmed as a legitimate and valuable component of a resilient landscape.

REDEFINING RIVERINE AESTHETICS

In the long negotiations surrounding the creation of floodscapes, landscape and urban planners have proven to be useful partners for local authorities and stakeholders, when actively involved in the planning process.

THE NEED FOR IMAGES AND VISIONS

Although hydrology remains the leading discipline in most flood mitigation projects, hydrology alone often appears insufficient for sharing, communicating, and articulating complex interventions. The Dutch tradition of 'research by design' and scenario-based studies, well established since the 1980s (Salewski, 2012), is visible in the development of the Room for the River programme. Sketch plans for the transformation of the Biesbosch were drawn up long before the programme itself was officially launched (see chapter 'From Policy to Proposal: Opening the Scope of Possible Measures'). Furthermore, in this context the elaboration and diffusion of images representing possible measures and landscape transformations functioned as a support for discussion rather than the communication of a finalized solution, and as an entrance for stakeholders to become involved in sensitive negotiations.

As the Isar controversy showed, images brought by landscape architects can have far-reaching consequences—in this case a two-year delay in the initial planning and a radical change in the final design. All the protagonists interviewed for this case, including landscape architects and city and water board representatives stressed the fact that the competition had revealed a latent conflict rather than generated it, and had allowed for an open and necessary discussion: 'People needed to see something to talk about.... You need design as a sort of a benchmark to have any discussion at all.... If people had started with an empty stage, with the thousands of people living around here, there would have been so many ideas, it just wouldn't have fitted together.' The delay was, in their opinion, more the result of confusing communication about pre-existing constraints, confusion exacerbated by political manoeuvres. Yet, 'At one point you just have to try and do a design, and then discuss it' (Engelmayer, 2013).

Rather than images, Valais' environmental associations first asked for the construction of a 'sample' on a section of the Rhône Correction, a

The new Isar banks in Munich, combining urbanity and naturalness (Kuenzel, 2012)

proposal that was rejected by the local authorities. Retrospectively, Pro Natura's director for Valais lamented the lack of a vision that could allow the population to become involved in the definition of the project. 'An engineering project was made, in engineering language with engineering concepts', instead of elaborating 'a political project; that is where a landscape vision would have been interesting: for people to connect on a global project' (Largey, 2013).

Local steering committees were set up with a similar ambition, but were retrospectively denounced as 'a way to divert attention while engineers worked out the project' (Largey, 2011), ultimately asking stakeholders to choose between predefined options. In Sion, however, the combination of urban renewal with the Third Correction, initially the ambition of the city urban planning department, eventually took off after the students from the ETH Zürich visualized the benefits that the city could gain from the transformation of the river. Tactically, the municipal authorities largely communicated the outcome of the studio, by displaying models, plans, sections, and perspectives, for five months in seven locations. The interest generated by the exhibition paved the way for a concrete project, and the city soon thereafter launched a landscape design competition for the project currently being developed.

The Munich and Sion design competitions showed how landscape architecture could bring to light latent expectations, and how plans and projects could trigger an open debate and cultural evolution—turning the distant, dangerous river into a motor of urban renewal and a platform for spatial negotiation, opening up the discussion around the space devoted to the different components of the landscape. In both cases, images and visions appeared to be powerful and necessary tools to develop shared and integrated projects.

SETTING AESTHETIC VALUES

At the end of the twentieth century, the reintroduction of spatial perception, aesthetics, and cultural values into flood prevention still appeared to be a multiple challenge. Could complex and relative variables, such as coherence or beauty, be defined and integrated into planning procedures? Could civil works reconcile these qualities with safety and efficiency? Should massive public means be invested in subjective, cultural productions? Should democratic representatives integrate new cultural values into public works production, or, on the contrary, defend civil engineering from irrational trends and volatile public taste?

With the reintroduction of the spatial dimension within the scope of flood management, (see chapter 'Embedding Spatial Quality into Flood Prevention'), qualitative questions arose together with more subjective criteria, engaging governments and citizens in a complex exercise of cultural redefinition. These issues emerged simultaneously within the scope of changing flood policy in Western European countries, although not in any coherent and legible order: qualitative criteria applied to river management today vary greatly from one country to the next. Key notions emerge, however, from a transnational perspective: freedom, authenticity and naturalness, robustness and openness, coherence and spatial quality.

FREEDOM, AUTHENTICITY, NATURALNESS

In the contemporary discourse surrounding river and floodplain restoration, authenticity is generally associated with the notion of freedom. Not unreminiscent of Rousseau's apologia for free streams and his critique of river channelling (see chapter 'Tame or Set Free: Nature and the Enlightenment'), a common idea suggests that the river should be 'set again', for ecological but also for cultural and aesthetic reasons.

The transformation of channelled rivers into 'free streams' was considered or promoted in a number of the studied projects, particularly in the Alpine context, where the role of dynamic and partly destructive fluctuation in the life of riparian habitats was known and acknowledged. The traditional compensation measures, focusing on the gross area dedicated to nature, were thus completed with a combination of aspects meant to shape a context for alluvial ecology: discharge minima and maxima, flood periodicity, stream velocity, input of alluvial materials, space available for fluctuation. Abandoning the idea of a natural climax or ideal stage, ecological restoration was to be directed towards 'the maintenance of processes, in particular hydrological processes', taking as a basis that 'flows of materials, water, and species must be renewed in order to ensure the sustainability of a landscape' (Girel, 2005: 58).

Paradoxically, the fluctuating nature of alpine streams is perceived both as a past ideal state and as a cause for public disinterest by some decision-makers. For Christiane Thalgott, Munich's former chief architect, the city 'has no urban waterfront because the Isar is a mountain river that brings pebbles, trees, so it was also and still is a bit wild. The water level is changing quite a lot: in the summer you have hardly enough water for anything, it is very low, and in the winter it is rather high. So it is not always a beautiful river' (Thalgott, 2013). A point of view echoed by Michel Pinhas, chairman of the Isère Water Board, about the Isère: 'Alpine rivers aren't full rivers; their level is often very low; waters aren't dirty but they are murky and full of alluvions; they don't add value to the areas they flow through' (Pinhas, 2013). Their water can be scary as well, according to the director of Valais Pro Natura organization: 'People are afraid of the Rhône, and it is frightening when you stand next [to the river banks]', a fear that was most present among parents of young children according to a survey made by the organization (Largey, 2011).

Changing streams and meandering rivers are however today again associated with ecological value and naturalness, although this dynamic character is difficult to translate into measurable goals. In Switzerland, despite the standards set by Pro Natura (Pro Natura, 1998) and by the Federation (OFEFP, et al., 2003), the 'optimal width' of the Rhône river in terms of ecological value remains controversial to this day. Michel Pinhas, dike master in Grenoble and associated to the design for Isère Amont, relates that during the planning process, 'there was a sort of a hesitation at some stage, because we had in mind this same old 'give the river its freedom back' story, and we could believe that by giving the Isère an extra 15 metres along its shores, we could give it back a bit of freedom. That is not freedom. Freedom has to be complete, or else you have only the downside of it' (Pinhas, 2013).

In Switzerland, the designers who were asked to define landscape guidelines for the Third Rhône Correction tried to combine freedom and constraint, envisioning a channel varied in width but bordered by fixed levees, between which 'the river's natural dynamic' would 'create its own landscape' (Veuve et al., 2008: 3), offering the spectacle of a contained but dynamic environment. The vocabulary used in the report often left the scientific and ecological scope: the Rhône would become a 'true life corridor in an increasingly anthropized valley', the 'Valais and Chablais' most beautiful brand image', and a 'true backbone' from which one could 'discover the plain's cultural and naturel assets' (idem).

In Bavaria, a cultural agenda implicitly supported the ecological ambitions within the Isar Plan. One of the main motifs of the plan from the start was the restoration of a lost 'character of Alpine wild river', as stated in the first municipal motion Nature in the City (CSU, 1984). This nostalgic rhetoric remained throughout the whole planning process, which promoted the plan as a restorative rather than transformative or creative enterprise. Yet this approach indirectly led to the rejection of the winning project of the 2003 design competition and its replacement by a more 'nature-like' (*naturnah*) proposal that would 'let the stream meander through the city' (*Süddeutsche Zeitung*, 5.06.2003). This freedom eventually proved unobtainable given the little space left for the river to play with; the meandering eventually had to be simulated by the designers and secured by steady foundations. As the landscape architect recalls, the Isar artificially acquired the image of a wild stream—something it 'never was and never could have been' within the urban constraints (Engelmayer, 2013).

By contrast, the restoration of an authentic natural riverine landscape remains rather absent from the Dutch debate. Dutch landscape architects prefer to stress the 'makeability' of nature. The possibility of actively creating nature areas was integrated into governmental policies as well already in the 1980s (Van Baalen, 1995). For the Room for the River programme, the past was sometimes used as an inspiration but never as a fixed and ideal state that should be reconstructed. More generally, a state of nature could scarcely be found in the Rhine-Meuse Delta, not even on historical maps, since the large-scale reclamation of the Dutch landscape preceded the development of precise topographic mapping. As Dirk Sijmons analyzed it, 'there is no memory and no nostalgia for a past state of nature' in the absence of physical or documented reference in the Netherlands. Besides the soft nature of estuarine grounds, this absence of preindustrial nostalgia can explain the fact that transformative projects were relatively easily accepted in the Netherlands, while restorative rhetoric dominated the Alpine projects.

ROBUSTNESS AND OPENNESS

Wilderness aesthetics and aesthetics of Wilderness in the Isar floodplain: artificial island/riverbanks after the 2013 flood (Künzel/Rossano, 2013)

Robustness is mentioned by several designers as a key value for the design of flood-ready landscapes, as a combination of formal simplicity and physical resistance. A 'robust landscape' being something 'different from a nice waterfront at the edge of the city. You don't need to design everything down to the paving stones'(De Koning, 2013). Some even mourn the times when limited means meant good and plain design: 'Technically, anything is possible [nowadays]. A century ago, technical options were limited, and therefore better integrated in the landscape—although enormous measures were taken... they somehow followed the morphology of the river' (Der Nederlanden, 2013). Promoters of robust river design might object multi-functional levees for similar reasons: levees are linear elements and should be legible in the landscape, rather than disappear under buildings or fade into shallow slopes. In order to combine multiple functions onto one levee, designers 'oversize them to be able to build or plant, turning slim, fitted levees into wide bodies where you can dump anything you like', remarks one of the Rijkswaterstaat experts regretfully about spatial quality (idem). 'Before you know it the river landscape is 'cut into pieces and fragmented into dull little things', some fear (Van der Grift, 2013).

This focus on simplicity was shared by the Q-team of Room for the River, which also wished to prevent the addition of useless elements meant only for beautification: 'In fact, the Q-team tried to safeguard sobriety and simplicity, and we calculated that this eventually saved taxpayers 50 to 60 million euros' (Sijmons, 2015). Experts at the Rijkswaterstaat agency saw in 'meaningful aesthetics' the third aspect attached to spatial quality, a motivation to avoid decorations: 'Meaningful [aesthetics] mainly means that you can explain why you should do it, and why it should look like this, not just as a way to give a nice tint or to cheer up the intervention' (Der Nederlanden, 2013).

Although the criteria weren't that explicit in Munich, the design of the Isar floodplain shared a similar ambition of simplicity and robustness, to cope with violent floods as well as overcrowding on sunny days. It is anticipated that the area might be flooded nearly every year, and the new Isarraum can be restored within a couple of weeks through 'a very low-key [restoration]: repair is basically bringing a bit of gravel somewhere' (Engelmayer, 2013). The hydraulic function of the floodplain implied that no buildings could be erected in the high-water channel, not even the basic amenities that would serve a public space hosting up to 30,000 visitors in a weekend. However, the limitations put on mass recreation are today valued by the inhabitants: 'You can't have events happening, and people like it for that; there is nothing you have to

Restoration of the Isar floodplain after a flood, and old 'enter at your own risk' warning sign (Rossano) • Isar beach reopened to the public (Kuenzel) • Design for a pedestrian promenade along the Rhône in Sion (Competition, Paysagestion 2012)

pay for, no beer gardens.... It's a nice thing to have, such a space that is... empty' (Engelmayer, 2013). This frugality, combining solidity to formal simplicity and economy of means, can thus be valued as liberating and relaxing, in contrast to fully-equipped and commercial urban spaces.

Public accessibility is indeed a crucial and complex issue in the design of floodscapes. Their recreational potential, especially when located near or within dense urban areas, was long identified but left unexploited for safety reasons. Their transformation within the frame of flood mitigation projects and/or ecological restoration has however allowed for a change in mentalities and in practices. Large areas were officially made accessible again to the public in Munich, Basel, Nijmegen, and Lyon. Everywhere, the restoration of floodplains into open and accessible public spaces, combined with softer embankments and slower streams shaped new opportunities for promenade and relaxation; flood protection measures offering an opportunity to bring people closer to water.

The first parameter that motivated openness was in all cases hydraulic efficiency: in order to secure enough storage and discharge capacity to mitigate flood events, floodplains should remain free from obstacles. A strong constraint for landscape architects, and frustrating for environmentalists, was the limitation of silting and vegetation growth—except when planted to protect sensitive structures, such as the Fort Steurgat in the former Noordwaard polder.

Besides hydraulic efficiency, the restoration of accessibility and openness can be motivated by various arguments: mobility, landscape restoration, or the development of a recreation economy, ambitions that are often combined but sometimes contradict each other. Projects of flood mitigation combine a wider accessibility for the public with the development of recreational routes through more attractive and more dynamic landscapes. Civil works on and along existing levees, whether heightened, reinforced, or lowered, allow for the creation of footpaths and cycling paths, or the connection of fragmented urban fabrics. This, however, may contradict ecological goals and generate anxiety among the inhabitants, leading to diversion strategies. Both the Isar Plan and Noordwaard projects tried to control and concentrate the expected flows of visitors.

In Noordwaard, much loved by fishermen and inhabitants for its quietness, public access and recreational amenities were concentrated in two peripheral areas. In Munich, possible conflicts emerged during the first rounds of public consultation: 'A lot of people said "we want true wilderness here in the city, and no one should go there".... Other people wanted to have parties all the time there, with mobile discos systems' (Engelmayer, 2013). Today the large crowds using the central areas clearly hinder the development of alluvial vegetation, but one should see 'ecological value in a different way: [The central meadow] keeps people here, and the areas upstream with more ecological value are not as heavily used as in the past' (idem). In both cases, accessibility to river and floodplain was thus not blocked but rather channelled in order to concentrate recreational facilities, motorized vehicles, and large groups, considered to bring more damage and trouble to plants, animals, and residents.

Besides physical accessibility, visual openness is often also a prerequisite in the elaboration of rivers and floodplain restoration projects. They should not only be usable; they should also be visible in metropolitan landscapes. In urbanized regions, floodplains were already appreciated for their visual openness in contrast to the built environment. The prerequisite of visual openness was thus a strong one, although often running contrary to the criteria of authenticity and ecological value. River landscapes are today visually open thanks to centuries of channelling, clearings and drainage, concreting of riverbanks that transformed these dense environments into open landscapes. Many, for instance, associated the Isère River with an open landscape, mistakenly considering this visual openness as a natural condition (Girel, 2013). In Munich, the Isar Competition brief required the competitor to design an open landscape and restore alluvial vegetation, two obviously contradictory prerequisites. In both Isère and Isar valleys, the designers had to make local choices, alternating local clearings so as to provide windows onto the horizon with new plantations to reinforce ecological value.

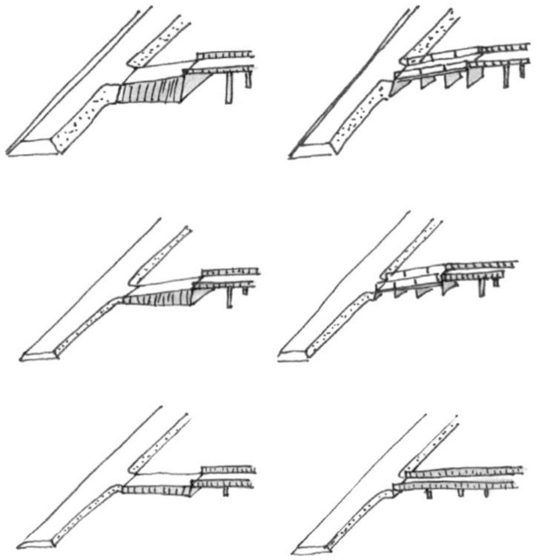

Noordwaard design guidelines for refuge platform and bridge over the creeks (De Koning, 2010: 47, 59). Design guidelines for the profile of future levees along the Rhône (Veuve et al., 2008)

COHERENCE AND SPATIAL QUALITY, A PLANNER'S SWISS ARMY KNIFE

The notions of coherence and spatial quality, although difficult to quantify, can set a general ambition for the project as a whole by its integration in the context and the coherence of its components. The need for coherence was affirmed for the complex programmes that involved multiple disciplines and numerous administrations, and concerned not only the rivers but also all components of the territory. Furthermore, the adaptation of rivers is often seen from the start as an opportunity to bring coherence to landscapes fragmented by infrastructures and urban development.

The external team hired to define guidelines for the landscape development of the Rhône and the plain reaffirmed that 'the transformation of the landscape of the Rhône [river] would go along with the transformation of the valley. This is an obvious demand of coherence' (Veuve et al., 2008: 4). Unfortunately, in the 2008 Project Synthesis, the 'global coherence' criterion was given a weighting of 1.5 percent in the evaluation process (Projet Rhône, 2008: 46). In the Netherlands, by contrast, spatial coherence was used early on as a criterion to evaluate the various measures of river widening and flood mitigation, and as a tool to assemble and articulate possible combinations into a global vision. A substantial part of the 2006 governmental Room for the River Key Decision was dedicated to spatial quality, and presented three possible strategies to connect and relate the proposed measures at the scale of the greater River Region: concentration of landscape measures around the cities; creation of flood corridors through rural areas; continuous widening of the large rivers (see chapter 'From Policy to Proposal: Opening the Scope of Possible Measures'). Typically, the Dutch law then decomposed and recomposed the three options into an optimal combination of 'urban developments, the development of robust nature core areas, and recreational possibilities' (Projectorganisatie Ruimte voor de Rivier, 2006, vol.4, Nota: 27). Paradoxically, coherence was thus sought in the hybridization of three different concepts.

Another tool used to reach coherent results are design guidelines, used as an intermediary step between spatial concept and landscape design. They provide a framework for the design, give information on profile measures and land use but also formal and aesthetic recommendations (for the design of the levees, embankments, bridges, paths, confluences). Although design guidelines can answer the need for coherence at an early stage, their efficiency is relative. If they are too precise, they uniform the design, if too vague, they miss their target. Their

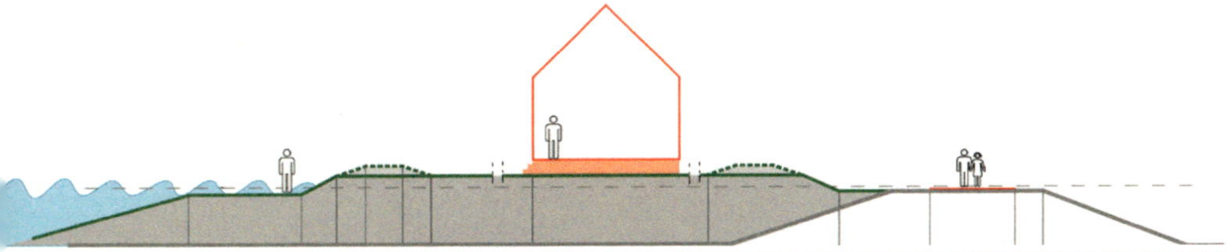

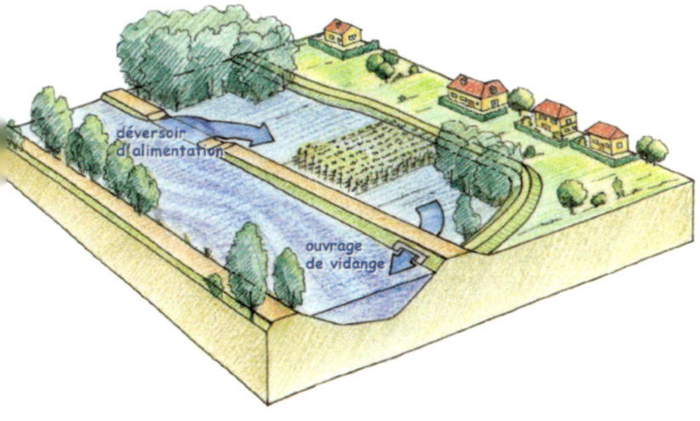

Diagrammatic cross-section of elevated platforms for the de-poldering of Noordwaard (De Koning, 2010) • Drawing of the future 'terpen' in the Overdiepse Polder, in normal and flood situations (Rudolf Das, 2008) • Schematic representation of a 'Controlled flood field' (CIC) along the Isère (SYMBHI, 2009)

effectiveness can also be weakened by long discussions and slow planning, as the original meaning might be lost and the contact between prescribers and designers becomes difficult (twelve years have passed between the publication of the guidelines for the Rhône embankments and the selection of a design team). Obviously, a shorter design process and a stable design team make guidelines more effective.

Spatial quality, less known as a general goal and a criterion for the evaluation of possible measures, became a key notion in Dutch planning policy in the last decades of the twentieth century (Reindorp et al., 1998), as well as within the departments in charge of water management and flood defence (see chapter 'Embedding Spatial Quality into Flood Prevention'). The 1998 Fourth Bill for Water Management stated that, for large projects of river widening, 'the promotion of spatial quality' was at least 'as important as legal and consultation instruments' in the development of large projects such as the Room for the River programme. The bill later highlighted the economic potential of spatial quality when integrated within water management together with ecological measures: 'Spatial quality can be an important economic factor, among other things, as a positive reason for business or inhabitants to settle' (Min. van Verkeer en Waterstaat et al., 1998: 107, 117). During the preliminary studies made for the Room for the River programme, spatial quality was increasingly used as a criterion to evaluate possible measures (RIZA and Bosch & Slabbers, 2000). It was eventually incorporated in the Governmental Guideline Room for the River in 2000, and became officially the second main goal of the Room for the River Programme, next to flood safety.

Spatial quality not only referred to the perception of a three-dimensional environment but was also developed as a catch-all tool of supervision. The Key Decision Room for the River presented it as a threefold concept that encompassed use, experience, and future value. Interestingly, the governmental document didn't provide a fixed definition of the notion but rather specified that 'the practical application of the spatial quality criterion is defined by the parties involved'. An example of this customization is the interpretation made by the architect of the Noordwaard project, who added a narrative dimension to spatial quality, insisting on the necessity for the designer to get to know the site and its past (De Koning, 2013), adding a temporal dimension to 'meaningful aesthetics'.

For Dirk Sijmons, first chairman of the programme's Q-Team, 'within the Room for the River programme, spatial quality included three dimensions: hydrological efficiency, robust ecology, and meaningful aesthetics', leaving room for negotiation and avoiding a focus on one isolated aspect of the living environment. Ecology was, for instance, present but 'hidden within spatial quality'. The Q-team was itself interdisciplinary: It included experts in hydrology, river management, ecology, urban planning, and landscape. For the water and civil engineering agency, it was important to rely on the definition set by the Q-Team in order to 'strive for coherence ..., and base your choices on integral considerations' (Der Nederlanden, 2013).

In the practice, this flexible concept allowed the various groups to stress different aspects: the environmental assessment reports summed it up in 'landscape and nature assets', municipalities stressed recreational possibilities, while the inhabitants linked the concept to the quality of their immediate living environment (Heems and Kothuis, 2012: 276). The polysemy of spatial quality could thus lead to frustration and conflict, and had to be reaffirmed and clarified by all parties concerned; but then again, this negotiation process was part of the strategy chosen by the national programme and its Q-team.

FROM WATERCOLOURS TO POINT CLOUDS: FLOODSCAPES EVOLVING DESIGN TOOLS

Water, terrain and natural dynamics form a shared culture for landscape architects. The representation of floodscape design however raises special problems, associated with the multiplicity of stakeholders, scales and temporalities of the projects. Visual communication, produced by professionals from various disciplines, each focusing on different aspects of the design, shows enormous differences in style, sophistication, and computerization. The articulation of design scales is also a crucial issue: Although the natural scale of flood management is intrinsically regional, with measures involving hectares or kilometres, its motivations are found

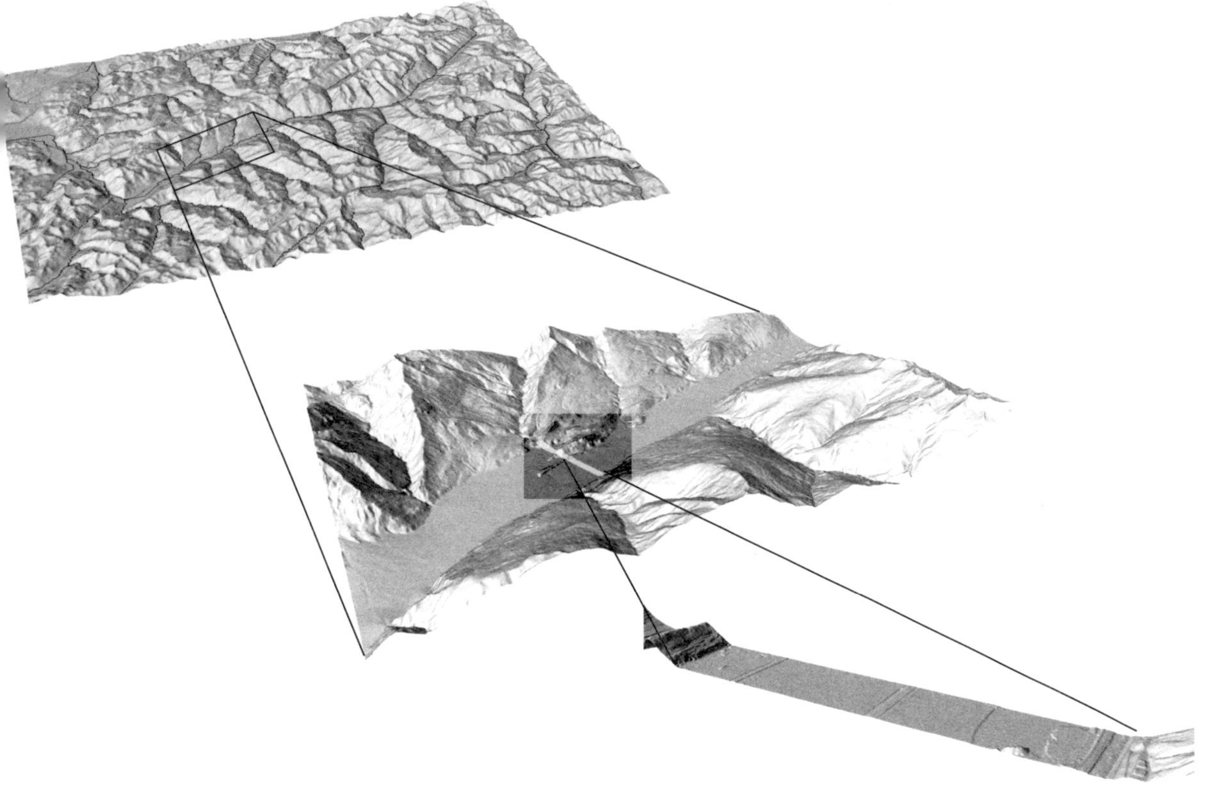

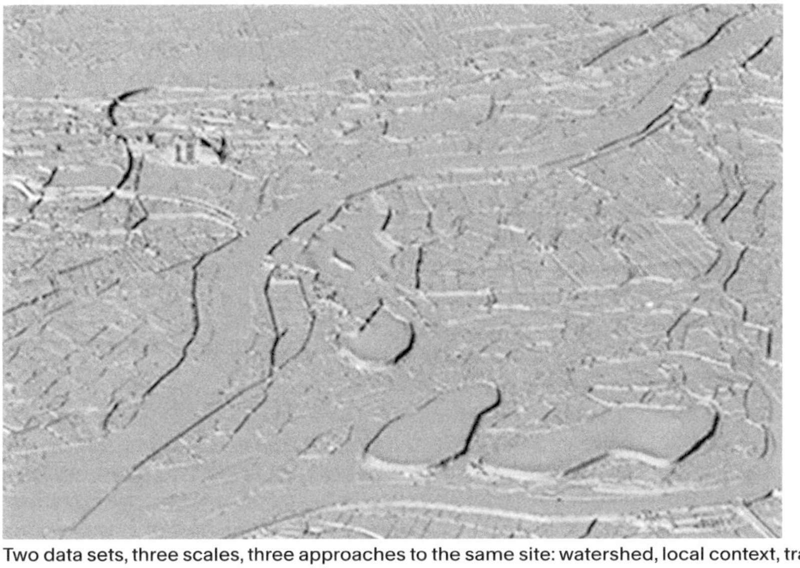

Two data sets, three scales, three approaches to the same site: watershed, local context, transect • The micro-topography of the Biesbosch revealed by a factor 10 on vertical axis

at the larger scale (state, watershed), while concrete measures take place at the smaller scale (city, district). It is yet crucial that all scales remain within the scope of the design work and influence each other during the whole planning process.

Representing, modelling and transforming the terrain elevation is another key aspect of water management, as it determines the course of water, the impact of rising levels and the vulnerability of an area. Fortunately, subtle irregularities can today be registered by high-precision terrain models, that can in their turn be manipulated and exaggerated to reveal the substratum of landscape design. Digital terrain models not only allow for modelling and visualization, they are also a source of precious information, often available but still rarely used by designers. Many elements of design can be retrieved through the use of GIS processing programs—such as the identification of potential flood areas, runoff patterns, slopes, and hydraulic compartments.

Although scenario-based design existed long before digital tools were developed, these tools enable designers to sketch multiple strategies based on the different variables, create numerous scenarios where drawing by hand would imply long and intensive labour, and make it possible for hydraulics engineers to evaluate the impact of possible interventions. This possibility was extensively used by the Dutch authorities during the Room for the River programme: Various options were translated into hydraulic and topographic digital models, so that open discussions could take place instead of reducing the design and decision process to a single channel.

With the help of historical maps, digital terrain models and hydraulic models, engineers and landscape architects can now develop a precise understanding of a site's past states, morphology, and dynamics. They can compare historical variations with present states and future options. They can easily visualize and put into context projects of impressive size, cost, duration, and complexity. They can integrate long and partly uncontrolled natural processes, and check their ideas against past and present dynamics, against regular and exceptional flood levels. Finally, they can communicate wider, and share the questions raised and the knowledge gained during the design process with a wider public. Combined together, hand-made sketches, watercolours, and digital modelling equally participate to a better embedding of floodscape design, and to a richer discussion needed to reach a broader consensus.

5

EPILOGUE: ADAPTING TO CHANGING TIMES

Kamagurka, *High Waters in Middelkerke*, 1987

> 'Nothing seems more difficult, not to say impossible, than to oppose the progressive cooling of the Earth and to raise the temperature of a climate; however, man can do this and has done this.' (Buffon, 1778)

More than two and a half century ago, naturalist Buffon praised the power of the human species, capable of warming the climate of entire regions (Buffon, 2018 ed.: 126). Studying Earth's climate, Buffon mentioned the then dominant climate cooling hypothesis, which he considered as correct but balanced by the human activities that warmed up populated regions. Basing himself on the observation that two cities located at the same latitude, namely Paris and Quebec, showed a difference in temperature relative to their difference in population, Buffon concluded that more human activity generated more heat, while forests had a cooling effect. 'To cleanse [or to drain, 'assainir' in the original French text], to reclaim and people a country, is to provide it with warmth for many thousands of years' (idem), he wrote in a striking statement. However, he added a visionary warning to his discovery: 'Now the power of man does not extend so far. He cannot make the cold descend like he can make heat rise.... But it is much easier to fell the forests in Guyana to warm the humid earth than to plant them in Arabia to refresh the arid sands' (id.: 129). Buffon remained a thinker of his time however, celebrating rather than lamenting the ability of humankind to exploit and transform nature in order to make it even more productive and hospitable.

PANIC AND CIRCUMSPECTION

Two hundred and forty years later, the conquering optimism of the Enlightenment has dramatically faded. Global warming has become a global concern, focusing the attention of scientists, governments, and increasingly worried citizens. The fight against climate change and against the inertia of political and economic powers is intensifying all around the world.

This global concern finds its echo in the media, particularly in movies which tend to overdramatize flood phenomena, and numerous films have exploited the fear of a new Flood since the turn of the century.

When information meets fiction. Cover of the German news magazine Der Spiegel (1986) • Imaginary view of Utrecht's Dome church in the Dutch news magazine Vrij Nederland (11 February 2019) • Icelandic poster for the movie '2012' featuring the Hallgrímskirkja church in Reykjavik (R. Emmerich, 2009)

A Japanese movie *Japan Sinks* (Shingi Higuchi, 2006) and an American one *2012* (Roland Emmerich, 2009), have exploited the theme of massive inundation, using very similar narratives. The overwhelming scale and the nature of the dangers depicted by both films did not shy away from reminding the viewer of the representations of Noah's Flood, combined with the real challenges raised by global warming, but both films remained silent on the phenomenon itself. By staging flood as a natural hazard, and heroes fighting back with a mix of technology and heroic action, they again opposed man's constructive rationality to nature's destructive instability, just like the thinkers of past centuries. Since the turn of the century, the 'natural disaster movie' flourishes on the fear of predicted floods ('We were warned', proclaimed the subtitle of the film *2012*), on the hopes of individual or collective salvation (through courage and love or through technology and innovation).

Sub-genres are in turn emerging, with various films inspired by the tsunami that have devastated the Asian coasts (*Tsunami*, a German TV film directed by Winfried Oelsner in 2005, *Deathwave*, a Thai film by Toranong Srichua (2009), *Tidal Wave*, a South Korean film by Youn Je-kyun (2009), *The Wave*, a Norwegian film by Roar Uthaug (2015)). Another diluvian subgenre is inspired by the great Atlantic storms: the North American films *Geostorm* by Dean Devlin (2017), or *Crawl* by Alexandre Aja (2019), among others 'superstorms' and television 'stormageddons'. Even John Rambo, the indefatigable hero who fought the Viet Cong, the Soviet army, the Burmese junta and his own government, had to face a torrential flood in the opening scene of the last and final episode of his saga *Rambo: Last Blood* (Adrian Grunberg, 2019). From a technical, if not bureaucratic subject, flooding has become a fertile and lucrative narrative engine, a source of intense and shared emotions.

More surprisingly, news media often cover the issue with a similar dramatic tension. Germany's largest news magazine, *Der Spiegel*, announced a 'Klima-Katastrophe' already in 1986, with a drawing of the Kölner Dom emerging from an endless ocean, therefore speculating on a sea level rise of at least fifty metres (Der Spiegel, 1986). The Dutch magazine

Heads of State inspect the areas devastated by floods: French Emperor Napoleon III (1856) • Queen Juliana and Queen Beatrix of the Netherlands (1953 and 1998) • U.S. Presidents G. W. Bush and B. Obama (2005 and 2012) • French President F. Hollande (2013) • German Chancellor A. Merkel (2013) • Charles, Prince of Wales (2014)

Vrij Nederland echoed *Der Spiegel* thirty years later with an article lamenting the lack of a 'plan B' in a country threatened by partial submersion (Vrij Nederland, 2019). This article, published in 2019, was illustrated with a similar view of the iconic Dom church of Utrecht, emerging from a vast, quiet sea. Frivolous and striking, both images match the poster made for the abovementioned movie *2012*, that showed the Hallgrímskirkja Church in Reykjavik thrown into the ocean by a gigantic wave. News media and works of fiction here used the same dramatization, evoking actual flood processes through images of an absolute and final Flood, that would not even spare national and religious icons. All three images actually use the same temporal reduction, as if slow, progressive and partially predictable water dynamics could only be grasped through an immediate and compressed version.

Information channels too tend to stress the devastating effects of floods and their emotional dimension, often with the active participation of elected representatives. Omnipresent, mighty, and paralyzing, more than any other catastrophe flood calls for cohesion in suffering and solidarity in reconstruction, the only remedies against the millenary ghosts of divine punishment and total annihilation. Unlike other natural catastrophes, flood is a lasting event, and its duration is as dangerous as its violence. Its aftermath calls for patient inventory and remedial work that leaders should experience on site, and not just supervise from a distance. With each flood, authorities are therefore expected to play their part, first of all in the realm of emotion. Carefully staged compassionate journeys remain a central element in information narratives, from Emperor Napoleon III inspecting the flooded Rhône Valley on a boat in 1856, to Prince Charles visiting the devastated Somerset Levels in 2014. The official visit not only shows that central authorities empathize with the fate of the victims, they also give politicians the opportunity to show their personal qualities through demonstrations of concern and compassion. In an article entitled 'From Schmidt to Merkel—politicians in high waters', German daily newspaper *Bild* attributed G. Schröder's re-election in 2002 to his rapid presence on the locations flooded by the Elbe river during the campaign (*Bild*, 05.06.2013). By contrast, George W. Bush's failure to engage physically, as he comfortably flew over the devastated city of New Orleans in August 2005, was widely commented upon by the US media and used by his political opponents to contest his credibility as head of state—more efficiently than criticism over his actual response to climate change. Donald Trump in his turn was widely mocked and criticized when he wrongly announced that hurricane Dorian would cross

Alabama, contradicting federal agencies, his critiques blaming both the president's lack of knowledge and lack of interest.

Since G.W. Bush's tactical mistake, no leader has skipped a dutiful visit to a damaged area. Still, although the compassionate visit is a requirement everywhere, it also reveals cultural and political specificities. While President Barack Obama was photographed warmly hugging a victim of hurricane Sandy in 2012, Chancellor Angela Merkel kept a dignified distance during her visit to Passau in 2013, avoiding any drama: 'The Chancellor deliberately avoided exaggerations when visiting Passau: instead of rubber boots and oilskins, she was wearing solid shoes and a robust blue jacket', reported *Bild* in the same article.

Mixing sentimentalism and dramatization, fiction and information media both tend to stress the devastating effects of flood, and their spectacular and exceptional nature, rather than their inscription into a predictable chain of events and actions. In this contradictory context, encouraging panic and circumspection if not a wait-and-see attitude, public communication around flood prevention and anticipation has become particularly delicate. Raising awareness of flood risk appears difficult even in Holland, as 'studying the debate on flood safety at the beginning of the 21st century made clear that most citizens showed little interest in becoming informed about flood safety nor in changing their behaviour' (Heems and Kothuis, 2012: 448). It remains difficult to communicate on a phenomenon that is both discreet (seas rising at a speed of a few millimetres a year) and spectacular (more violent storms and floods), a phenomenon both ancient (the natural variability of climate and river regimes) and current (accelerated global warming), progressive in its development but immediate in its particular manifestations. This last opposition is particularly significant for flood management and adaptation of rivers to climate change.

The difficulty of grasping the planet's slow changes isn't new in itself: Aristotle noted already that 'the whole vital process of the earth takes place so gradually and in periods of time which are so immense compared to the length of our life, that these changes are not observed, and before their course can be recorded from beginning to end whole nations perish and are destroyed' (Aristotle, 1931, I, 14). In ancient Mesopotamian, Greek, Hebrew, or Christian traditions, the Flood myth remained a short and a singular episode, however, hiding the reality of long-term geological processes and hindering a continuous vision of the earth history—a history that was thought to be only a few millennia long until recently. The main difficulty for river adaptation projects lies precisely in the gap that separates the different temporalities of reflection and action that condition these projects: the history of climate and its long episodes, measured in millennia for the Holocene epoch; the centuries of accelerated global warming (whether called as Buffon did the 'Seventh Epoch', When the Power of Man Has Assisted That of Nature, or currently the 'Anthropocene'); the time of a human life, which determines the perception of risk and the acceptability of planned measures; finally, political time, which determines the scale of these measures and the continuity of investments.

CLIMATE UNCERTAINTY, A CONSTRAINT AND A CHALLENGE FOR DESIGNERS

This necessary synchronization of Earth, water, and human life time is made even more complex by an accelerated climate change, yet whose future development remains largely unpredictable.

For the twentieth century, the Fifth IPCC reports showed a significant rise of sea levels, a temperature increase in the whole of Europe, with a clear increase in precipitation in North-western Europe and a decrease around the Mediterranean basin (IPCC, 2014: 41), aggravating risks of river floods and the devastating effects of flash floods.

Upstream of the great European rivers, Switzerland expects temperatures to rise between 0.5 to 3.6 degrees Celsius in 2060, much more than the predicted global values. The Alpine regions should see their temperatures rise even more than the national average (Meteosuisse, 2014). With less snow stored in upper altitudes, more frequent heat waves and more frequent torrential rains, Alpine rivers should be under stress, and their regime should evolve significantly.

Downstream, the forecast for the Rhine estuary is worrying as well. In its 2014 report, the Dutch Royal Meteorological Institute KNMI drew various climate scenarios, forecasting a sea level rise of 20 to 90 centimetres in 2100 (KNMI, 2014), while by 2300 the sea level could be fifty centimetres to several metres higher than it is today. The same report forecasts more frequent extreme rainfalls in summer and winter, while the volume of precipitation should increase by 10 to 20 percent on the Dutch estuary (EEA, 2017: 81). If greenhouse gases emissions are not significantly reduced, the recent IPCC report on Oceans and Cryosphere forecasts a sea level rise of half a metre up to five metres in 2300 (IPCC, 2019), a crucial difference for the population of the estuary, living for the most part below the current sea level. On the basis of clear trends in the medium term but with crucial uncertainties beyond the end of the century, how to conceive and plan regional response and adaptation measures? Should one panic, worry, or rejoice? The coverage of the projections published by the KNMI in 2014 illustrated the country's ambivalence: temperatures were expected to rise by 1 to 2.3 °C around 2050, possibly by 3.7 °C in 2085, and Dutch winters were expected to become warmer and stormier, the report stated. The online summary illustrated these projections with a comparison of future Dutch winters to those now seen in Bordeaux (KNMI, 26.05.2014). But while the original article was entitled 'More extreme rains caused by global warming', the warning was echoed by ironically cheerful titles in national newspapers ('Dutch winters will look like Bordeaux's winters', *Volkskrant*, 26.05.14, *NRC*, 26.05.2014), illustrated for the *NRC* with an image of a joyful, crowded, and sunny sidewalk café. Regarding the country's cold, grey and humid winters, this seemed to announce free and unlimited holidays for all.

In his *Cultural History of Climate*, Wolfgang Behringer pleaded for more relativism towards an ongoing change that had until then been no more extreme than the previous climate episodes survived by humankind. Behringer wrote: 'The Neolithic revolution and the rise of ancient civilization became possible in periods when it was somewhat warmer than it is today. If the IPCC's latest predictions are accurate, those levels will be reached again somewhere in the twenty-first century' (Behringer, 2010: 216). As the historian noted, past glaciations and deglaciations had been a blessing for some regions, a drama for others, and he pointed out that the Sahara used to be a rich and fertile region before it turned into a desert.

This relativism could be linked to the statement made by Prof. Richard Tol, who left the IPCC in 2014, accusing other authors of systematically anticipating the worst, and missing the chance to stress that 'through adaptation and clever development these were manageable risks'. (McGrath, BBC, 25.03.2014; Volkskrant, 1.04.2014). This relativist and more positive approach is also promoted by the Global Commission on Adaptation chaired by Ban Ki-moon, Bill Gates and Kristalina Georgieva, a commission initiated in 2018 by the Netherlands to 'share its knowledge on how it has managed to adopt innovative water management solutions as sea levels rise' (GCA, 2019).

Humankind did indeed survive various episodes of glaciation and warming, which each time drastically transformed our living environment, starting with the layout of coastlines and rivers. The current transition, however, differs from the previous in key aspects: the size, the sprawl and the fixity of today's human settlement, which make adaptation measures much heavier, slower, and more complex. Compared to the hundreds or thousands of Scandinavians who escaped a cooling Greenland, today's global warming could affect billions of people living in regions threatened by sea level rise, river flood or desertification. Contrary to Neolithic settlements, contemporary cities are nor transportable, nor compostable nor recyclable, and adaptation measures should be found on the spot, within existing infrastructures that were built for stable conditions. Conversely, human resources and knowledge are now much greater than in previous transitions, and adaptation measures, although necessarily local, can benefit from an unprecedented pool of experience and knowledge.

In terms of water management, the disparity of global warming scenarios beyond 2050, and the uncertainty concerning their local effects, neither 'lasting' solutions nor 'unsinkable' defence structures to combat yet unknown discharge levels are conceivable. The opposite approach, 'set waters free' and wait for a hypothetical natural balance, cannot ensure a credible flood protection for densely exploited and inhabited regions either. Lacking measurable objectives, adaptation policies can only focus on restoring margins for future fluctuations. This makes the shift from fixed flood-containment measures to flexible spatial strategies even more legitimate: more than sustainability (in the sense of a lasting state), elasticity should be ambitioned today for the development of river regions and estuaries.

FROM FULL SAFETY TO CONTROLLED DISASTER: THE NECESSITY OF FEAR

Today, high waters are predictable thanks to a complex networks of information providers, from satellite pictures to physical inspections still widely used. Imminent danger will eventually lead to preventive action such as targeted messaging, the removal of potential obstacles from the riverbed, or gradual access restrictions to the river dikes, shores, and floodplains. Serious threats may also activate mitigation measures, such as the controlled flooding of predefined areas, in order to lower risks in the most sensitive areas, sometimes located kilometres away.

These costly and complex flood management systems, set at a territorial scale, contrast sharply with the lack of actual and practical anticipation that generally applies to local planning authorities as well as individuals. Flood defence infrastructure plays an ambiguous role here, being both protection and entrapment, as it often encourages risky behaviour in the protective shadow of the levee.

The case of the Netherlands water policy is particularly revealing in that regard. Planning policies and regulations have long downsized the question to a binary issue: one lives 'within-the-dike' or 'outside-the-dike'. Within the dike ring, the safety level is defined by law, guaranteed by the state, and provided for by its powerful water administration. The concentration of knowledge and responsibility within one specialized authority leads to a paradoxical result: are the Netherlands suffering from excessive self-confidence due to their excellent flood defence system? The construction of the new town Westergouwe in a polder located six metres below sea-level, close to a branch of the Rhine river, has recently raised controversy about the level of risk awareness found among developers and local authorities, as well as doubts about the sustainability of neighbourhoods built in the Green Heart region, located between Rotterdam, Utrecht and Amsterdam. The consequences this denial has for spatial planning in the Netherlands was recently confirmed by a comparative study held on the localization of sensitive facilities in England and the Netherlands (Van den Hurk et al., 2012). It showed that in the specific case of a regional hospital, exposure to flood played a minor role in the Dutch planning process, in contrast to its English equivalent, and despite the Dutch compulsory 'water test', that requires the recommendation of the water board for new building plans since 2003.

A form of collective denial seems to be at work, possibly created by cognitive dissonance—after all, would one rather be constantly reminded of the fact that one lives, works, and invests in a place exposed to destruction by flood, or rather trust blindly in world-famous technological know-how? Another factor in this denial is the absence of any alternative. When the short-term growth of a city, a region, or even a country, in the case of the Netherlands, is in direct contradiction with its long-term safety, the former might prevail over the latter. As Alfred van Hall, former Dike Master of the Hunze-en-Aa's region in the north of the Netherlands, pointed out about the Meerstad project: 'It can happen, and it happened here, that you build on the worst spot for reasons other than water-technical ones', reasons that Harm Küpers, Director of the same Water Board, expressed in a few words in the same conversation: 'We are in one of the world's most densely populated regions. If we build only where it's reasonable in terms of hydrology, we simply couldn't fit all the people into the Netherlands.'

This perception of exposure as a forced choice can be seen in the past development of large European cities such as Bordeaux in France, or Hamburg in Germany, which planned and built massive urban

Island in the Flaucher area, Munich (2013, June 24 and September 7, Rossano/Künzel)

developments within known flood zones, having no 'safe' ground left to offer to developers and being bound to choose between risky growth or no growth at all. Flood events have a limited effect on risk awareness, if at all, as a Swiss study showed for Valais, reporting that 'practical effects onto land value and land management practices are relatively weak' (Barbisch et al., 2006: 6).

The lack of awareness or the downplaying of flood risks equally affect the reaction capacity. Holland here again combines exposure and denial at a high level. An investigation published in 2011 by the magazine *HP de Tijd* revealed the total lack of preparation of the Randstad, even though the then Prince of Orange and now King Willem-Alexander had pointed out, on several occasions, the enormous damage to be expected in case of dike breach. 'A breach in dike ring 14 [which protects most of Amsterdam, Rotterdam and the Hague] could today cause 37 billion euros in damage' (Prince Willem-Alexander, quoted in *HP de Tijd*, 2012). 'And with a lead time of eight to ten hours, it is out of the question that all 4.5 million inhabitants of this dike ring can be evacuated in an orderly way' (ibid.). Bas Kolen, from the consulting firm HKV, explained this situation as an excess of confidence: 'A specific, prepared planning for the evacuation of dike ring 14 is currently lacking', he explained. 'The confidence placed in water managers and in the quality of the Dutch dikes has contributed to the fact that the preparation of the Netherlands in terms of aid has received little attention' (id.). And so, thanks to the efficiency of the Dutch flood protection system, flood has become a foremost technological issue to be dealt with at the regional or national level, and flood risk is today as invisible as the rivers themselves, hidden behind massive levees.

FLOODSCAPES AS REALITY CHECK

Navigating between mobilization and a reassuring discourse, official communication and policies must today create space for anticipation, adaptation, and mitigation, rather than celebrate ambition or a pseudo-permanent safety. Rather than combative ardour, cautious care and measured adjustments are required to adapt our coasts and rivers.

Significantly, evacuation exercises are no longer taboo. In the Swiss Rhône Valley, the canton's capital informs its inhabitants about the closest dry shelter reachable in case of a dam-breach alert. Some Dutch municipalities are developing disaster management strategies, and the local Red Cross has since 2015 become involved in exercises for flood rescue using a network of civilian volunteers. In case of high water, the municipality of Munich has set up a seven-step scheme of action in order to protect the population living along the Isar. Numerous examples show that the adaptation of spaces and minds go together, and follow a path closer to citizens reactivity and self-organization capacity, but also closer to their desires regarding the accessibility and public use of water courses.

As the rivers 'portrayed' in this book show, the transformation of riverine landscapes is a highly specific enterprise, historically, geographically, and culturally, and it remains dependent on economic and political ups and downs. Yet the pioneering work done in the past twenty years can be studied and compared to reflect on the modus operandi that has led to successful transformations. They allow us to build combinatory and consensual adaptation strategies, and they encourage stakeholders and architects of regional planning to play an active role in this historical enterprise.

More than just an answer to flood risks, these new floodscapes participate in the repositioning of natural processes within the living environment, and therefore within the field of spatial design, with respect for the habitats shaped by our predecessors over time. Creating a space that emphasizes fluctuation rather than neutralizing all its effects means a radical breakthrough, with regard to coercive flood defence that has turned most European rivers into drainage channels, hiding risks and flood processes behind ever-higher levees. These new 'elastic landscapes' reintroduce physicality and positivism into environments that have been literally drained by intensive exploitation and petrified by urbanization. They form instead a permanent display of natural variation, offering all citizens the experience of change, stimulating observation, concern, and anticipation. Witnesses to a flood episode along the Isar, Meuse, Rhône, Rhine or Isère Rivers can now observe the river's seasonal changes, both affecting and protecting their daily environment, and giving meaning and presence to the term 'floodplain'. Floodscapes serve as a reality check, more efficiently than communication campaigns, dramatized news, or apocalyptic fiction. In that regard, floodscapes not only mitigate the direct effects of the flood but, by making it visible, tangible, and acceptable, they also help foster stronger resilience on the part of individuals, communities, and territories.

Like yesterday's massive dams, today's floodscapes are the expression of a new cosmogony, which sees a logic of negotiated cohabitation with natural processes replace the old narratives of heroic fight, hypothetical conquest, and absolute safety. Technology, in particular hydro-engineering, is no longer perceived as the final remedy against natural threats, but is used as one ingredient in the delicate balance that needs to be found, in each specific situation, between natural fluctuations and human needs. From that perspective, floodscapes are a collective achievement: they require a consensus in terms of risk redistribution, a joint and lasting effort in terms of investment, design, and care. They are also, as the examples presented show, the shared success of supra-national, national and local institutions, which were able to articulate the various scales of reflection and intervention.

Judging by the first landscapes transformed since the turn of the millennium, which are now a source of curiosity and pride, all relevant actors in the territories concerned are ready and willing to play an active role in this historical turn, and together search for combinatory benefits. There is every reason to believe that many more floodscapes will emerge in the coming decades, as necessary adaptation measures and as positive and flexible answers to an uncertain future. A clever mix of Mesopotamian fatalism, biblical fear, Enlightenment optimism and Frisian adaptability could still inspire new ideas and agile projects, in order to carry out an inevitable and desirable transition.

BIBLIOGRAPHY

Quotes in the text originally written and published in other languages than English were translated by the author.

Quotes from the Bible are retrieved from the New International Version (2008–2011), retrieved from https://www.biblegateway.com.

Quotes from the Koran are retrieved from The Holy Qur'an (1983), trans. Shakir, Tahrike Tarsile Qur'an, Inc., text retrieved from https://quod.lib.umich.edu/k/koran/.

1 GOD, THE RIVER AND THE ENGINEER

Agard, Maurice (1942), 'L'endiguement de l'Isère en Grésivaudan', *Revue de Géographie Alpine*, 30 (4), 701–71.

Alexander, David (1984), 'The Reclamation of Val-di-Chiana (Tuscany)', *Annals of the Association of American Geographers*, 74 (4), 527–50.

Allard, Paul and Labeur, Christine (2009), 'Évolution des vulnérabilités des riverains du Bas-Rhône (France) du XVIIIe à nos jours', Le Rhône: Dynamique, Histoire et Société (*Cahiers de Vallesia* 21; Sion: Archives de l'Etat du Valais), 177–96.

Apollodorus (1921), *The Library*, Trans. James George Frazer (Cambridge, MA: Harvard University Press; London: William Heinemann).

Apollonius Rhodius, *The Argonautica* (1919), trans. R. C. Seaton (London: William Heinemann, New York: G. P. Putnam's Sons).

Aristotle (1931 ed.), *Meteorology*, trans. E. W. Webster (London: Oxford Unversity Press), retrieved from http://classics.mit.edu.

Attinger, Victor, Godet, Marcel, and Turler, Henri (1930), *Dictionnaire historique et biographique de la Suisse* (Neuchatel: Administration du Dictionnaire historique et biographique de la Suisse).

Bender, Gabriel (2001), 'Où est passée la Camargue valaisanne?', in Pierre Dubuis (ed.), *La mémoire dans la vie* (Sion), 87–103.

Bender, Gabriel (2004), 'Corriger le Rhône et les Valaisans: trois siècles de travaux et de débats', *Revue de Géographie Alpine*, 92 (92–3), 51–61.

Behringer, Wolfgang (2010), *A Cultural History of Climate* (Cambridge: Polity Press).

Bink, Kees, et al. (2011), *De Overdiepse Polder. 750 jaar geschiedenis van een polder in de delta* (Waspik: Heemkundekring Op't Goede Spoor).

Blackbourn, David (2000), 'Conquests from Barbarism: Interpreting Land Reclamation in 18th Century Prussia', International Congress of Historical Sciences (Oslo).

Blackbourn, David (2007), *The Conquest of Nature. Water, Landscape and the Making of Modern Germany* (London: WW Norton & Co).

Blanc, François, known as La Goutte (1733), Grenoblo malhérou (Grenoble: René Faure, ed. 1733; Dardelet, reed. 1860 illustrated by D. Rahoult and engraved by E. Dardelet).

Boersema, J. J. (2011), 'Deugt de Natuur ?', in Bettine Siertsema and Bert Musschenga (eds), *Het kwaad. Reflectie op de zwarte zijde van ons bestaan* (Vught: Skandalon).

Boersma, J. W., et al. (1972), Terpen, mens en milieu, ed. J. W. Boersma (Haren: Knoop & Niemeijer).

Bonaparte, Louis Napoleon (1856), 'Lettre de l'Empereur à son Ministre des Travaux Publics', (2S 99; Plombières: Archives Départementales du Loiret).

Bonaparte, Louis Napoleon (1857) 'Discours de S. M. Napoléon III à l'ouverture de la session législative, le 17 février 1857' quoted in Méjean, 1996.

Bonaparte, Louis Napoleon (1860), 'Lettre de l'Empereur au ministre d'Etat', published in Journal d'Agriculture Pratique, yr. 24 (1).

Borgeat-Theler, Muriel, Scheurer, Alexandre, and Dubuis, Pierre (2011), 'Le Rhône et ses riverains entre Riddes et Martigny (1400–1860)' *Cahiers de Vallesia*, (66; Sion: Archives de l'Etat du Valais), 1-106.

Boulanger, Nicolas-Antoine (1772), L'Antiquité dévoilée par ses usages, ou Examen critique des principales opinions, cérémonies et institutions religieuses et politiques des différents peuples de la terre (Amsterdam: Marc Michel Rey).

Buffon, Georges-Louis Leclerc, comte de Buffon (1778, ed. 2018), The epochs of nature, trans. J. Zalasiewicz, S. Sverker, L. Robin, J. Grinevald (Chicago, London: Chicago University Press).

Buffon, Georges Louis Leclerc, comte de Buffon (1797), *Natural History, Containing a Theory of the Earth, a General History of Man, of the Brute Creation, and of Vegetables, Mineral, etc*, Smith Barr, James trans., Ebook by Project Gutenberg, Release Date: January 29, 2014.

Cadell, W. A. (1820), A journey in Carniola, Italy, and France in the Years 1817, 1818 (Edinburgh: A. Constable).

Champion, Maurice (1858-64), Les inondations en France depuis le VIe siècle jusqu'à nos jours, volumes 3 and 4 (Paris: Dunod, 1861 and 1862).

Chepalyga, A. L. (2007), 'Late glacial great flood in the Ponto-Caspian basin', The Black Sea Flood Question: Changes in

Coastline, Climate, and Human Settlement (Dordrecht: Springer), 119-48.
Choron, M. (1871), 'Mémoire sur le colmatage des terrains de la vallée de l'Isère, partie comprise dans la Savoie entre le pont de Grésy et la limite du département', *Annales des ponts et chaussées. Mémoires et documents relatifs à l'art des constructions et au service de l'ingénieur* (5e série, 2e semestre), 311-60.
Cœur, Denis (2008), *La plaine de Grenoble face aux inondations. Genèse d'une politique publique du XVIIe au XXe siècle* (Versailles: Quae).
Confédéré du Valais, 28.02.1864, 'A la rédaction du Confédéré du Valais', sec. Canton du Valais p. 1.
Conseil Fédéral Suisse (1860), 'Rapport du Conseil fédéral sur sa gestion en 1860', Archives fédérales suisses.
Cook, Hadrian, Stearne, Kathy, and Williamson, Tom (2003), 'The origins of water meadows in England', The Agricultural History Review, 51.
Cooke, Alice M. (1893), 'The Settlement of the Cistercians in England', The English Historical Review, 8 (32).
Courrier du Valais, 10.06.1856, 'Toujours les inondations !', sec. Nouvelles étrangères, p. 4.
Courrier du Valais, 03.07.1856, 'De l'utilite d'une école d'agriculture en valais', p. 1-2.
Creyke, Ralph (1845), 'Some Account of the Process of Warping', The Bristish Farmer's Magazine, 9, 28-36.
Curtis, Daniel R. and Campopiano, Michele (2013), 'Medieval land reclamation and the creation of new societies: comparing Holland and the Po Valley, ca. 800 – ca. 1500', *Journal of Historical Geography*, 30, 1-16.
Dalley, Stephanie (2000), *Myths from Mesopotamia. Creation, The Flood, Gilgamesh, and others* (Oxford: Oxford World's Classics).
Dapples, et al. (1861), 'Rapport de la Commission du Conseil des Etats sur la correction du Rhin (Du 11 juillet 1861.)', *Feuille Fédérale Suisse* (13, 2, 45; Bern), 663-71.
Dauphiné Libéré, 02.02.1968, 'Au Versoud, l'Isère a changé de lit'.
De Torrenté, Charles and de Kalbermatten, Arnold (1964), *La correction du Rhône en amont du lac Léman* (Bern: Service fédéral des routes et des digues).
Diodorus Siculus (1935), *The Library of History* (Cambridge, MA: Loeb Classical Library edition, Vol. II), available at http://penelope.uchicago.edu/Thayer/E/Roman/Texts/Diodorus_Siculus/4B*.html.
Dion, Roger (1934), *Le val de Loire: Etude de géographie régionale* (Tours: Arrault).
Dion, Roger (1961), *Histoire des levées de la Loire* (Paris: chez l'auteur).
Drizard, Antoine (1868), 'Mémoire sur le colmatage des terrains de la vallée de l'Isère', in *Annales des Ponts et Chaussées*, tome XVI, 4e serie, 2d sem., 593-632 (Paris: Dunod).
Du Buat, Pierre L. G. (1786), *Principes D'Hydraulique Vérifiés par un grand nombre d'Expériences faites par ordre du Gouvernement*, 2 vol. (Paris: Imprimerie de Monsieur).
Duhem, Pierre (1958), *Le Système du monde : histoire des doctrines cosmologiques de Platon à Copernic*, vol. 9 (Paris: Hermann).
Etat du Valais, (1907), 'Constitution du Canton du Valais du 8 mars 1907', http://www.admin.ch/opc/fr/classified-compilation/19070006/index.html.
Evéquoz-Dayen, Myriam (2009), 'Des sources pour l'étude du Rhône valaisan', Le Rhône: Dynamique, Histoire et Société (*Cahiers de Vallesia* 21; Sion: Archives de l'Etat du Valais), 47-62.
Feuille Fédérale Suisse, 14.02.1863, y. XV, vol. I, nr. 7, pp 273-372 (Bern).
Folland, C. K., T. R. Karl, J. R. Christy, R. A. Clarke, G. V. Gruza, J. Jouzel, M. E. Mann, J. Oerlemans, M. J. Salinger and S.-W. Wang (2001), '2001: Observed Climate Variability and Change', in R. K. Pachauri and L. A. Meyer Core Writing Team (ed.), *Climate Change 2001: The Scientific Basis. Contribution of Working Group I to the Third Assessment Report of the Intergovernmental Panel on Climate Change* (Houghton, J. T. Y. Ding, D. J. Griggs, M. Noguer, P. J. Van der Linden, X. Dai, K. Maskell, and C. A. Johnson ed.; Cambridge, United Kingdom and New York, NY, USA: Cambridge University Press).
Fossombroni, Vittorio (1835), Memorie idraulico-storiche sopra la Valdi-Chiana (Montepulciano: Angiolo Fumi), available online at http://hdl.handle.net/2027/uiuo.ark:/13960/t2m62n609.
Fournier, Marie (2008), *Quelle place pour les riverains dans la gestion des inondations de la Loire? Les leçons du passé* (Orléans: Etablissement Public Loire).
Frederick II, King of Prussia and Voltaire, François Arouet (1889 ed.) *Correspondance de Voltaire avec le Roi de Prusse, 1736-1778* (Paris: Librairie de la Bibliothèque Nationale), available online at https://gallica.bnf.fr/ark:/12148/bpt6k75628n.texteImage.
Gazette du Valais, 24.02.1859, sec. Canton du Valais, p. 1-2.
Girel, Jacky (2005), *Les paysages alluviaux alpins (histoire, diagnostic, gestion, restauration et protection)* (Université Joseph Fourier Grenoble, Laboratoire d'Ecologie Alpine).
Girel, Jacky (2008), 'River Diking and Reclamation in the Alpine Piedmont', in Christof Mauch and Thomas Zeller (eds.), *Rivers in History. Perspectives on Waterways in Europe and North America* (Pittsburgh: University of Pittsburgh Press), 78-88.
Girel, Jacky (2010), 'Histoire de l'endiguement de l'Isère en Savoie: conséquences sur l'organisation du paysage et la biodiversité actuelle', *Géocarrefour*, Vol. 85 (1).
Gornitz, Vivien (2009), 'Sea Level Change, Post-Glacial', Encyclopedia of Paleoclimatology and Ancient Environments (*Encyclopedia of Earth Sciences Series*: Springer), 887-92.
Guillou, Samuel and Maurin, Jean (2005), 'Les levées de la Loire: des turcies au Plan Loire, huit siècles d'évolution', *Ingénieries EAT*, Spécial Sécurité des digues fluviales et de navigation, 31-39.
Haenni, William (1929), 'Le Rhône Valaisan et ses corrections', 3e congrès du Rhône (Geneva), 37-45.
Holzhauser, H. and Zumbühl, H. J. (1999), 'Glacier Fluctuations in the Western Swiss and French Alps in the 16th century', *Climatic Change* 43 (1), 223-37.
Hydronat (2000) *Rapport de synthèse juin 2000*, Canton du Valais. Département des Transports, de l'Equipement et de l'Environnement, Service des Routes et des Cours d'Eau. [online text retrieved from http://www.vs.ch/Navig/navig.asp?MenuID=5256&Language=fr.
Laval (1845), 'Rapport sur un mémoire de M. Puvis relatif à l'endiguement de la Saône', *Annales des sciences physiques et naturelles, d'agriculture et d'industrie*, 8, 217-31.
Lebecq, Stéphane (1997), 'Entre terre et mer: la mise en valeur des contrées littorales de l'ancienne Frise', *Histoire, économie et société*, year 16 (3), 361-76.
Lehouck, Alexander and van Acker, Jan (2011), 'Vorming van monastieke landschappen? Domeinbeheer door religieuze instellingen in Noordwest-Europa van de 12de tot de 18de eeuw', *Novi Monasterii*. Jaarboek Abdijmuseum Ten Duinen, 1-9.
Lendering, Jona, (2002), 'The Edges of the Earth' [online text], http://www.livius.org/concept/the-edges-of-the-earth-1/.
Lendering, Jona and Bosman, Arjen (2012), *Edge of empire: Rome's frontier on the Lower Rhine* (Rotterdam: Karwansaray).
Lenthéric, Charles (1892). *Le Rhône: du Saint-Gothard à la mer: histoire d'un fleuve* (Paris: Plon).

Lesher, James (2014) *Xenophanes* [online text], The Stanford Encyclopedia of Philosophy (Fall 2014 Edition), http://plato.stanford.edu/archives/fall2014/entries/xenophanes/.

'Lex Frisionum' (ca. 785) (Basel: transcription by Joannis Basilius Herold, 1557, trans. Kees Nieuwenhuijsen) [online text], http://www.keesn.nl/lex/.

Locke, John (1690) *Second Treatise of Government*, original text available online at http://www.gutenberg.org.

Lortet (1846), 'Rapport fait à la Société Royale d'Agriculture, etc, de Lyon, dans sa séance du 5 juin 1846, par M. le Dr Lortet, sur un ouvrage intitulé: Atterrissements artificiels dans la vallée de l'Isère supérieure, cause des fièvres endémiques et périodiques de cette vallée et moyens de les prévenir; par Ant. Chiron, Architecte de la Ville, à Chambéry', *Annales des sciences physiques et naturelles, d'agriculture et d'industrie*, 9, 729-36.

Loudon, J. C. (1871), *An Encyclopedia of Agriculture* (London: Longmans, Green and co.).

Mac Donald, David (1988), 'The Flood: Mesopotamian Archaeological Evidence', *Creation Evolution Journal*, 8 (2), 14-20. http://ncse.com/cej/8/2/flood-mesopotamian-archaeological-evidence.

Mallowan, M. E. L. (1964), 'Noah's Flood Reconsidered', *Iraq*, 26 (2), 62-82.

Meyer, Han, Bobbink, Inge, and Nijhuis, Steffen (2010), *Delta Urbanism: The Netherlands* (Chicago and Washington: American Planning Association).

Origen (1982 edn), *Homilies on Genesis and Exodus*, trans. Ronald E. Heine (Washington: Catholic University of America Press).

Oudot, D. (1989), 'Les Îles de la Loire angevine armoricaine', *Norois*, n°142, avril-juin 1989, 205-223.

Ovid (1893), *The Metamorphoses*, trans. Henry Thomas Riley (London: Bell).

Pautou, Guy, et al. (1995), 'Hydrologie, paludisme et démoustication. (L'exemple de la Région Rhône-Alpes)', *Revue de géographie alpine*, 83 (1), 33-52.

Pichard, Georges (1995), 'Les crues sur le bas Rhône de 1500 à nos jours. Pour une histoire hydro-climatique', *Méditerranée*, 82 (03.04.1995), 105-16.

Picon, Bernard and Allard, Paul (2006), *Gestion du risque inondation et changement social dans le delta du Rhône* (Cemagref).

Pigeon, et al. (1845), 'Rapport sur un Mémoire de M. Puvis, concernant l'endiguement des rivières en général et celui de la Saône en particulier, par M. PIGEON', *Annales des sciences physiques et naturelles, d'agriculture et d'industrie*, 8, 233-44.

Plato, *Laws*, in *Plato in Twelve Volumes*, Vols. 10 and 11, trans. R. G. Bury 1967 and 1968 edition (Cambridge: Harvard University Press, William Heinemann Ltd.). Available at http://www.perseus.tufts.edu/hopper/text?doc=Perseus:text:1999.01.0166.

Pliny the Elder (1855 edn), *Natural History*, trans. John Bostock et al. (London: Taylor and Francis), available online at http://data.perseus.org/citations/urn:cts:latinLit:phi0978.phi001.perseus-eng1:16.1.

Polyaenus (1793), Stratagems of war, trans. R. Shepherd (London: Georg Nicol), available on http://www.attalus.org/info/polyaenus.html.

Puvis, Marc-Antoine (1844), *De l'Endiguement des fleuves, des rivières et des torrens* (Paris: Mme Huzard).

Puvis, Marc-Antoine (1847), *Über die Eindämmung flissender Gewässer*, trans. Carl Müller (Vienna: Carl Gerold).

Rigord, Breton, G. (1882 and 1885), Œuvres de Rigord et de Guillaume le Breton, historiens de Philippe-Auguste, vol. 1: Chroniques de Rigord et de Guillaume le Breton ; vol. 2: Philippide de Guillaume Le Breton (Paris: Librairie Renouard).

Rooijendijk, Cordula (2009), *Waterwolven: een geschiedenis van stormvloeden, dijkenbouwers en droogmakers* (Amsterdam: Atlas Contact).

Rose, Jeffrey I. (2010), 'New Light on Human Prehistory in the Arabo-Persian Gulf Oasis', *Current Anthropology*, 51 (6), 849-83.

Rousseau, Jean-Jacques (1754), *A Dissertation on the origin and foundation of the inequality of mankind*, (1923 ed.; London and Toronto: J. M. Dent and Sons).

Rousseau, Jean-Jacques (1762), *Emile, ou de l'éducation*, tome II, livre V, in *Collection complète des œuvres*, Genève, 1780-1789, vol. 5, in-4°, available online at www.rousseauonline.ch.

Rousseau, Jean-Jacques (1763a), 'Lettre à M. Le Maréchal de Luxembourg, le 28 janvier 1763', (Môtiers).

Rousseau, Jean-Jacques (1763b), 'Letter to Beaumont', in Roger D. Masters and Christopher Kelly (eds.), *Letters written from the mountains and related things. The collected writings of Rousseau* (2001 ed., 9; Hanover & London: University Press of New England).

Ryan, William B. F., et al. (1997), 'An abrupt drowning of the Black Sea shelf', *Marine Geology*, Volume 138 (1-2), 119-26.

Sandars, N. K. (trans), *The Epic of Gilgamesh*, anonymous, 1972 ed. (London: Penguin).

Sartre, Jean-Paul (1938), *Nausea*, trans. Lloyd Alexander (1969 ed.; New York: New Directions Publishing).

Schenk, et al. (1863), 'Rapport de la Commission du Conseil des Etats sur la correction du Rhône (Du 24 juin 1863.)', *Feuille Fédérale Suisse* (15, 3, 29; Bern), 33-65.

Schoeneich, Philippe (2009), 'Histoire des aménagements du Rhône et de la plaine dans le Chablais', *Cahiers de Vallesia* (21; Sion: Archives de l'Etat du Valais), 151-66.

Smart, P. J., Wheeler, B. D., and Willis, A. J. (1986), 'Plants and peat cutting: Historical ecology of a much exploited peatland—Thorne Waste, Yorkshire, UK', *New Phytologist*, 104 (4), 731-48.

Smith, Thomas M. (2012), 'Enclosure & agricultural improvement in north-west Lincolnshire from circa 1600 to 1850', PhD thesis (University of Nottingham).

Speich, Daniel (2002a), *Linth Kanal. Die korrigierte Landschaft—*200 Jahre Geschichte (Glarus: Baeschlin).

Speich, Daniel (2002b), 'Draining the Marshlands, Disciplining the Masses: The Linth Valley Hydro Engineering Scheme (1807-1823) and the Genesis of Swiss National Unity', *Environment and History*, Volume 8, Number 4, 429-47.

Staüble, Sabine, and Reynard, Emmanuel (2005), Evolution du paysage de la plaine du Rhône dans la région de Conthey depuis 1850. Les apports de l'analyse de cartes historiques, in *Vallesia: bulletin annuel de la Bibliothèque et des Archives cantonales du Valais, des Musées de Valère et de la Majorie*, p. 433-456.

Staüble, Sabine (2009), 'Evolution de la plaine du Rhône suisse du début du XIXe siècle à nos jours: étude cartographique dans le Valais central', Le Rhône: Dynamique, Histoire et Société (*Cahiers de Vallesia* 21; Sion: Archives de l'Etat du Valais), 167-76.

Steno, Nicolaus (1916 ed.), *The prodromus of Nicolaus Steno's dissertation concerning a solid body enclosed by process of nature within a solid*, trans. J. G. Winter (Macmillan and Company; New York, London).

Sully, Maximilien de Béthune (1820 ed.), *Economies Royales*, Chailley-Bert, Joseph ed. (Paris: Guillaumin & Cie).

Talon, Philippe and Anthonioz, Stéphanie (2019), *Enuma elis—Lorsqu'en haut*, collection Littératures anciennes du Proche-Orient, n°22, avril 2019 (Paris: éditions du Cerf).

Ter Haar, Gerard and polhuis, Peter L. (2004), *De loop van het Friese water. Geschiedenis van het waterbeheer en de waterschappen in Friesland* (Franeker: Van Wijnen).

Tulla, Johann G. (1825), *Über die Rektifikation des Rheins von seinem Austritt aus der Schweitz bis zu seinem Eintritt in das Großherzogthum Hessen* (Karlsruhe: Müller).
Van Acker, J. (2014) *800 Jaar Cîteaux in Zuid-West-Vlaanderen* [online text], https://www.academia.edu/.
Van Wijk, Wim (2012), *Historische Atlas van de Biesbosch* (Zwolle: W Books).
Varenius (1734 ed.), *A compleat system of general geography*, eds Isaac Newton and Dr Jurin, trans. Dugdale (London: Stephen Austen).
Venetz, Ignace (1851), *Mémoire sur les digues insubmersibles, sur les écluses à cheminées pour le colmatage et sur les principes à suivre dans les corrections des cours d'eau, avec leur application spéciale aux redressements de l'Orbe et de la Broye et à l'assainissement des marais adjacents* (Genève: Bibliothèque Universelle).
Vierlingh, Andries (1579), *Tractaet van Dyckagie* (1920 ed.; Den Haag: Martinus Nijhoff).
Vischer, Daniel L. (2003), 'Histoire de la protection contre les crues en Suisse', *Rapports de l'OFEG*, Série Eaux (5; Bern: Office fédéral des eaux et de la géologie).
Walckenaer, Charles A. (1798), *Essai sur l'histoire de l'espèce humaine* (Paris: Du Pont).
Walter, François (1990), *Les Suisses et l'environnement. Une histoire du rapport à la nature du 18e siècle à nos jours* (Geneva: Zoé).
Weiss, Johann Heinrich (1797), *Atlas Suisse*, Partie du canton de Berne, du Vallais et canton de Fribourg (Aarau: J. R. Meyer). ETH-Bibliothek Zurich, http://doi.org/10.3931/e-rara-23379, Public Domain Mark.
Williams, M. (1963), 'The Draining and Reclamation of the Somerset Levels, 1770-1833', Transactions and Papers (Institute of British Geographers), 33 (Dec., 1963), pp. 163-79.
Wuhrmann, Sylvie and Cariel, Rémi (2006), 'Diluvio e sua dimostrazione in pittura': les enjeux artistiques du déluge', *Visions du déluge de la Renaissance au XIXe siècle* (Paris: Réunion des Musées Nationaux).
Yanko-Hombach, Valentina, et al. (eds.) (2007), *The Black Sea Flood Question: Changes in Coastline, Climate, and Human Settlement* (Dordrecht: Springer).
Zuercher, Christoph (2007), 'Jan Lelewel', in *Dictionnaire historique de la Suisse*, online dictionnary, version 30.11.2007, https://hls-dhs-dss.ch/fr/articles/014783/2007-11-30/.

2 IN SEARCH OF COMBINATORY LANDSCAPES

Bonaparte, Louis Napoleon (1856), 'Lettre de l'Empereur à son Ministre des Travaux Publics', (2S 99; Plombières: Archives Départementales du Loiret).
Bruijn, D., Hamhuis, D., Nieuwenhuijze, L. van, Overmars, W., Sijmons, D., & Vera, F. (1987). Ooievaar: De toekomst van het Rivierengebied (Arnhem: Gelderse Milieufederatie).
Prominski, Martin, Stokman, Antje, Zeller, Susanne, Stimberg, Daniel, Voermanek, Hinnerk (2012), *River.Space.Design. Planning Strategies, Methods and Projects for Urban Rivers* (Basel: Birkhäuser).
Warner, Jeroen Frank, Buuren, Arwin Van, and Edelenbos, Jurian (2013), 'Making Space for the River. Governance challenges', in Arwin Van Buuren, Jeroen Frank Warner, Jurian Edelenbos (ed.), *Making Space for the River. Governance experiences with multifunctional river flood management in the US and Europe* (London: IWA), 1-13.

The Floodplain as Urban Park: Isar Plan, Munich

Armonat, Thomas (2010), 'Isar—drei Ziele, ein plan', *Garten+ Landschaft* (120), 20-24.
Baumeister, Nicolette (2003), 'Neues Leben für die Isar. Ergebnisse des internationalen Wettbewerbs zwischen Braunauer Eisenbahnbrücke und Deutschem Museum in München' (Munich: Landshaupstadt München, Wasserwirtschaftsamt München).
Blasy + Mader (1998), 'Isar-Plan. Verbesserung des Hochwasserschutzes und naturnahe Umgestaltung der Isar unter Berücksichtigung der Erholungsnutzung zwischen der südlichen Stadtgrenze und der Corneliusbrücke in München. Kurzfassung zur Vorplannung' (Munich: Landsehauptstadt München, Wasserwirtschaftsamt München).
Böhm, O. and Wetzel, K.-F. (2006), 'Flood history of the Danube tributaries Lech and Isar in the Alpine foreland of Germany', *Hydrological Sciences Journal* (51: 5), 784-98.
CSU-Stadtratsfraktion München (1984), 'Natur in der Stadt. Antrag Nr. 282 der Stadtratsfraktion der CSU vom 30. Juli 1984' (Munich: Landeshauptstadt München).
Düchs, Julia (2014), *Wann wird's an der Isar wieder schön? Die Renaturierung der Isar in München. Über das Verständnis von Natür in der Großstadt*, ed. Institut für Volkskunde/Europäische Ethnologie der Ludwig-Maximilians-Universität München (Münchner Ethnographische Schriften; München: Herbert Utz Verlag).
Jerney, Winfrid (2003), 'Rasen betreten erlaubt—Günther Grzimek: Planung mit den Bürgern', *Garten+Landschaft* (113) 13-16.
Oppermann, Bettina and Pauleit, Stephan (2005), 'The Green structure of Munich: The need for and risk of regional cooperation', *Green Structures and Urban Planning*. Final Report (European Cooperation in the field of Scientific and Technical Research–COST Action C11), 72-79.
Rädlinger, Christine (2012), *Geschichte der Isar in München*, ed. Stadtarchiv München (Munich: Franz Schiermeier Verlag).
Referat für Stadtplanung und Bauordnung (1983), 'Stadtentwicklungsplan' (München: Landeshauptstadt München).
Referat für Stadtplanung und Bauordnung, Landeshauptstadt München (1986), 'Beschluß des Ausschusses für Stadtplanung und Bauordnung vom 4. Juni 1986' (Munich).
SPD-Stadtratsfraktion München (1985), 'Isar-Plan. Konzept zum Schutz und Wiederherstellung, zur Erhaltung und Erschließung des Isarraumes, der Auen und der Hangkanten. Antrag Nr. 636 der Stadtratsfraktion der SPD vom 3. Mai 1985' (Munich: Landeshauptstadt München).
Statistisches Amt München (2012), 'Die Bevölkerung seit 1900'. Landeshauptstadt München http://www.muenchen.de/rathaus/Stadtinfos/Statistik/Bev-lkerung/Bev-lkerungsbestand.html.
Süddeutsche Zeitung, 'Die Isar soll wieder wild fließen. Bagger beseitigen Uferbefestigung / Fluß wird aus dem Betonkorsett befreit' (24.02.1998), 39.
Süddeutsche Zeitung, 'Banales Betonmonster. Isar-Renaturierung: Bürger lehnen Siegerentwurf ab' (05.06.2003), 42.
Süddeutsche Zeitung, 'Kompromiss für Isar-Renaturierung gefunden' (12.01.2005), p. 38.
Ude, Christian (2008) Burgerfragen: Warum ist der Riemer Park so hässlich?, weblog: https://www.christian-ude.de/politisch/buergerfragen/495270.html, version 28.06.2008.
Zinsser, Tilmann (2003), 'Neues Leben für die Isar', *Garten+ Landschaft* (113), 12-15.

Flood as Investment: Meerstad Groningen

Abrahamse, Jan (1999), 'Woldmeer op de website', *Noorderbreedte*, year 23 (2), http://www.noorderbreedte.nl/nog_niet_toegekend/woldmeer-op-de-website/.

Abrahamse, Jan (2002), 'Groningen, de compacte stad', *Noorderbreedte*, Themanummer 'Ruimtelijke ordening in Noord-Nederland', no. 5a, 36-38, https://www.noorderbreedte.nl/2001/11/30/groningen-de-compacte-stad/?highlight=Groningen%2C%20de%20compacte%20stad.
Alle Hosper landschapsarchitecten and KCAP architects and planners (2005), 'Masterplan Meerstad' (Haarlem: Gemeente Groningen, Gemeente Slochteren, Provincie Groningen, Ministerie van Landbouw, Natuurbeheer en Visserij, Dienst Landelijk Gebied, Waterschap Hunze-en-Aa's, Grondbank Meerstad Groningen).
Archined (1999), 'Het Woldmeer gelanceerd', Archined.nl, https://www.archined.nl/1999/03/het-woldmeer-gelanceerd/.
Archined (2000), 'Groningen boven water', Archined.nl, https://www.archined.nl/woldmeer/index.html, https://web.archive.org/web/20000413173620/http://www.archined.nl:80/woldmeer/index.html.
Bijnsdorp, Rob, et al. (2002), 'Meerstad: meer water en meer stad', *De Water*, waterbeleid en -beheer in uitvoering, 103 (nov. 2004), 7-8.
Das, Marjolijn and van Daalen, Gelske (2012), 'Dynamiek op de woningmarkt', *De Nederlandse economie 2011* (The Hague: Centraal Bureau voor de Statistiek).
De Jonge, Jannemarie (2009), *Landscape Architecture between Politics and Science*. (Wageningen: Blauwdruk).
Gemeente Groningen, gemeente Slochteren, Dienst Landelijk Gebied (1998), ontwerpstudie Maximum Laadvermogen
Gemeente Groningen, dienst RO/EZ (1994), 'De stad van straks: Groningen in 2005: concept structuurplan' (Groningen).
Gemeente Groningen, Onderzoek Informatie en Statistiek. Statistische jaaroverzichten. Online document: http://www.os-groningen.nl/images/stories/rapport/sjb/thema02_bevolking.pdf.
Gemeente Slochteren (2012). De gemeente Slochteren in cijfers, Statistisch onderzoek. Online document: https://www.raadslochteren.nl/Vergaderingen/Gemeenteraad/2012/12-juli/20:00/.Bestuurlijke-agenda/04-Slochteren-in-cijfers-20121.pdf.
Grontmij (2003), 'Milieueffectrapport Masterplan Meerstad Groningen' (Groningen: Waterschap Hunze-en-Aa's, Gemeente Slochteren, Gemeente Groningen, Provincie Groningen).
Keuter, Mark (2007), 'Samenwerken: waar, wanneer en met wie?', Master Thesis (University of Twente).
Ministerie van Volkshuisvesting, Ruimtelijke Ordening en Milieubeheer (1988), 'Vierde Nota ruimtelijke ordening' (The Hague).
Ministerie van Volkshuisvesting, Ruimtelijke Ordening en Milieubeheer (1990), 'Vierde Nota ruimtelijke ordening Extra' (The Hague).
Ministerie van Volkshuisvesting, Ruimtelijke Ordening en Milieubeheer (2006), 'Nota Ruimte. Ruimte voor ontwikkeling. Katern over nieuw ruimtelijk beleid in 2006' (The Hague).
Notten, Frank (2012), 'Hypotheekschuld in Nederland', *De Nederlandse economie 2011* (9; The Hague: Centraal Bureau voor de Statistiek), 206-24.
Salewski, Christian (2012), *Dutch New Worlds: Scenarios in Physical Planning and Design in the Netherlands, 1970-2000* (Rotterdam: 010 Publishers).
Wildeman, Klaas (1997), 'De Wolden Landschap tussen Fievel en Siep', *Noorderbreedte* (1997-1), 40-43.

Farmers as Risk-Catchers: Flood Fields in the Isère Valley

Blanic, M., Verdet, G. (1975) Quelques travaux de correction sur le cours de l'Isère, *La Houille Blanche* (2/3, 1975) 191-197.
Bravard, Jean-Paul (1989), La métamorphose des rivières des Alpes Françaises à la fin du Moyen-Age et à l'époque moderne. *Bulletin de la Société Géographique de Liège* (25), 151-154.
Castex, Joëlle, Belleudy, Philippe (2005) L'Isère endiguée dans le Grésivaudan. Du risque zéro à la rivière libérée: entre ces deux utopies, quelles attente sociale? Programme de recherche Risque Décision Territoire, Laboratoire d'études des transferts en hydrologie et environnement. Université Joseph Fourier, Grenoble.
Coeur, Denis (2008), *La plaine de Grenoble face aux inondations. Genèse d'une politique publique du XVIIe au XXe siècle*. (Versailles: Quae).
Conseil Général de l'Isère, Syndicat mixte des bassins hydrauliques de l'Isère (SYMBHI), and Chambre d'agriculture de l'Isère (2010), 'Projet Isère amont. Accord cadre portant sur les incidences agricoles' (Grenoble).
Dauphiné Libéré, 02.02.1968, 'Au Versoud, l'Isère a changé de lit'.
Girel, Jacky (2010), 'Histoire de l'endiguement de l'Isère en Savoie: conséquences sur l'organisation du paysage et la biodiversité actuelle', *Géocarrefour* (Vol. 85-1), 41-54.
Pinhas, Michel (2009), 'Rétrospective historique du système d'endiguement de l'Isère dans le Grésivaudan', Colloque ISERE 1859-2009. 150ème anniversaire de la Crue de Référence de l'Isère (Grenoble).
Pinhas, Michel (2009-2011), 'Histoire des digues de l'Isère amont au 20e siècle'. Online text: http://www.isere-drac-romanche.fr/?Histoire-des-digues-de-l-Isere.
Le Progrès, 02.02 1968, 'À 11 h 17, heure de Grenoble, l'Isère a obéi aux hommes des Ponts & Chaussées et changé de lit'.
SYMBHI, Département de l'Isère (2016), 'Projet Isère amont 2016-2021, Les tranches de travaux 2 & 3', Dossier de presse novembre 2016 (Grenoble).
SYMBHI, Syndicat Mixte des Bassins Hydrauliques de l'Isère (2009), 'Rapport de présentation synthétique du projet Isère Amont' (Grenoble: Conseil Général de l'Isère).

3 UPSCALING FLOODSCAPES DESIGN

Bender, Gabriel (2004), 'Correcting the Rhône and the Valaisans: three centuries of works and debate', Revue de Géographie Alpine (92-3), 62-72.
Ministerie van Infrastructuur en Waterstaat (2018), '31e Voortgangsrapportage Ruimte voor de Rivier' (The Hague).
Ministerie van Landbouw, Natuur en Voedselkwaliteit, Ministerie van Verkeer en Waterstaat, VROM (2006), 'Spatial Planning Key Decision 'Room for the River', Investing in the safety and vitality of the Dutch river basin region', information brochure (sept. 2006).
Ministerie van Verkeer en Waterstaat and Ministerie van Volkshuisvesting, Ruimtelijke Ordening en Milieubeheer (1996), 'Beleidslijn ruimte voor de rivier' (The Hague).
Projet Rhône (2008), 'Troisième correction du Rhône. Sécurité pour le futur. Rapport de Synthèse du Plan d'aménagement' (Sion: Canton du Valais, Département des transports, de l'équipement et de l'environnement, Service des routes et des cours d'eau).
Projet Rhône (2011), 'Zones de danger d'inondation du Rhône. Rapport technique, 26 mai 2011' (Sion: Canton du Valais, Département des transports, de l'équipement et de l'environnement, Service des routes et des cours d'eau).

The River as *Gesamtkunstwerk*? The Third Rhône Correction

ADSA, Association de défense du sol agricole, et al. (2008), 'Remarques sur le Plan d'aménagement Troisième correction du Rhône' (Conthey).

Arborino, Tony (2019), Le fleuve retrouvé, *Tracés (*16-17, 30.08. 2019), 6-11.
Assemblée fédérale de la Confédération suisse (1991), 'Loi fédérale sur l'aménagement des cours d'eau du 21 juin 1991'.
Bender, Gabriel (2004), 'Correcting the Rhône and the Valaisans: three centuries of works and debate', *Revue de Géographie Alpine* (92-3), 62-72.
Bloetzer, Werner, et al. (1998), Klimaänderungen und Naturgefahren in der Raumplanung. Methodische Ansätze und Fallbeispiele (Nationales Forschungsprogramm 'Klimaänderungen und Naturkatastrophen', NFP 31; Zurich: VDF Hochschulverlag).
Blotnitzki, L. and Hartmann, A. (1861), 'Rapport d'expert sur la Correction du Rhône au Département fédéral de l'Intérieur et des Travaux publics. (Du 6 avril 1861.)', *Feuille Fédérale Suisse*, 14 Février 1863 (15, 1, 7; Bern), 294-306.
Canal 9, Le débat. Face à Face: Rhône 3: camps opposés? With Thierry Largey (Pro Natura Valais) and Olivier Schupbach (Association de Défense des Surfaces Agricoles), television programme, Canal 9 (08.06.2011).
Confédéré, 06.09.1948, 'Les frasques du Rhône', p. 2.
Chablais Agglo, COPIL (2016), Chablais Agglo 3, 'Projet d'agglomération du Chablais 3e Génération, Rapport final décembre 2016', Canton de Vaud et Canton du Valais.
DFI. Département fédéral de l'intérieur, Service fédéral des routes et des digues (1964), 'La correction du Rhône en amont du lac Léman', ed. Département Fédéral de l'Intérieur (Bern: Office fédéral des imprimés et du matériel).
Dupont, Cédric and Pigeon, Patrick (2008), 'Le Haut-Rhône et son bassin versant montagneux: pour une gestion intégrée de territoires transfrontaliers. Projet INTERREG IIIA', Report (Institut de la Montagne, Environnements DYnamiques et Territoires de Montagne, UMR Univ. Savoie-CNRS, Chambéry, Institut de Hautes Etudes internationales et du Développement, Genève, Institut Universitaire Kurt Bösch, Sion).
Girot, Christophe, Rossano, Frédéric, and Duner, Isabelle (2010), *Sion-sur-Rhône*, pamphlet 13 (Zurich: GTA verlag).
Hydronat (2000) 'Rapport de synthèse juin 2000', Canton du Valais. Département des Transports, de l'Equipement et de l'Environnement, Service des Routes et des Cours d'Eau. Online text retrieved from http://www.vs.ch/Navig/navig.asp?MenuID=5256&Language=fr.
Nomad architectes (2016), 'Troisième correction du Rhône, Lignes directrices paysagères / Synthèse' (Sion: Canton du Valais, DTEE, SRTCE, Section de protection contre les crues du Rhône).
Nouvelliste, 07.09.1948, 'Le cycle d'épouvante', p. 1.
OFEE, SHGN (1991), 'Analyse des causes des crues de l'année 1987: rapport final' (Bern: Département Fédéral des Transports, des Communications et de l'Energie, Office Fédéral de l'Economie des Eaux, Département Fédéral de l'Intérieur, Office Fédéral de l'Environnement, des Forêts et du Paysage, Service Hydrologique et Géologique National).
OFEFP, et al. (2003), 'Idées directrices—Cours d'eau suisses. Pour une politique de gestion durable de nos eaux' (Bern: Office fédéral de l'environnement).
OFEG (2002), 'La protection contre les crues au fil du temps' (Bern: Office fédéral des eaux et de la géologie).
Office Cantonal de la construction du Rhône (2018), Aménagement des espaces publics du Rhône, Appel à candidatures—procédure sélective, Cahier des charges (Sion: Canton du Valais).
Pro Natura, Frossard, Pierre-André, et al. (1998), 'Mehr Raum für unsere Fliessgewässer. Ein Gewinn für Mensch und Natur' / 'Davantage d'espace pour nos cours d'eau. Pour l'homme et la nature' (Beiträge zum Naturschutz in der Schweiz / Contributions à la protection de la nature en Suisse, 19-20/1998; Basle: Pro Natura).
Projet Rhône, Département des transports, de l'équipement et de l'environnement, Service des routes et des cours d'eau, Canton du Valais (2006), 'Plan sectoriel 3ème correction du Rhône, juin 2006' (Sion: Canton du Valais).
Projet Rhône (2008), 'Troisième correction du Rhône. Sécurité pour le futur. Rapport de Synthèse du Plan d'aménagement' (Sion: Canton du Valais, Département des transports, de l'équipement et de l'environnement, Service des routes et des cours d'eau).
Projet Rhône (2011), 'Zones de danger d'inondation du Rhône. Rapport technique, 26 mai 2011' (Sion: Canton du Valais, Département des transports, de l'équipement et de l'environnement, Service des routes et des cours d'eau).
Projet Rhône, Département des transports, de l'équipement et de l'environnement, Service des routes et des cours d'eau, Canton du Valais (2016), 'Plan d'Aménagement (PA-R3), Rapport de synthèse Février 2014', mise à jour septembre 2015.
Reynard, Emmanuel, et al. (2006), 'Apprendre par les catastrophes naturelles: le cas des inondations récentes en Suisse', Actes du colloque L'eau en montagne (Megève).
Rhone.vs, Magazine d'information sur la troisième correction du Rhône, Canton du Valais, n°17, septembre 2010.
Staüble, Sabine (2009), 'Evolution de la plaine du Rhône suisse du début du XIXe siècle à nos jours: étude cartographique dans le Valais central', Le Rhône: Dynamique, Histoire et Société (*Cahiers de Vallesia* 21; Sion: Archives de l'Etat du Valais), 167-76.
Le Temps, Parvex, Marie, 6 may 2015, 'La correction du Rhône déchire la droite valaisanne', https://www.letemps.ch/suisse/correction-rhone-dechire-droite-valaisanne.
Veuve, Léopold, Feddersen, Pierre, and Paysagestion (2008), 'Lignes directrices paysagères pour le Rhône et pour la plaine' (Sion: Projet Rhône, Département des transports, de l'équipement et de l'environnement, Service des routes et des cours d'eau, Canton du Valais).
WWF World Wide Fund for Nature, Pro Natura, SLFP Fondation Suisse pour la Protection et l'Aménagement du Paysage, ATE Association Transport et Environnement (2005), 'Plan sectoriel 3ème correction du Rhône (PS-R3). La position des organisations de protection de l'environnement'.
Zaugg, Marc (2002), 'More Space for running waters: Negotiating institutional change in the Swiss flood protection system', *GeoJournal*, 58, 275-84.

From Raising Dikes to Making 'Room for the River'

Bink, Kees, et al. (2011), *De Overdiepse Polder* (Waspik: Heemkundekring Op't Goede Spoor).
Bosch & Slabbers (2005), 'Noordwaard. Ontwerpen met cultuurhistorie, Erfgoed en water, voor nu en later', Projectbureau Belvedere, Rijkswaterstaat (RIZA), Rijksdienst voor de Monumentenzorg, Rijksdienst voor de Oudheidkundig Bodemonderzoek.
Dauvellier, Peter (1991), 'Ruimtelijke kwaliteit: de oorsprong en de toepassing van een begrip', Symposium Werkgroep Landelijke Gebieden / LU (Wageningen).
De Gast, Cees (2004), *De macht van het water. Leven met water tussen Maas en Merwede* (Raamsdonkveer: Verse Hoeven).
De Jong, S. A., et al. (2000), 'Vergroting van de afvoercapaciteit en berging in de benedenloop van Rijn en Maas. Bestuurlijk advies aangeboden aan de staatssecretaris van Verkeer en Waterstaat door de Stuurgroep Integrale Verkenning Benedenrivieren' (Rotterdam: Rijkswaterstaat, Directie Zuid-Holland).
de Hullu, Elk (1995), 'Natuurontwikkeling bij Staatsbosbeheer', *De Levende Natuur*, 96 (5), 141-47.

De Jonge, Jannemarie (2009), *Landscape Architecture between Politics and Science* (Wageningen: Blauwdruk).
Eo Wijers Stichting (1986), 'Juryrapport ideeënprijsvraag Nederland rivierenland' (The Hague: Eo Wijers Stichting).
Feddes, Yttje and Halenbeek, Frans (1988), *Een scherpe Grens, Ontwerpstudie naar de ruimtelijke kwaliteit van verzwaarde rivierdijken* (Utrecht: Landschapsarchitectuur Staatsbosbeheer).
Heezik, Alex van (2007), *De strijd om water* (HNT Historische producties, Rijkswaterstaat: The Hague).
Kappers, F. I. (2000), 'Overlaten: (g)een oplossing van de problemen bij te veel water? Een bijdrage aan de geschiedenis van de overlaten in het Nederlandse rivierengebied in de 18e en 19e eeuw' (Lelystad: RIZA).
Klijn, Frans, et al. (2013), 'Design quality of room-for-the-river measures in the Netherlands: role and assessment of the quality team (Q-team)', *International Journal of River Basin Management*, 11 (3), 287-99.
Kors, A. G., 'The DSS Planning Kit' and its application in the spankracht study', *Lowland Technology International*, Vol.6, N. 2, 67-73, December 2004, international Association of Lowland Technology.
Leenders, K. A. H. W. (1999/2005), 'Landschapsgeschiedenis van de Hoekse Waard', Tekst van een voordracht op 27 mei 1999, gehouden voor de KNMG afdeling Zuid-Hollandse Eilanden. http://users.bart.nl/~leenders/txt/hoeksew1.html.
Ministerie van Verkeer en Waterstaat and Ministerie van Volkshuisvesting, Ruimtelijke Ordening en Milieubeheer (1996), 'Beleidslijn ruimte voor de rivier' (The Hague).
Ministerie van Verkeer en Waterstaat (1998), 'Vierde Nota waterhuishouding, regeringsbeslissing' (The Hague).
Ministerie van Verkeer en Waterstaat (1997b), 'Deltaplan Grote Rivieren, Voortgangsrapport 4, Stand van zaken per 1-1-1997' (The Hague).
Ministerie van Verkeer en Waterstaat, Directoraat-Generaal Rijkswaterstaat (2000), 'Ruimte voor de rivier. Kabinetstandpunt' (The Hague).
Ministerie van Verkeer en Waterstaat, Rijkswaterstaat, Directie Zuid-Holland (2000b), 'Integrale Verkenning Benedenrivieren' (Rotterdam).
Nienhuis, Piet H. (2008), Environmental History of the Rhine–Meuse Delta. An ecological story on evolving human–environmental relations coping with climate change and sea-level rise (Springer).
Pierlinck, Jacob (1757) *De verschrikkelyke watersnood, langs de rivieren de Waal en de Maas, voorgevallen in de maand February des jaars 1757 met derzelver droevige gevolgen, uit geloofbaare berigten en eigen bevindinge by een verzameld en beschreven: Waar by gevoegd is een vertoog over het eenig middel van redres, voor het toekomende* (Amsterdam: Isaak Tirion).
Projectbureau Noordwaard (2006), 'Ontpoldering Noordwaard. Ontwerpvisie' (Rotterdam: Provincie Noord-Brabant, Waterschap rivierenland, Gemeente Werkendam, Ministerie van Landbouw Natuur en Voedselkwaliteit, Ruimte voor de Rivier, Nederland leeft met water).
Projectbureau Noordwaard (2007), 'Ontpoldering Noordwaard. Ontwerpvisie: Het regio-alternatief' (Rotterdam: Provincie Noord-Brabant, Waterschap rivierenland, Gemeente Werkendam, Ministerie van Landbouw Natuur en Voedselkwaliteit, Ruimte voor de Rivier, Nederland leeft met water).
Projectorganisatie Ruimte voor de Rivier (2003), '2e Voortgangsrapportage 1 januari 2003 – 30 juni 2003' (The Hague).
Projectorganisatie Ruimte voor de Rivier (2006), 'Planologische Kernbeslissing Ruimte voor de Rivier', Ministerie van Verkeer en Waterstaat, Ministerie van Volkshuisvesting, Ruimtelijke Ordening en Milieubeheer, Ministerie van Landbouw, Natuur en Voedselkwaliteit, in 4 volumes (The Hague).
Provincie Noord-Brabant (2008), 'Rivierverruiming Overdiepse Polder. Projectnota/MER' ('s-Hertogenbosch).
RIZA and Bosch & Slabbers (1999), 'Landschappelijke verkenning rivierverruimende maatregelen' (Lelystad: Rijksinstituut voor Integraal Zoetwaterbeheer en Afvalwaterbehandeling, Projectteam Integrale Verkenning Benedenrivieren IVB, Rijkswaterstaat, Ministerie van Verkeer en Waterstaat).
RIZA and Bosch & Slabbers (1999b), 'Integrale verkenning Benedenrivieren, Deelstudie Landschap, Rivierverruiming en landschapsontwikkeling in de Biesbosch'. (Lelystad: Rijksinstituut voor Integraal Zoetwaterbeheer en Afvalwaterbehandeling, Projectteam Integrale Verkenning Benedenrivieren IVB, Rijkswaterstaat, Ministerie van Verkeer en Waterstaat).
RIZA and Bosch & Slabbers (2000), 'Landschappelijke verkenning rivierverruimende maatregelen' (Lelystad: Ministerie van Verkeer en Waterstaat Rijkswaterstaat, Rijksinstituut voor Integraal Zoetwaterbeheer en Afvalwaterbehandeling (RWS, RIZA), Projectteam Integrale Verkenning Benedenrivieren IVB).
RIZA, Rijksinstituut voor Integraal Zoetwaterbeheer en Afvalwaterbehandeling (2002), 'De spankracht van ons rivierenland. Eindrapport Spankracht' (Lelystad: Ministerie van Verkeer en Waterstaat, Ministerie van Volkshuisvesting, Ruimtelijke Ordening en Milieu. Ministerie van Landbouw, Natuurbeheer en Visserij, Provincies Utrecht, Gelderland, Overijssel, Zuid-Holland, Noord-Brabant, Vereniging Nederlandse Riviergemeenten, Unie van Waterschappen).
Sanderse, Wouter, et al. (2004), 'Voor de watersnoodramp van 1953'. Online text: http://www.deltawerken.com/Voor-de-ramp-van-1953/24.html.
Sijmons, Dirk (2008), 'Het deskundigenoordeel als instrument voor ruimtelijke kwaliteit', Jaarverslag Kwaliteitsteam Ruimte voor de Rivier 2006-2007 (Utrecht: Rijkswaterstaat).
Sijmons, Dirk (2009), 'Room for the River', *Topos* (68), 61-63.
Stolwijk, Herman and Verrips, Annemiek (2000), 'Ruimte voor water. Kosten en baten van zes projecten en enige alternatieven' (Den Haag: Centraal Planbureau).
Van de Vorstenbosch, Harry (2011), 'De verlegging van de Maasmond', Brabants Historisch Informatie Centrum. Online text: https://www.bhic.nl/ontdekken/verhalen/de-verlegging-van-de-maasmond.
Van Herwijnen, M., et al. (2003), 'Succes- en faalfactoren van natuurontwikkeling in en langs het water' (Amsterdam: IVM Instituut voor Milieuvraagstukken, Vrije Universiteit).
Van Rooy, Peter and Slootweg, Roel (2003), 'Spiegelproject Overdiepse polder: rapportage verkenning' (Gouda: Habiforum, Vereniging 'Belangengroep Overdiepse polder', Bezinningsgroep Water, Stuurgroep Benedenrivieren).
Van Veen, Johan (1948), *Dredge, drain, reclaim: the art of a nation* (The Hague: Nijhoff).
Van Wijk, Wim (2012), *Historische Atlas van de Biesbosch* (Zwolle: W Books).
Vandersmissen, Hans (1998), *Het woelige water. Watermanagement in Nederland* (Teleacnot / INMERC).
Verschelling, Eelco (2003), 'Rivierkundige analyses maatregelen Benedenrivierengebied ten behoeve van PKB Ruimte voor de Rivier' (Lelystad: DG Rijkswaterstaat RIZA).
Vis, Joh A. (1936), *De inpoldering van den Zuidhollandschen Biesbosch* (Dordrecht: Vereeniging De Biesbosch).
Warner, Jeroen Frank, Buuren, Arwin Van, and Edelenbos, Jurian (2013), 'Making Space for the River. Governance challenges', in Arwin Van Buuren, Jeroen Frank Warner, Jurian Edelenbos (ed.), *Making Space for the River. Governance experiences with multifunctional river flood management in the US and Europe* (London: IWA), 1-13.

4 DESIGNING ELASTIC LANDSCAPES FOR AN UNCERTAIN FUTURE

Arborino, Tony and Perret, Jacques (2008), 'Corriger le Rhône', Tracés 6 (09.04.2008), 7-10.
Arborino, Tony (2011), 'Prise en compte du changement climatique dans la gestion du risque inondation: l'exemple du Rhône en Valais', Actes du séminaire international d'experts 'Adaptation de la gestion des risques naturels face au changement climatique' (Domancy, France).
CSU-Stadtratsfraktion München (1984), 'Natur in der Stadt. Antrag Nr. 282 der Stadtratsfraktion der CSU vom 30. Juli 1984' (Munich: Landeshauptstadt München).
De Koning, Robbert (2010), and Projectbureau Noordwaard, Ambitiedocument Ontpoldering Noordwaard (Arnhem).
EEA, European Environnement Agency (2019), 'Land and soil in Europe. Ever-sprawling urban concrete?', online document, https://www.eea.europa.eu/signals/signals-2019-content-list/articles/land-and-soil-in-europe.
Girel, Jacky (2005), 'Les paysages alluviaux alpins (histoire, diagnostic, gestion, restauration et protection)' (Université Joseph Fourier Grenoble, Laboratoire d'Ecologie Alpine).
Hajer, Maarten, et al. (2006), *Een plan dat werkt. Ontwerp en politiek in de regionale planvorming*, eds. Maarten Hajer, Dirk Sijmons, and Fred Feddes (Rotterdam: Nai).
Heems, Geertruida C. and Kothuis, Baukje L. M. (2012), 'Waterveiligheid; Managen van Kwetsbaarheid. Voorbij de Mythe van Droge Voeten. De Nederlandse omgang met overstromingsdreiging in sociaal-cultureel perspectief', PhD (University of Maastricht).
Hydronat (2000) 'Rapport de synthèse juin 2000', Canton du Valais. Département des Transports, de l'Equipement et de l'Environnement, Service des Routes et des Cours d'Eau. Online text retrieved from http://www.vs.ch/Navig/navig.asp?MenuID=5256&Language=fr.
Ministerie van Infrastructuur en Milieu (2015), 'Ons Water in Nederland. Nieuw Nationaal Waterplan 2016-2021' (The Hague).
NRC, Wie woont er graag in de calamiteitenpolder?, Rotterdam, 02.12.2002, Schreuder, A., http://vorige.nrc.nl/binnenland/article1554360.ece.
OFEFP, et al. (2003), 'Idées directrices—Cours d'eau suisses. Pour une politique de gestion durable de nos eaux' (Bern: Office fédéral de l'environnement).
Pellaud, Marc, et al. (2005), 'Objectifs environnementaux et paysagers d'un aménagement hydraulique à buts multiples. Nouveaux développements dans la gestion des crues', Conférence sur la recherche appliquée en relation avec la troisième correction du Rhône, 9 juin 2005 (Martigny).
Projectorganisatie Ruimte voor de Rivier (2006), 'Planologische Kernbeslissing Ruimte voor de Rivier', Ministerie van Verkeer en Waterstaat, Ministerie van Volkshuisvesting, Ruimtelijke Ordening en Milieubeheer, Ministerie van Landbouw, Natuur en Voedselkwaliteit, in 4 volumes (The Hague).
Projet Rhône (2008), 'Troisième correction du Rhône. Sécurité pour le futur. Rapport de Synthèse du Plan d'aménagement' (Sion: Canton du Valais, Département des transports, de l'équipement et de l'environnement, Service des routes et des cours d'eau).
RIZA, Bosch & Slabbers (1999), 'Integrale verkenning Benedenrivieren, Deelstudie Landschap, Rivierverruiming en Landschapsontwikkeling in de Biesbosch'. (Lelystad: Rijksinstituut voor Integraal Zoetwaterbeheer en Afvalwaterbehandeling, Projectteam Integrale Verkenning Benedenrivieren IVB, Rijkswaterstaat, Ministerie van Verkeer en Waterstaat).
Rossano, Frédéric (2012), 'Dynamic Designs for Fluctuating Landscapes', in Girot et al., *Rising Waters, Shifting Lands*. Pamphlet 16, Zurich: GTA verlag).
Salewski, Christian (2012), *Dutch New Worlds: Scenarios in Physical Planning and Design in the Netherlands*, 1970-2000 (Rotterdam: 010 Publishers).
Süddeutsche Zeitung, 'Banales Betonmonster. Isar-Renaturierung: Bürger lehnen Siegerentwurf ab' (05.06.2003), 42.
SYMBHI, Syndicat Mixte des Bassins Hydrauliques de l'Isère (2009), 'Rapport de présentation synthétique du projet Isère Amont' (Grenoble: Conseil Général de l'Isère).
Van Baalen, Jieles (1995), 'Natuurontwikkeling en natuurbeleid', De Levende Natuur, 96, 6, pp 130-133.
Van Rooy, Peter and Slootweg, Roel (2003), 'Spiegelproject Overdiepse polder: rapportage verkenning' (Gouda: Habiforum, Vereniging 'Belangengroep Overdiepse polder', Bezinningsgroep Water, Stuurgroep Benedenrivieren).
Veuve, Léopold, Feddersen, Pierre, and Paysagestion (2008), 'Lignes directrices paysagères pour le Rhône et pour la plaine' (Sion: Projet Rhône, Département des transports, de l'équipement et de l'environnement, Service des routes et des cours d'eau, Canton du Valais).

5 EPILOGUE: ADAPTING TO CHANGING TIMES

Aristotle (1931 ed.), *Meteorology*, trans. E. W. Webster (London: Oxford Unversity Press), http://classics.mit.edu.
Barbisch, Caroline, et al. (2006), 'Vulnérabilité des infrastructures urbaines et gestion de crise. Impacts et enseignements de cas d'inondation en Suisse', Cost C19-Proactive Crisis Management of Urban Infrastructure (Université de Genève, Université de Lausanne, Université de Fribourg, Universität Zürich).
Bild.de (2013, June 5), 'Von Schmidt bis Merkel—Politiker im Hochwasser. Wahlkampf in Gummistiefeln', http://www.bild.de/politik/inland/flut/entscheidet-das-hochwasser-die-bundestagswahl-30691242.bild.html.
Behringer, Wolfgang (2010), *A Cultural History of Climate* (Cambridge: Polity Press).
Buffon, Georges-Louis Leclerc, comte de Buffon (1778, ed. 2018), *The epochs of nature*, trans. J. Zalasiewicz, S. Sverker, L. Robin, J. Grinevald (Chicago, London: Chicago University Press).
Der Spiegel (1986, August 11), 'Die Klima-Katastrophe, Das Weltklima gerät aus den Fugen' (33-1986).
EEA, European Environment Agency (2017), Climate change, impacts and vulnerability in Europe 2016, An indicator-based report.
GCA, Global Commission on adaptation (2019), 'about us'. Online document: https://gca.org/global-commission-on-adaptation/about-us.
Heems, Geertruida C. and Kothuis, Baukje L. M. (2012), 'Waterveiligheid; Managen van Kwetsbaarheid. Voorbij de Mythe van Droge Voeten. De Nederlandse omgang met overstromingsdreiging in sociaal-cultureel perspectief', PhD (University of Maastricht).
HP de Tijd, Traa, Mark, 'Wat gebeurt er eigenlijk als de Randstad overstroomt?' (30.10.2012), www.hpdetijd.nl/2012-10-30/wat-gebeurt-er-eigenlijk-als-de-randstad-overstroomt/
IPCC (2014), 'Climate Change 2014: Synthesis Report. Contribution of Working Groups I, II and III to the Fifth Assessment Report of the Intergovernmental Panel on Climate Change' [Core Writing Team, R. K. Pachauri and L. A. Meyer (eds.)] (Geneva, Switzerland: IPCC).
KNMI (2014), 'KNMI'14, klimaatscenario's voor Nederland; Leidraad voor professionals in klimaatadaptatie' (De Bilt: KNMI, Koninklijk Nederlands Meteorologisch Instituut).
KNMI (2014, May 26), 'Meer extreme buien door opwarming', Koninklijk Nederlands Meteorologisch Instituut. Online text:

http://www.knmi.nl/cms/content/119655/meer_extreme_buien_door_opwarming.

McGrath, Math (2014), 'Dissent among scientists over key climate impact report', BBC news, Sciences and environment (25.03.2014), http://www.bbc.com/news/science-environment-26655779?print=true.

Météosuisse, Office Fédéral de Métorologie et de Climatologie (2014), Rapport technique n°243, 'Scénarios climatiques Suisse, un aperçu régional' (Bern: Office fédéral de l'environnement).

NRC (2014, May 26), Extremer weer: Nederlandse winter gaat lijken op winter in Bordeaux, Schreuder, Arjen, http://www.nrc.nl/nieuws/2014/05/26/knmi-verwacht-extremer-weer-komende-decennia/.

Van het Hurk, Martin, and van Ark, Donald (2012), 'Meerlagsveiligheid in Nederland en Engeland. Zonder overstromingen geen urgentiegevoel', *ROm, Ruimtelijke ontwikkeling, infrastructuur en milieu* (30/4, april 2012), 33-35.

Volkskrant (2014, May 26), 'Nederlandse winters gaan op die van Bordeaux lijken', http://www.volkskrant.nl/vk/nl/9104/2014/article/print/detail/3661910/Nederlandse-winters-gaan-op-die-van-Bordeaux-lijken.dhtml.

Volkskrant (2014, April 1st), 'Reële klimaatscenario's dramatisch samengevat. Analyse IPCC rapport', p. 3-4.

Vrij Nederland (2019, February 19), 'De zeespiegelstijging is een groter probleem dan we denken. En Nederland heeft geen plan B'. Schuttenhelm, Rolf and Limarev, Nina.

INTERVIEWS

France

Buisson, Morgane, environment engineer for Isère-Amont, Conseil Général de l'Isère (Grenoble, 20 December 2012).
Girel, Jacky, engineer-researcher, Laboratoire d'écologie Alpine, CNRS, Grenoble. (Chambéry, 19 February 2013).
Jardin, Paul, hydrogeologist, former engineer of SOGREAH/Artelia (Grenoble, 26 December 2012).
Manin, Olivier, Engineer and Project leader, SYMBHI (Grenoble, 19 December 2012).
Pinhas, Michel, Director Association des Digues Isère-Drac-Romanche (telephone interview, Zurich/Grenoble, 14 May 2013).
Vignal, Bertrand, co-director BASE landscape/urban design (telephone interview, Strasbourg/Lyon, 5 may 2020).

Germany

Engelmayer, Oliver, project leader, Irene Burckhardt Landschaftarchitekt (Munich, 24 June 2013).
Mahida, Nivedita, Management and communication, Water Board of Bavaria, State of Bavaria (Munich, 26 June 2013).
Schaufuß, Daniela, Director Water Works, City of Munich (Munich, 25 June 2013).
Segern, Hille von, Professor Leibniz Universität Hannover (Hanover, 21 June 2013).
Thalgott, Christiane, chief architect of the City of Munich from 1992 to 2007 (Munich, 25 June 2013).

Switzerland

Gross, Damien, Planner, Department of Architecture and Urban Planning, City of Sion (Sion, 24 July 2013).
Mettan, Nicolas, Planner, Department of Roads and Rivers, Direction Rhône 3, Canton of Valais (Sion, 23 July 2013).
Michelet, Paul, Department of Agriculture, Canton of Valais (Sion, 24 July 2013).
Largey, Thierry, manager Pro Natura Valais (telephone interview, Zurich/Sion, 13 August 2013).
Tille, Didier, Engineer hydrologist, Project leader Central Valais, Direction Rhône 3, Department of Roads and Rivers, Canton of Valais. (Sion, 17 April 2014).

The Netherlands

De Bont, Jos, historian, former farmer and inhabitant Overdiepse Polder (Waspik, 23 September 2013).
De Jonge, Jannemarie, Landscape Architect, guest lecturer at Wageningen University (Wageningen, 22 March 2013).
De Koning, Robbert, Landscape Architect Noordwaard (Arnhem, 24 April 2013).
Houwing, Erik-Jan, Technical Manager Room for the River / Noordwaard, Group West (Werkendam, 26 November 2012).
Sijmons, Dirk, former chairman of the Q-Team Room for the River (15 September 2015, Delft, and 20 October 2015, Amsterdam).
Van der Grift, Jan and Der Nederlanden, Hermine, experts Spatial Quality, Department of Knowledge Management, Rijkswaterstaat, Room for the River. (Utrecht, 13 August 2013).
Leverman, Robert, Project Manager, and De Vries, Aldert, Civil engineering, Bureau Meerstad (Harkstede, 12 March 2013).
Rolvink, Remco, Project leader and former partner Alle Hosper Landschapsarchitectuur (Haarlem, 12 December 2012).
Van de Bospoort, Jan, landscape architect, City of Groningen (Groningen, 11 March 2013).
Van Hall, Albert, Chairman, and Küpers, Harm, Director-Secretary, Water Board Hunze-en-Aa's (Veendam, 13 March 2013).

SOURCES TOPOGRAPHIC AND TERRAIN DATA

France: Institut Géographique National
Netherlands: Actueel Hoogtebestand Nederland, Kadaster
Switzerland: Federal Office of Topography – Swisstopo
Germany: Bayerisches Landesamt für Vermessung und Geoinformation
OpenStreetMap
Eurogeographics

This book was made possible by people and institutions that I especially want to thank for their support:

The Swiss National Science Foundation, who funded my PhD thesis from 2012 to 2015, as well as the team of the Institute of Landscape Architecture/ETH Zurich who supported this project at its beginning, Pr. Christophe Girot, Anette Freytag and Suzanne Kríženecký; those who assisted this research with talent and dedication: Georgios Orfanopoulos and Marcel Goethals.

Professors Christophe Girot, Dirk Sijmons and Kelly Shannon, who shared their expertise and vision during this investigation, as well as Marc Bedarida, Pierre Chabard and Marcel Witvoet, who helped me clarify my ideas and this manuscript.

Tom Leighton, Stéphanie Grégoire and Leo Reijnen, who patiently streamlined the text resulting from this nomadic and polyglot investigation.

All the professionals who kindly shared their knowledge during the field trips and interviews held in Switzerland, Germany, France, and the Netherlands: Damien Gross, Thierry Largey, Nicolas Mettan, Paul Michelet, Didier Tille; Oliver Engelmayer, Nivetida Mahida, Daniela Schaufuss, Christiane Thalgott; Morgane Buisson, Jacky Girel, Paul Jardin, Olivier Manin, Michel Pinhas, Bertrand Vignal; Jan van de Bospoort, Jan van der Grift, Alfred van Hall, Erik Jan Houwing, Robbert de Koning, Harm Küpers, Jannemarie de Jonge, Robert Leverman, Remco Rolvink, Aaldert de Vrieze, Hermine van der Nederlanden.

The institutions that freely provided the geographical data needed for modelling and visualization: the Federal Office of Topography and the Institute of Cartography and Geoinformation for Switzerland, the Institut Géographique National for France, Eurogeographics; to the programmers who create the geographic tools to process this data, and share their creation.

My partner, family and close friends, who stood by me during these exciting years: thank you Tim, and Andreia, Audrey, Christophe, Gilbert, Maryline, Monique, Muriel, Nathaly, Philine, Susanne.

To my grandfather Marcel Lebègue, for his communicative dedication and many contributions to the harsh but generous Alpine landscape, to keep it cultivated and inhabitable.

Texts and drawings (unless otherwise specified): Frédéric Rossano
Editing: Stéphanie Grégoire
Copy editing: Leo Reijnen
Copy editing 2015 PhD manuscript: Tom Leighton

Design: SJG / Joost Grootens, Dimitri Jeannottat, Julie da Silva
Printing: NPN Drukkers, Breda
Publisher: Marcel Witvoet, nai010 publishers, Rotterdam

Printed and bound in the Netherlands

ISBN 978-94-6208-525-1
NUR 648, 945
BISAC ARC008000, ARC010000

Floodscapes is also available as e-book:
ISBN 978-94-6208-536-7 (e-book)

This book was published with the support of the AMUP Research Unit, a joint unit of the National School of Architecture in Strasbourg and Insa in Strasbourg, and with the support of the Ministry of Culture (bureau de la recherche architecturale, urbaine et paysagère).

RÉPUBLIQUE FRANÇAISE
Liberté Égalité Fraternité

Strasbourg, école d'architecture

amup
Architecture, Morphologie / Morphogenèse Urbaine et Projet

MINISTÈRE DE LA CULTURE
Liberté Égalité Fraternité

nai010 publishers is an internationally orientated publisher specialized in developing, producing and distributing books in the fields of architecture, urbanism, art and design. www.nai010.com

nai010 books are available internationally at selected bookstores and from the following distribution partners:
North, Central and South America: Artbook | D.A.P., New York, USA, dap@dapinc.com
Rest of the world: Idea Books, Amsterdam, the Netherlands, idea@ideabooks.nl
For general questions, please contact nai010 publishers directly at sales@nai010.com or visit our website www.nai010.com for further information.